The BEST WRITING on MATHEMATICS

2011

The BEST WRITING on MATHEMATICS

2011

Mircea Pitici, Editor

PRINCETON UNIVERSITY PRESS
PRINCETON AND OXFORD

Published by Princeton University Press, 41 William Street, Princeton, New Jersey 08540

In the United Kingdom: Princeton University Press, 6 Oxford Street, Woodstock, Oxfordshire OX20 1TW

press.princeton.edu

ISBN 978-0-691-15315-5

This book has been composed in Perpetua Std

Printed on acid-free paper. ∞

Printed in the United States of America

1 3 5 7 9 10 8 6 4 2

To my teacher, Ioan Candrea

In memoriam

Contents

Foreword: Recreational Mathematics

FREEMAN DYSON

Hobbies are the spice of life. Recreational mathematics is a splendid hobby which young and old can equally enjoy. The popularity of Sudoku shows that an aptitude for recreational mathematics is widespread in the population. From Sudoku it is easy to ascend to mathematical pursuits that offer more scope for imagination and originality. To enjoy recreational mathematics you do not need to be an expert. You do not need to know the modern abstract mathematical jargon. You do not need to know the difference between homology and homotopy. You need only the good old nineteenth-century mathematics that is taught in high schools, arithmetic and algebra and a little geometry. Luckily for me, the same nineteenth-century mathematics was all that I needed to do useful calculations in theoretical physics. So, when I decided to become a professional physicist, I remained a recreational mathematician. This foreword gives me a chance to share a few of my adventures in recreational mathematics.

The articles in this collection, *The Best Writing on Mathematics 2011*, do not say much about recreational mathematics. Many of them describe the interactions of mathematics with the serious worlds of education and finance and politics and history and philosophy. They are mostly looking at mathematics from the outside rather than from the inside. Three of the articles, Doris Schattschneider's piece about Maurits Escher, the Fergusons' piece on mathematical sculpture, and Dana Mackenzie's piece about packing a circle with circles, come closest to being recreational. I particularly enjoyed those pieces, but I recommend the others too, whether they are recreational or not. I hope they will get you interested and excited about mathematics. I hope they will tempt a few of you to take up recreational mathematics as a hobby.

I began my addiction to recreational mathematics in high school with the fifty-nine icosahedra. *The Fifty-Nine Icosahedra* is a little book published in 1938 by the University of Toronto Press with four authors, H.S.M. Coxeter, P. DuVal, H. T. Flather, and J. F. Petrie. I saw the title in a catalog and ordered the book from a local bookstore. Coxeter was the world expert on polyhedra, and Flather was the amateur who made models of them. The book contains a complete description of the fifty-nine stellations of the icosahedron. The icosahedron is the Platonic solid with twenty equilateral triangular faces. A stellation is a symmetrical solid figure obtained by continuing the planes of the twenty faces outside the original triangles. I joined my school friends Christopher and Michael Longuet-Higgins in a campaign to build as many as we could of the fifty-nine icosahedra out of cardboard and glue, with brightly colored coats of enamel paint to enhance their beauty. Christopher and Michael both went on later to become distinguished scientists. Christopher, now deceased, was a theoretical chemist. Michael is an oceanographer. In 1952, Michael took a holiday from oceanography and wrote a paper with Coxeter giving a complete enumeration of higher-dimensional polytopes. Today, if you visit the senior mathematics classroom at our old school in England, you will see the fruits of our teenage labors grandly displayed in a glass case, looking as bright and new as they did seventy years ago.

My favorites among the stellations are the twin figures consisting of five regular tetrahedra with the twenty vertices of a regular dodecahedron. The twins are mirror images of each other, one right-handed and the other left-handed. These models give to anyone who looks at them a vivid introduction to symmetry groups. They show in a dramatic fashion how the symmetry group of the icosahedron is the same as the group of 120 permutations of the five tetrahedra, and the subgroup of rotations without reflections is the same as the subgroup of 60 even permutations of the tetrahedra. Each of the twins has the symmetry of the even permutation subgroup, and any odd permutation changes one twin into the other.

Another book which I acquired in high-school was *An Introduction to the Theory of Numbers* by G. H. Hardy and E. M. Wright, a wonderful cornucopia of recreational mathematics published in 1938. Chapter 2 contains the history of the Fermat numbers, $F_n = 2^{2^n} + 1$, which Fermat conjectured to be all prime. Fermat was famously wrong. The first four Fermat numbers are prime, but Euler discovered in 1732 that $F_5 = 2^{32} + 1$ is divisible by 641, and Landry discovered in 1880 that

$F_6 = 2^{64} + 1$ is divisible by 274,177. Since fast computers became available, many larger Fermat numbers have been tested for primality, and not one has been found to be prime.

Hardy and Wright provide a simple argument to explain why F_5 is divisible by 641. Since $641 = 1 + 5a = 2^4 + 5^4$ with $a = 2^7$, we have

$$F_5 = 2^4 a^4 + 1 = (1 + 5a - 5^4)a^4 + 1 = (1 + 5a)a^4 + (1 - (5a)^4),$$

which is obviously divisible by $1 + 5a$. I was always intrigued by the question of whether a similarly elementary argument could be found to explain the factorization of F_6. Sixty years later, I found the answer. This was another joyful piece of recreational mathematics. The answer turned out to depend on a theorem concerning palindromic continued fractions. If a and q are positive integers with $a < q$, the fraction a/q can be expressed in two ways as a continued fraction:

$$a/q = 1/(p_1 + 1/(p_2 + 1/\ldots(p_{n-1} + 1/p_n)\ldots)),$$

where the partial quotients p_j are positive integers. The fraction is palindromic if $p_j = p_{n+1-j}$ for each j. The theorem says that the fraction is palindromic if and only if $a^2 + (-1)^n$ is divisible by q.

Landry's factor of F_6 has the structure

$$q = 274177 = 1 + 2^8 f, \quad f = (2^6 - 1)(2^4 + 1),$$

where f is a factor of $2^{24} - 1$, so that

$$2^{24} - 1 = fg, \quad g = (2^6 + 1)(2^8 - 2^4 + 1),$$

and

$$2^{32} = 2^8(1 + fg) = gq - a, \quad a = g - 2^8 = 15409.$$

The partial quotients of the fraction a/q are (17, 1, 3, 1, 5, 5, 1, 3, 1, 17). A beautiful palindrome, and the palindrome theorem tells us that $a^2 + 1 = qu$ with u integer. Therefore $F_6 = 1 + (gq - a)^2 = q(g^2 q - 2ga + u)$ is divisible by q. I was particularly proud of having discovered the palindrome theorem, until I learned that I had been scooped by the French mathematician Joseph Alfred Serret, who published it in 1848. To be scooped by one of the lesser-known luminaries of the nineteenth century is part of the game of recreational mathematics. On another occasion I was scooped by Riemann, but that is a long story and I do not have space to tell it here.

When I was an undergraduate at Cambridge University, I was intrigued by a famous discovery of the Indian prodigy Ramanujan concerning the arithmetical properties of the partition function $p(n)$. This discovery was lovingly described in Hardy and Wright's *An Introduction to the Theory of Numbers*. For any positive integer n, $p(n)$ is the number of ways of expressing n as a sum of positive integer parts. Ramanujan discovered that $p(5k + 4)$ is always divisible by 5, $p(7k + 5)$ is divisible by 7, and $p(11k + 6)$ is divisible by 11. I wanted to find a way of actually dividing the partitions of $5k + 4$ into 5 equal classes, and similarly for 7 and 11. I found a simple way to do the equal division. The "rank" of any partition is defined as the biggest part minus the number of parts. The partitions of any n can be divided into 5 rank classes, putting into class m the partitions that have rank of the form $5j + m$ for $m = 1, 2, 3, 4, 5$. I found to my delight that the 5 rank classes of partitions of $5k + 4$ are exactly equal. The same trick works for 7 but not for 11. It was easy to check numerically that the rank classes were equal for the partitions of $5k + 4$ and $7k + 5$ all the way up to 100, but I failed to find a proof. I also conjectured the existence of another property of a partition that would do the same job for the partitions of $11k + 6$, and I called that hypothetical property the "crank."

Ten years later, Oliver Atkin and Peter Swinnerton-Dyer succeeded in proving the equality of the rank-classes for $5k + 4$ and $7k + 5$, and 45 years later, Frank Garvan and George Andrews identified the crank. Garvan and Andrews not only found the crank, but also proved that it provides an equal division of the crank classes for all three cases, 5, 7, and 11. More recently, in 2008, Andrews made another discovery as beautiful as Ramanujan's original discovery. Andrews was looking at another function $S(n)$ enumerating the smallest parts of partitions of n. $S(n)$ is defined as the sum, over all partitions of n, of the number of smallest parts in that partition. Andrews discovered and also proved that $S(5k + 4)$ is divisible by 5, $S(7k + 5)$ is divisible by 7, and $S(13k + 6)$ is divisible by 13. The appearance of 13 instead of 11 in this statement is not a typographical error. It is a big surprise and adds a new mystery to the mysteries discovered by Ramanujan.

The question now arises whether there exists another property of partitions with their smallest parts, like the rank and the crank, allowing us to divide the partitions with their smallest parts into 5 or 7 or 13 equal classes. I conjecture that such a property exists, and I offer a challenge to readers of this volume to find it. To find it requires no expert knowledge. All that you have to do is to study the partitions and smallest parts for a few small values of n, and make an inspired guess at the property that divides them equally. A second challenge is to prove that the guess actually works. To succeed with the second challenge probably requires some expert knowledge, since I am asking you to beat George Andrews at his own game.

My most recent adventure in recreational mathematics is concerned with the hypothesis of Decadactylic Divinity. Decadactylic is Greek for ten-fingered. In days gone by, serious mathematicians were seriously concerned with theology. Famous examples were Pythagoras and Descartes. Each of them applied his analytical abilities to the elucidation of the attributes of God. I recently found myself unexpectedly following in their footsteps, applying elementary number theory to answer a theological question. The question is whether God has ten fingers. The evidence in favor of a ten-fingered God was brought to my attention by Norman Frankel and Lawrence Glasser. I hasten to add that Frankel and Glasser were only concerned with the mathematics, and I am solely responsible for the theological interpretation. Frankel and Glasser were studying a sequence of rational approximations to π discovered by Derek Lehmer. For each integer k, there is a rational approximation $[R_1(k)/R_2(k)]$ to π, with numerator and denominator defined by the identity

$$R_1(k) + R_2(k)\pi = \sum_{m=1}^{\infty} [(m!)^2/(2m)!]2^m m^k.$$

The right-hand side of this identity has interesting analytic properties which Frankel and Glasser explored. The approximations to π that it generates are remarkably accurate, beginning with 3, 22/7, 22/7, 335/113, for $k = 1, 2, 3, 4$. Frankel and Glasser calculated the first hundred approximations to high accuracy, and found to their astonishment that the kth approximation agrees with the exact value of π to roughly k places of decimals. I deduced from this discovery that God must calculate as we do, using arithmetic to base ten, and it was then easy to conclude that He has ten fingers. It seemed obvious that no other theological hypothesis could account for the appearance of powers of ten in the approximations to such a transcendental quantity as π.

Unfortunately, I soon found out that my theological breakthrough was illusory. I calculated precisely the magnitude of the error of the kth Lehmer approximation for large k, and it turned out that the error does not go like 10^{-k} but like Q^{-k}, where $Q = 9.1197 \ldots$ is a little smaller than ten. For large k, the approximation is in fact only accurate to $0.96k$ places of decimals, where 0.96 is the logarithm of Q to base ten. Q is defined as the absolute value of the complex number $q = 1 + (2\pi i/\ln(2))$. When we are dealing with complex numbers, the logarithm is a many-valued function. The logarithm of 2 to the base 2 has many values, beginning with the trivial value 1. The first nontrivial value of $\log_2(2)$ is q. This is the reason why q determines the accuracy of the Lehmer approximations. This calculation demonstrates that God does not use arithmetic to base ten. He uses only fundamental units such as π and $\ln(2)$ in the design of His mathematical sensorium. The number of His fingers remains an open question.

Two of these recreational adventures were from my boyhood and two from my old age. In between, I was doing mathematics in a more professional style, finding problems in the understanding of nature where elegant nineteenth-century mathematics could be usefully applied. Mathematics can be highly enjoyable even when it is not recreational. I hope that the articles in this volume will spark readers' interest in digging deeper into some aspect of mathematics, whether it is puzzles and games, history of mathematics, mathematics education, or perhaps studying for a professional degree in mathematics. The joys of mathematics are to be found at all levels of the game.

Introduction

Mircea Pitici

This new volume in the series of *The Best Writing on Mathematics* brings together a collection of remarkable texts, originally printed during 2010 in publications from several countries. A few exceptions from the strict timeframe are inevitable, due to the time required for the distribution, surveying, reading, and selection of a vast literature, part of it coming from afar.

Over the past decade or so, writing about mathematics has become a genre, with its own professional practitioners—some highly talented, some struggling to be relevant, some well established, some newcomers. Every year these authors, considered together, publish many books. This abundance is welcome, since writing on mathematics realizes the semantic component of a mental activity too often identified to its syntactic-procedural mode of operation. The appropriation to the natural language of meaningful intricacies latent in symbolic formulas opens up paths toward comprehending the abstraction that characterizes mathematical thinking and some mathematical notions; it also offers unlimited expressive, imaginative, and cognitive possibilities. In the second part of the introduction, I mention the books on mathematics that came to my attention last year; the selection in this volume concerns mostly pieces that are not yet available in book form—either articles from academic journals or good writing in the media that goes unobserved or is forgotten after a little while. *The Best Writing on Mathematics* reflects the literature on mathematics available out there in myriad publications, some difficult to consult even for people who have access to exceptional academic resources. In editing this series I see my task as restitution to the public, in convenient form, of excellent writing on mathematics that deserves enhanced reception beyond the initial publication. By editing this series I also want to make widely available good texts about mathematics that otherwise would be lost in the deluge of information that surrounds us. The content of each volume builds itself up to a point; I only give it a coherent structure and present it to the reader. That means that every year some prevailing themes will be new, others will reappear.

Most readers of this book are likely to be engaged with mathematics in some way, at least by being curious about it. But most of them are inevitably engaged with only a (small) part of mathematics. That is true even for professional mathematicians, with rare exceptions. Mathematics has far-reaching tentacles, in pure research branches as well as in mundane applications and in instructional contexts. No wonder the stakeholders in the metamorphosis of mathematics as a social phenomenon can hardly be well informed about the main ideas and developments in all the different aspects connected to mathematics. Solipsism among mathematicians is surely not as common as the general public assumes it is; yet specialization is widespread, with many professionals finding it difficult to keep abreast of developments beyond their strict areas of interest. By making this volume intentionally eclectic, I aim to break some of the barriers laid by intense specialization. I hope that the enterprise makes it easier for readers, insiders and outsiders, to identify the main trends in thinking about mathematics in areas unfamiliar to them.

Anthologies of writings on mathematics have a long—if sparse and irregular—history. Countless volumes of contributed collections in particular fields of mathematics exist but, to my knowledge, only a handful of anthologies that include panoramic selections across multiple fields. Soon after the Second World War, William Schaaf edited *Mathematics, Our Great Heritage*, which included writings by G. H. Hardy, George Sarton, D. J. Struik, Carl G. Hempel, and others. A few years later James Newman edited *The World of Mathematics* in four massive tomes, a collection widely read for decades by many mathematicians still active today. In parallel, in francophone countries circulated Le Lionnaise's *Les Grand Courants de la Pensée Mathématique*, translated into English only several years ago. During the 1950s and 1960s a synthesis in three volumes edited by A. D. Aleksandrov, A. N. Kolmogorov, and M. A. Lavrentiev, including contributions by Soviet authors, was translated and widely circulated. Very few similar books appeared during the last three decades of the twentieth century; notable was *Mathematics Today,* edited by Lynn Arthur Steen in the late 1970s. In the present century the pace quickened; several excellent volumes were published, starting with

Mathematics: Frontiers and Perspectives edited in 2000 by Vladimir Arnold, Michael Atiyah, Peter Lax, and Barry Mazur, followed by the voluminous tomes edited by Björn Engquist and Wilfried Schmid in 2001, Timothy Gowers in 2008, as well as the smaller collections edited by Raymond Ayoub in 2004 and Reuben Hersh in 2006 (for complete references, see the list of works mentioned at the end of this introduction). *The Best Writing on Mathematics* builds on this illustrious tradition, capitalizing on an ever more interconnected world of ideas and benefiting from the regularity of yearly serialization.

Overview of the Volume

The texts included in this volume touch on many topics related to mathematics. I gave up the thematic organization adopted in the first volume, since some of the texts are not easy to categorize and some themes would have been represented by only one or very few pieces.

Underwood Dudley argues that mathematics beyond the elementary notions is the best preparation for reasoning in general and that most people value it primarily for that purpose, not for its immediate practicality.

Dana Mackenzie describes the overt and the hidden properties of the Apollonian gasket, a configuration of infinitely nested tangent circles akin to a fractal.

Rik van Grol tells the story of finding the optimal number of steps that solve scrambled Rubik's cubes of different sizes—starting with the easy cases and going to the still unsolved ones.

Andrew Schultz writes on the friendly professional interactions that shape the career of a mathematician, from learning the ropes as a graduate student to becoming an accomplished academic.

In a polemical reply to a text we selected in last year's volume of this series (Gowers and Nielsen), Melvyn Nathanson argues that the most original mathematical achievements are distinctively individual, rather than results of collaboration.

Martin Campbell-Kelly meditates on the long flourishing popularity and recent demise of mathematical tables.

Reuben Hersh ponders on the post–World War II abundance of Jewish mathematicians at American universities, in contrast to the pale prewar representation of Jews among American mathematicians.

David Hand points out that attributing the 2008 financial crisis to the widespread reliance on mathematical models of investment and risk oversimplifies causality in complex economies.

Jordan Ellenberg describes how compressed sensing, a process based on a mathematical algorithm, leads to the reconstruction of big amounts of information starting from smaller samples.

After reviewing the broad stages that led to computing as we know it (from a blend of applied mathematics, science, and engineering in the 1930s and continuing into the present) Peter Denning characterizes the interactions between computing and the other domains of science—physical, social, and life.

James Hamlin and Carlo Séquin identify the mathematical properties of Charles Perry's airy sculptures and extend those ideas by suggesting other ways of constructing transparent surfaces in a similar vein.

Barry Cipra describes the Lorenz system of three simple differential equations, a versatile system with solutions of aesthetic interest but whose dynamic properties are only partly understood by mathematicians.

Doris Schattschneider identifies the mathematical inspirations of the Dutch artist M. C. Escher—most of them fruits of long-term preoccupation with geometry and of his interaction with mathematicians.

Counterbalancing the impression of fragility conveyed by the light structures presented in the Hamlin and Séquin text, Helaman and Claire Ferguson detail the mathematical notions underlying Helaman's massive bronze sculptures.

John Mason recounts the main events, people, and trends that influenced his views of mathematics education over a half century of activity.

Douglas Fisher, Nancy Frey, and Heather Anderson discuss modeling, vocabulary enrichment, and group work as ways for engaging students in the mathematics classroom.

Francis Edward Su asks what makes a learner become an inquirer—and further, a discoverer—and offers refreshing answers, contradicting some of the widely held assumptions concerning the effectiveness of instructional approaches in secondary and undergraduate mathematics.

Echoing the autobiographical tone in John Mason's text, Alan Schoenfeld reviews the ideas that guided him in researching mathematical problem solving and anchoring the role of theory in mathematics education research.

Hans Niels Jahnke combines etymological considerations with the analysis of scientific thinking in ancient Greece, to discern the sources and the changing meaning of the notion of mathematical proof.

Ian Hacking asks what unifies and gives a common "identity" to varieties of mathematics—and concludes that the answer has changed through time, depending on the contingent historical conditions.

Feng Ye formulates several challenges that must be overcome by philosophers who claim that mathematical entities reside in imagination, in order to adopt a viable position—and provides grounds for reconstructing such a philosophy of mathematics.

Ivan Havel speculates on the possible mental imagery of number representation, in connection with the polysemy of the concept of number and the unusual abilities displayed by people able to perform exceptional computational abilities.

Ioan James reviews recent research on possible connections between characteristics of the autistic personality and salient features of mathematical talent visible in people affected by the syndrome.

Chris Budd and Rob Eastaway take on the conundrum of communicating mathematics well to the nonspecialized public—and suggest balancing the symbolic language of mathematics with good examples of the imprint left by mathematics on everyday life.

Marianne Freiberger explores connections between physics and the geometry of space, starting from conversations she has had with Shing-Tung Yau, one of the prominent geometers who studies them.

To conclude this collection of articles on mathematics, Erica Klarreich explains how a computerized algorithm helps place in residency couples of medical school graduates and how the process has changed over several decades to accommodate the transformation of the marketplace for doctors.

Other 2010 Writings on Mathematics

I selected the texts in this volume from a much larger group. At the end of the book the reader finds a list of other remarkable pieces, which I initially considered in a wider selection but I did not include for reasons of space or related to copyright. In this section of the introduction I mention a string of books on mathematics (as different from mathematics books proper) that came to my attention over the past year.

A captivating first-person account of research in pure mathematics, with insights into the collaborative work of mathematicians, is given by Shing-Tung Yau (with Steve Nadis) in *The Shape of Inner Space*. Alexandre Borovik gives a rich exploration of cognitive aspects of mathematical intuition, based on many biographical accounts, in *Mathematics under the Microscope*. A book difficult to categorize, touching on psychological, cognitive, and historical aspects of mathematical thinking, in an original manner, is *Mathematical Reasoning* by Raymond Nickerson. Also interdisciplinary is *Grammar, Geometry, and Brain* by Jens Erik Fenstad. Two recent books of interviews also offer glimpses into the rich life experiences of mathematicians: *Recountings: Conversations with MIT Mathematicians* edited by Joel Segel and *Creative Minds, Charmed Lives* edited by Yu Kiang Leong—as does Reuben Hersh and Vera John-Steiner's *Loving and Hating Mathematics*.

Among several books that combine excellent expository writing with mathematics proper are *Charming Proofs* by Claudi Alsina and Roger Nelsen, *Creative Mathematics* by Alan Beardon, *Making Mathematics Come to Life* by O. A. Ivanov, *A Mathematical Medley* by George Szpiro, and the latest volume (number 8) of *What's Happening in the Mathematical Sciences* by Dana Mackenzie.

Many notable books on the history of mathematics are being published these days; some of the most recent titles and their authors are the following. Benjamin Wardhaugh has written a brief but badly needed guide called *How to Read Historical Mathematics*. Two excellent books on the history of geometry are *Revolutions of Geometry* by Michael O'Leary and *Geometry from Euclid to Knots* by Saul Stahl. Other accounts of particular mathematical branches, times, or personalities are *The Birth of Numerical Analysis* edited by Adhemar Bultheel and Ronald Cools, *The Babylonian Theorem* by Peter Rudman, *The Pythagorean Theorem: The Story of Its Power and Beauty* by Alfred Posamentier, *Hidden Harmonies: The Lives and Times of the Pythagorean Theorem* by Robert and Ellen Kaplan, *Galileo* by J. L. Heilbron, *Defending Hypatia* by Robert Goulding, *Voltaire's Riddle* by Andrew Simoson, *The Scientific Legacy of Poincare* edited by Éric Charpentier, Étienne Ghys, and Annick Lesne, *Emmy Noether's Wonderful Theorem* by Dwight Neuenschwander, and *Studies in the History of Indian Mathematics* edited by C. S. Seshadri. Thematically more encompassing are *Mathematics and Its History* by John Stillwell, *An Episodic History of Mathematics* by Steven Krantz, *Duel at Dawn* by Amir Alexander, and *History of Mathematics: High-*

ways and Byways by Amy Dahan-Dalmédico and Jeanne Peiffer. The mathematical avatars of representative politics since Ancient Greece are wonderfully narrated by George Szpiro in *Numbers Rule*. Joseph Mazur, in *What's Luck Got to Do with It?* also takes a historical perspective, by telling the story of the mathematics involved in gambling. Alex Bellos gives a refreshingly informal tour of the history of elementary mathematics in a new edition of *Here's Looking at Euclid*. Two histories of sciences with substantial material dedicated to mathematics are *Kinematics: The Lost Origins of Einstein's Relativity* by Alberto Martinez and *Technology and Science in Ancient Civilizations* by Richard Olson. Among editions of old mathematical writings are *Pappus of Alexandria: Book 4 of the Collection* edited by Heike Sefrin-Weis and the fourth volume of Lewis Carroll's pamphlets, containing *The Logic Pamphlets*, an edition by Francine Abeles.

Among recent books of philosophical issues in mathematics I note *There's Something about Gödel* by Francesco Berto, *Roads to Infinity* by John Stillwell, the reprinting of the *Philosophy of Mathematics* by Charles S. Peirce, and several new volumes on logic: *The Evolution of Logic* by W. D. Hart, *Logic and Philosophy of Mathematics in the Early Husserl* by Stefania Centrone, *Logic and How It Gets that Way* by Dale Jacquette, and the collection of classic essays *Thinking about Logic* edited by Steven Cahn, Robert Talisse, and Scott Aikin. In *Mathematics and Reality*, Mary Leng is concerned with the ontology of mathematical objects and its implications for the status of mathematical practice.

Several recent books on mathematics education that came to my attention are worth mentioning. *Theories of Mathematics Education* edited by Bharath Sriraman and Lyn English is a synthesis of contributions by foremost researchers in the field. Others, concerned with particular themes, are *Proof in Mathematics Education* by David Reid and Christine Knipping, *Mathematical Action and Structures of Noticing* edited by Stephen Lerman and Brent Davis (honoring the work of John Mason, a contributor to this volume), *Culturally Responsive Mathematics Education* edited by Brian Greer and collaborators, and *Winning the Math Wars* by Martin Abbott and collaborators. A detailed and authoritative presentation of inquiry-based teaching of mathematics is *The Moore Method* by Charles Coppin, Ted Mahavier, Lee May, and Edgar Parker. A new volume (number VII) in the series of *Research in Collegiate Mathematics* was edited recently by Fernando Hitt, Derek Holton, and Patrick Thompson. Volumes concerned with education generally but with excellent chapters on mathematical aspects are *Instructional Explanations in the Disciplines* edited by Mary Kay Stein and Linda Kucan, *Beauty and Education* by Joe Winston, and *Visualization in Mathematics, Reading, and Science Education* edited by Linda Phillips, Stephen Norris, and John Macnab. The National Council of Teachers of Mathematics has continued the publication of several series, listed below under the authors Marian Small, Michael Shaughnessy, Mark Saul, Jane Schielack, and Frank Lester.

In *Numbers Rule Your World*, Kaiser Fung (who also maintains the Junk Charts blog: http://junkcharts.typepad.com/junk_charts/) presents in detail several cases of sensible statistical thinking in engineering, health, finance, sports, and other walks of life. Related recent titles are *The Pleasures of Statistics: The Autobiography of Frederick Mosteller*, *What Is p-Value anyway?* by Andrew Vickers, and *Probabilities* by Peter Olofsson.

In applied mathematics several remarkable volumes became available or were reissued lately. Colin Clark's *Mathematical Bioeconomics* is now in print in its third edition, offering a wide range of examples in resource management and environmental studies, all in an easy-to-read presentation. Mohammed Farid edited a massive volume covering 37 major topics concerning properties of food, under the title *Mathematical Modeling of Food Processing*. A group including Warren Hare and collaborators wrote *Modelling in Healthcare*, a substantial report on data collection, mathematical modeling, and interpretation in healthcare institutions. Robert Keidel authored a refreshing visual-conceptual view of strategic management in *The Geometry of Strategy*. Closer to mathematics proper is *Mrs. Perkins's Electric Quilt* by Paul Nahin, an introduction to mathematical physics written with much verve. *Once Before Time* by Martin Bojowald and *What If the Earth Had Two Moons?* by Neil Comins underlie the crucial role of mathematics in understanding the meaning of physical laws governing the observable universe. Highly readable, appealing to basic notions of randomness and complexity, is *Biology's First Law* by Daniel McShea and Robert Brandon. In *The Mathematics of Sex*, Stephen Ceci and Wendy Williams zoom in on gender aspects of education, social inequality, and public policy. Two books with a wide span of applications are *Critical Transitions in Nature and Society* by Marten Scheffer and *Disrupted Networks* by Bruce West and Nicola Scafetta. David Easley and Jon Kleinberg take an interdisciplinary approach in *Networks, Crowds, and Markets*. And a lively account of the use, overuse, and abuse of mathematical methods in the financial industry is given by Scott Patterson in *The Quants*.

A splendid illustration of mathematical methods in charting complex data sets is the *Atlas of Science* by Katy Börner. Also visual, at the intersection between mathematics, arts, and philosophy, is *Quadrivium*, edited by John Martineau and others.

Popularizing mathematics is well represented by several titles. In *Our Days Are Numbered,* Jason Brown writes on common occurrences of many elementary (and a few less-than-elementary) mathematical notions—as do Ian Stewart in *Cows in the Maze and Other Mathematical Explorations* (pointing to a wealth of online resources), John D. Barrow in *100 Essential Things You Didn't Know You Didn't Know*, Marcus du Sautoy in *Symmetry: A Journey into the Patterns of Nature* (an attractive blend of history, games, and storytelling), and Jamie Buchan in *Easy as Pi*. A sort of dictionary of number properties is *Number Freak* by Derrick Niederman. First-hand accounts of learning mathematics are *Hot X: Algebra Exposed* by Danica McKellar, *The Calculus Diaries* by Jennifer Ouellette, and *Dude, Can You Count?* by Christian Constanda. Finally in this category, a must-read antidote to blindly taking for granted numerical arguments in public discourse is *Proofiness* by Charles Seife.

In a separate register, it is worth mentioning a moving novel of love and loneliness on a mathematical metaphor, *The Solitude of Prime Numbers* by the young Italian physicist Paolo Giordano.

I conclude this summary overview of the vast number of books on mathematics published last year by mentioning the recent publication of the first issue (January 2011) of a new periodical, the *Journal of Humanistic Mathematics* (thanks to Fernando Gouvêa for drawing my attention to this journal).

This is not a critical review of the recent literature on mathematics and surely not a comprehensive list. Other books, not mentioned here, can be found on the list of notable texts at the end of the volume; perhaps still others have escaped my survey. Authors and publishers can make sure I know about a certain title by contacting me at the address provided just before the list of works mentioned in this introduction.

A Few Internet Resources

The sheer number of excellent websites on mathematics (including those hosted by educational institutions and individual mathematicians) makes any attempt to compile a comprehensive inventory look quixotic. Here

I only suggest a handful of online mathematical resources that caught my attention over the past year that I did not mention in the introduction to the previous volume.

I begin with a few specialized sites. The *Why Do Math* website (http://www.whydomath.org/) is particularly original in highlighting the success stories of applied mathematics in science, society, and everyday life. A multilanguage site profiled on number sequences is the On-Line Encyclopedia of Integer Sequences (http://oeis.org/). The Geometric Dissections site (http://home.btconnect.com/GavinTheobald/Index.html) obviously needs no further description. Neither does the World Federation of National Mathematics Competitions (http://www.amt.edu.au/wfnmc/). And an excellent website for questions and answers is the Math Overflow site (http://mathoverflow.net/).

Among the many good blogs on mathematics, I mention a few very active ones: the Math-blog (http://math-blog.com/), the Computational Complexity blog (http://blog.computationalcomplexity.org/), Wild about Math! blog (http://wildaboutmath.com/), and the Thinkfinity blog (http://www.thinkfinity.org/). A number of prestigious mathematicians maintain active blogs, with a rich network of links to other blogs—including Timothy Gowers (http://gowers.wordpress.com/), Terence Tao (http://terrytao.wordpress.com/), and Richard Lipton (http://rjlipton.wordpress.com/), to mention just a few.

Some good instructional/educational sites, among many, are the Khan Academy (http://www.khanacademy.org/), Mathematically Sane (http://mathematicallysane.com/), and the Reasoning Mind (http://www.reasoningmind.org/).

Works Mentioned

Abbott, Martin, Duane Baker, Karen Smith, and Thomas Trzyna. *Winning the Math Wars: No Teacher Left Behind*. Seattle, WA: University of Washington Press, 2010.

Aleksandrov, A. D., A. N. Kolmogorov, and M. A. Lavrentiev. (Eds.) *Mathematics: Its Content, Method, and Meaning.* Three volumes. Cambridge, Mass.: MIT Press, 1963. (Originally published in Moscow, 1956.)

Alexander, Amir. *Duel at Dawn: Heroes, Martyrs, and the Rise of Modern Mathematics*. Cambridge, Mass.: Harvard University Press, 2010.

Alsina, Claudi, and Roger B. Nelsen. *Charming Proofs: A Journey into Elegant Mathematics*. Washington, D.C.: Mathematical Association of America, 2010.

Arnold, V. I., M. Atiyah, P. Lax, and B. Mazur. (Eds.) *Mathematics: Frontiers and Perspectives.* Providence, R.I.: American Mathematical Society, 2000.

Ayoub, Raymond G. (Ed.) *Musings of the Masters: An Anthology of Mathematical Reflections.* Washington, D.C.: Mathematical Association of America, 2004.

Barrow, John D. *100 Things You Didn't Know You Didn't Know: Math Explains Your World*. New York: W. W. Norton, 2010.

Beardon, Alan F. *Creative Mathematics.* Cape Town, South Africa: Cambridge University Press, 2009.

Bellos, Alex. *Here's Looking at Euclid: A Surprising Excursion through the Astonishing World of Mathematics*. New York: Free Press, 2010.

Berto, Francesco. *There's Something about Gödel: A Complete Guide to the Incompleteness Theorem.* Oxford, U.K.: Wiley & Sons, 2009.

Bojowald, Martin. *Once Before Time: A Whole Story of the Universe*. New York: Alfred A. Knopf, 2010.

Börner, Katy. *Atlas of Science: Visualizing What We Know*. Cambridge, Mass.: MIT Press, 2010.

Borovik, Alexandre V. *Mathematics under the Microscope: Notes on Cognitive Aspects of Mathematical Practice*. Providence, R.I.: American Mathematical Society, 2010.

Brown, Jason I. *Our Days Are Numbered: How Mathematics Orders Our Lives.* Toronto, Canada: McClelland & Stewart, 2009.

Buchan, Jamie. *Easy as Pi: The Countless Ways We Use Numbers Every Day*. Pleasantville, N.Y.: Reader's Digest, 2010.

Bultheel, Adhemar, and Ronald Cools. (Eds.) *The Birth of Numerical Analysis*. Singapore: World Scientific, 2010.

Cahn, Steven M., Robert B. Talisse, and Scott F. Aikin. (Eds.) *Thinking about Logic: Classic Essays.* Boulder, Colo.: Westview Press, 2010.

Carroll, Lewis. *The Pamphlets of Lewis Carroll: The Mathematical Pamphlets of Charles Lutwidge Dodgson and Related Pieces*, edited by Francine F. Abeles. Charlottesville, Va.: University of Virginia Press, 2010.

Ceci, Stephen J., and Wendy M. Williams. *The Mathematics of Sex: How Biology and Society Conspire to Limit Talented Women and Girls*. Oxford, U.K.: Oxford University Press, 2010.

Centrone, Stefania. *Logic and Philosophy of Mathematics in the Early Husserl*. New York: Springer Science + Business Media, 2010.

Charpentier, Éric, Étienne Ghys, and Annick Lesne. (Eds.) *The Scienctific Legacy of Poincare*. Providence, R.I.: American Mathematical Society, 2010.

Clark, Colin W. *Mathematical Bioeconomics: The Mathematics of Conservation*. New York: John Wiley and Sons, 2010.

Comins, Neil F. *What If the Earth Had Two Moons?* New York: St. Martin's, 2010.

Constanda, Christian. *Dude, Can You Count? Stories, Challenges, and Adventures in Mathematics.* London, U.K.: Springer Verlag, 2009.

Coppin, Charles A., Ted W. Mahavier, Lee E. May, and Edgar G. Parker. *The Moore Method: A Pathway to Learner-Centered Instruction*. Washington, D.C.: Mathematical Association of America, 2010.

Dahan-Dalmédico, Amy, and Jeanne Peiffer. *History of Mathematics: Highways and Byways.* Washington, D.C.: Mathematical Association of America, 2010.

du Sautoy, Marcus. *Symmetry: A Journey into the Patterns of Nature*. New York: HarperCollins, 2009.

Easley, David, and Jon Kleinberg. *Networks, Crowds, and Markets: Reasoning about a Highly Connected World*. New York: Cambridge University Press, 2010.

Engquist, Björn, and Wilfried Schmid. (Eds.) *Mathematics Unlimited: 2001 and Beyond*. New York: Springer, 2001.

Farid, Mohammed M. (Ed.) *Mathematical Modeling of Food Processing*. Boca Raton, Fla.: Taylor and Francis, 2010.

Fenstad, Jens Erik. *Grammar, Geometry, and Brain*. Stanford, Calif.: Center for the Study of Language, 2010.

Fung, Kaiser. *Numbers Rule Your World: The Hidden Influence of Probabilities and Statistics on Everything You Do*. New York: McGraw Hill, 2010.

Giordano, Paolo. *The Solitude of Prime Numbers*. New York: Viking, 2010.

Goulding, Robert. *Defending Hypatia: Ramus, Savile, and the Renaissance Rediscovery of Mathematical History*. New York: Springer Science+Business Media, 2010.

Gowers, Timothy. (Ed.) *Princeton Companion to Mathematics*. Princeton, N.J.: Princeton University Press, 2008.

Greer, Brian, et al. (Eds.) *Culturally Responsive Mathematics Education*. New York: Routledge, 2010.

Hare, Warren, et al. *Modelling in Healthcare*. Providence, R.I.: American Mathematical Society, 2010.

Hart, W. D. *The Evolution of Logic*. Cambridge, U.K.: Cambridge University Press, 2010.

Heilbron, J. L. *Galileo*. Oxford, U.K.: Oxford University Press, 2010.

Hersh, Reuben. (Ed.) *18 Unconventional Essays on the Nature of Mathematics*. New York: Springer, 2006.

Hersh, Reuben, and Vera John-Steiner. *Loving and Hating Mathematics: Challenging the Myths of Mathematical Life*. Princeton, N.J.: Princeton University Press, 2010.

Hitt, Fernando, Derek Holton, and Patrick W. Thompson. (Eds.) *Research in Collegiate Mathematics Education, VII*. Providence, R.I.: American Mathematical Society, 2010.

Huber, Mark, and Gizem Karaali. *Journal of Humanistic Mathematics*, online only, Vol. 1, January 2011. http://scholarship.Claremont.edu/jhm/.

Ivanov, Oleg A. *Making Mathematics Come to Life: A Guide for Teachers and Students*. Providence, R.I.: American Mathematical Society, 2010.

Jacquette, Dale. *Logic and How It Gets That Way*. Montreal: McGill-Queen's University Press, 2010.

Kaplan, Robert, and Ellen Kaplan. *Hidden Harmonies: The Lives and Times of the Pythagorean Theorem*. New York: Bloomsbury Press, 2011.

Keidel, Robert W. *The Geometry of Strategy: Concepts for Strategic Management*. New York: Routledge, 2010.

Krantz, Steven G. *An Episodic History of Mathematics: Mathematical Culture through Problem Solving*. Washington, D.C.: Mathematical Association of America, 2010.

Le Lionnaise, F. (Ed.) *Great Currents of Mathematical Thought*. Two volumes translating the 1962 French original *Les Grand Courants de la Pensée Mathématique*. Mineola, N.Y.: Dover Publications, 2004.

Leng, Mary. *Mathematics and Reality*. Oxford, U.K.: Oxford University Press, 2010.

Leong, Yu Kiang. *Creative Minds, Charmed Lives: Interviews at the Institute for Mathematical Sciences, National University of Singapore*. Singapore: World Scientific, 2010.

Lerman, Stephen, and Brent Davis. (Eds.) *Mathematical Action and Structures of Noticing*. Rotterdam, Netherlands: Sense Publishers, 2009.

Lester, Frank. (Series Editor) *Teaching and Learning Mathematics*. Several volumes. Reston, Va.: National Council of Teachers of Mathematics, 2010.

Mackenzie, Dana. *What's Happening in the Mathematical Sciences*. Vol 8. Providence, R.I.: American Mathematical Society, 2011.

Martineau, John. (Ed.) *Quadrivium: Number, Geometry, Music, Heaven*. New York: Bloomsbury Publishing, 2010.

Martinez, Alberto A. *Kinematics: The Lost Origins of Einstein's Relativity*. Baltimore, Md.: Johns Hopkins University Press, 2009.

Mazur, Joseph. *What's Luck Got to Do with It? The History, Mathematics, and Psychology of the Gambler's Illusion*. Princeton, N.J.: Princeton University Press, 2010.

McKellar, Danica. *Hot X: Algebra Exposed*. New York: Penguin Books, 2010.

McShea, Daniel, and Robert N. Brandon. *Biology's First Law: The Tendency for Diversity and Complexity to Increase in Evolutionary Systems*. Chicago: University of Chicago Press, 2010.

Mosteller, Frederick. *The Pleasures of Statistics. The Autobiography of Frederick Mosteller*. New York: Springer Science+Business Media, 2010.

Nahin, Paul J. *Mrs. Perkins's Electric Quilt and Other Intriguing Stories of Mathematical Physics*. Princeton, N.J.: Princeton University Press, 2010.

Neuenschwander, Dwight E. *Emmy Noether's Wonderful Theorem*. Baltimore, Md.: Johns Hopkins University Press, 2010.

Newman, James R. (Ed.) *The World of Mathematics: A Small Library of the Literature of Mathematics*, 4 volumes. New York: Simon and Schuster, 1956.

Nickerson, Raymond S. *Mathematical Reasoning: Patterns, Problems, Conjectures, and Proofs*. New York: Taylor and Francis Group, 2010.

Niederman, Derrick. *Number Freak: From 1 to 200, the Hidden Language of Numbers Revealed*. New York: Perigee, 2009.

O'Leary, Michael. *Revolutions of Geometry*. Hoboken, N.J.: John Wiley and Sons, 2010.

Olofsson, Peter. *Probabilities: The Little Numbers that Rule Our Lives*. New York: Wiley & Sons, 2010.

Olson, Richard G. *Technology and Science in Ancient Civilizations*. Santa Barbara, Calif.: ABC-CLIO, 2010.

Ouellette, Jennifer. *The Calculus Diaries: How Math Can Help You Lose Weight, Win in Vegas, and Survive a Zombie Apocalypse*. London, U.K.: Penguin, 2010.

Pappus of Alexandria. *Book 4 of the Collection*. Edited with translation and commentary by Heike Sefrin-Weis. Heidelberd, Germany: Springer Verlag, 2010.

Patterson, Scott. *The Quants: How a Small Band of Math Wizards Took Over Wall Street and Nearly Destroyed It*. New York: Crown, 2010.

Peirce, Charles S. *Philosophy of Mathematics: Selected Writings*. Edited by Matthew E. Moore. Bloomington, Ind.: Indiana University Press, 2010.

Phillips, Linda M., Stephen P. Norris, and John S. Macnab. (Eds.) *Visualization in Mathematics, Reading, and Science Education*. New York: Springer Verlag, 2010.

Posamentier, Alfred S. *The Pythagorean Theorem: The Story of Its Power and Beauty*. Amherst, N.Y.: Prometheus Books, 2010.

Reid, David A., and Christine Knipping. *Proof in Mathematics Education*. Rotterdam, Netherlands: Sense Publishers, 2010.

Rudman, Peter S. *The Babylonian Theorem: The Mathematical Journey to Pythagoras and Euclid*. Amherst, N.Y.: Prometheus Books, 2010.

Saul, Mark, Susan Assouline, and Linda Jensen Sheffield. *The Peak in the Middle: Developing Mathematically Gifted Students in the Middle Grades*. Reston, Va.: National Council of Teachers of Mathematics, 2010.

Schaaf, William L. (Ed.) *Mathematics, Our Great Heritage: Essays on the Nature and Cultural Significance of Mathematics*. New York: Harper, 1948.

Scheffer, Marten. *Critical Transitions in Nature and Society*. Princeton, N.J.: Princeton University Press, 2009.

Schielack, Jane F. (series adviser) *Teaching with Curriculum Focal Points*, various grades. Reston, Va.: National Council of Teachers of Mathematics, 2010.

Segel, Joel. (Ed.) *Recountings: Conversations with MIT Mathematicians*. Wellesley, Mass.: A. K. Peters, 2009.

Seife, Charles. *Proofiness: The Dark Arts of Mathematical Deception*. New York: Viking, 2010.

Seshadri, C. S. (Ed.) *Studies in the History of Indian Mathematics*. New Delhi, India: Hindustan Book Agency, 2010.

Shaughnessy, Michael J., Beth Chance, and Henry Kranendonk. *Focus in High School Mathematics: Reasoning and Sense Making*. Reston, Va.: National Council of Teachers of Mathematics, 2010.

Simoson, Andrew. *Voltaire's Riddle: Micromegas and the Measure of All Things*. Washington, D.C.: Mathematical Association of America, 2010.

Small, Marian. *Good Questions: Great Ways to Differentiate Mathematics Instruction*. New York: Teachers College Press, 2010.

Sriraman, Bharath, and Lyn English. (Eds.) *Theories of Mathematics Education: Seeking New Frontiers*. Berlin, Germany: Springer-Verlag Business + Media, 2010.

Stahl, Saul. *Geometry from Euclid to Knots*. Mineola, N.Y.: Dover Publications, 2010.

Steen, Lynn Arthur. (Ed.) *Mathematics Today: Twelve Informal Essays*. New York: Springer Verlag, 1978.

Stein, Mary Kay, and Linda Kucan. (Eds.) *Instructional Explanations in the Disciplines*. New York: Springer Science+Business Media, 2010.

Stewart, Ian. *Cows in the Maze and Other Mathematical Explorations*. Oxford, U.K.: Oxford University Press, 2010.

Stillwell, John. *Mathematics and Its History*, 3rd edition. New York: Springer Science+Business Media, 2010.

______. *Roads to Infinity: The Mathematics of Truth and Proof*. Natick, Mass.: A. K. Peters, 2010.

Szpiro, George G. *A Mathematical Medley: Fifty Easy Pieces on Mathematics*. Providence, R.I.: American Mathematical Society, 2010.

______. *Numbers Rule: The Vexing Mathematics of Democracy, from Plato to the Present*. Princeton, N.J.: Princeton University Press, 2010.

Vickers, Andrew. *What Is a p-Value Anyway? 34 Stories to Help You Actually Understand Statistics*. Boston: Addison-Wesley, 2010.

Wardhaugh, Benjamin. *How to Read Historical Mathematics*. Princeton, N.J.: Princeton University Press, 2010.

West, Bruce J., and Nicola Scafetta. *Disrupted Networks: From Physics to Climate Change*. Singapore: World Scientific, 2010.

Winston, Joe. *Beauty and Education*. New York: Routledge, 2010.

Yau, Shing-Tung, and Steve Nadis. *The Shape of Inner Space: String Theory and the Geometry of the Universe's Hidden Dimensions*. New York: Basic Books, 2010.

What Is Mathematics For?

Underwood Dudley

A more accurate title is "What is mathematics education for?" but the shorter one is more attention-getting and allows me more generality. My answer will become apparent soon, as will my answer to the subquestion of why the public supports mathematics education as much as it does.

So that there is no confusion, let me say that by "mathematics" I mean algebra, trigonometry, calculus, linear algebra, and so on: all those subjects beyond arithmetic. There is no question about what arithmetic is for or why it is supported. Society cannot proceed without it. Addition, subtraction, multiplication, division, percentages: though not all citizens can deal fluently with all of them, we make the assumption that they can when necessary. Those who cannot are sometimes at a disadvantage.

Algebra, though, is another matter. Almost all citizens can and do get through life very well without it, after their schooling is over. Nevertheless it becomes more and more pervasive, seeping down into more and more eighth-grade classrooms and being required by more and more states for graduation from high school. There is unspoken agreement that everyone should be exposed to algebra. We live in an era of universal mathematical education.

This is something new in the world. Mathematics has not always loomed so large in the education of the rising generation. There is no telling how many children in ancient Egypt and Babylon received training in numbers, but there were not many. Of course, in ancient civilizations education was not for everyone, much less mathematical education. Literacy was not universal, and I suspect that many who could read and write could not subtract or multiply numbers. The ancient Greeks, to their glory, originated real mathematics, but they did not do it to fill classrooms with students learning how to prove theorems. Compared to them, the ancient Romans were a mathematical blank. The Arab scholars who started to develop algebra after the fall of Rome were doing it for their own pleasure and not as something intended for the masses. When Brahmagupta was solving Pell's equation a millennium before Pell was born, he did not have students in mind.

Of course, you may think, those were the ancients; in modern times we have learned better, and arithmetic at least has always been part of everyone's schooling. Not so. It may come as a surprise to you, as it did to me, that arithmetic was not part of elementary education in the United States in the colonial period. In *A History of Mathematics Education in the United States and Canada* (National Council of Teachers of Mathematics, 1970), we read

> Until within a few years no studies have been permitted in the day school but spelling, reading, and writing. Arithmetic was taught by a few instructors one or two evenings a week. But in spite of the most determined opposition, arithmetic is now being permitted in the day school.

Opposition to arithmetic! *Determined* opposition! How could such a thing be? How could society function without a population competent in arithmetic? Well, it did, and it even thrived. Arithmetic was indeed needed in many occupations, but those who needed it learned it on the job. It was a system that worked with arithmetic then and that can work with algebra today.

Arithmetic did make it into the curriculum, but, then as now, employers were not happy with what the schools were turning out. Patricia Cline Cohen, in her estimable *A Calculating People: The Spread of Numeracy in Early America* (U. of Chicago Press, 1983; Routledge paperback, 1999) tells us that

> Prior to this act [1789] arithmetic had not been required in the Boston schools at all. Within a few years a group of Boston businessmen protested to the School committee that the pupils taught by the method of arithmetic instruction then in use were totally unprepared for business. Unfortunately, the educators in this case insisted that they were doing an adequate job and refused to make changes in the program.

Both sides were right. It is impossible to prepare everyone for every possible occupation, and it is foolish to try. Hence many school leavers will be unprepared for many businesses. But mathematics teachers, then as now, were doing an adequate job.

A few years ago I was at a meeting that had on its program a talk on the mathematics used by the Florida Department of Transportation. There is quite a bit. For example, the Florida DoT uses Riemann sums to determine the area of irregular plots of land, though it does not call the sums that. After the talk I asked the speaker what mathematical preparation the DoT expects in its new hires. The answer was, none at all. The DoT has determined that it is best for all concerned to assume that the background of its employees includes nothing beyond elementary arithmetic. What employees need, they can learn on the job.

There seems to be abroad in the land the delusion that skill in algebra is necessary in the world of work and in everyday life. In *Moving Beyond Myths* (National Academy of Sciences, 1991) we see

> *Myth:* Most jobs require little mathematics.
> *Reality:* The truth is just the opposite.

I looked very hard in the publication for evidence for that assertion, but found none. Perhaps the NAS was equating mathematics with arithmetic. Many people do this, as I have found in asking them about how, or if, they use mathematics. Almost always, the "mathematics" they tell me about is material that appears in the first eight grades of school.

Algebra, though, is mentioned explicitly in *Everybody Counts* (National Research Council, 1989):

> Over 75 percent of all jobs require proficiency in simple algebra and geometry, either as a prerequisite to a training program or as part of a licensure examination.

I find that statement extraordinary. I will take my telephone Yellow Pages, open it at random, and list in order the first eight categories that I see:

> Janitor service, Janitors' equipment and supplies, Jewelers, Karate and other martial arts, Kennels, Labeling, Labor organizations, Lamps and lamp shades.

In which six is algebra required, even for training or license? I again looked very hard for evidence in the NRC's publication but couldn't find any.

It may be that no evidence is presented because none is needed: everybody knows that algebra is needed for all sorts of jobs. For example, there was an algebra book whose publisher advertised that it contained

> "Career Applications"—Includes explanations, examples, exercises, and answers for work in electronics; civil/chemical engineering; law enforcement; nursing; teaching; and more. Shows students the relationship of chapter concepts and job skills—with applications developed through interviews and market research in the workplace that ensure relevance.

Of course I requested an examination copy, and the publisher graciously sent me one. To return the favor, I will refrain from naming the publisher or the author. The career applications were along the lines of

> In preparation for the 2002 Winter Olympic Games in Salt Lake City, several people decide to pool their money and share equally the $12,000 expense of renting a four-bedroom house in Salt Lake City for two weeks. The original number of people who agreed to share the house changed after two people dropped out of the deal because they thought the house was too small. Those left in the deal must now pay an additional $300 each for the rental. How many people were left?

Exactly what career this applied to was not specified. Nor was it mentioned that the best way to solve this problem is to find a member of the group and ask. The answer should be forthcoming. If the person's reply is the conundrum in the text, the member of the group should be beaten about the head until he or she promises to behave in a more civilized manner.

This is not to say that the problem is not a good one. It is a good one, a very good one, and one that students should try to solve. Students should be made to solve many word problems, the more the better. The reason for solving them, though, is not that they will arise in their careers.

Another text, whose author and publisher I will not name—alas, still in print in its third edition—asserts

> This text aims to show that mathematics is useful to virtually everyone. I hope that users will complete the course with greater confidence in their ability to solve practical problems.

Here is one of the practical problems:

> An investment club decided to buy \$9,000 worth of stock with each member paying an equal share. But two members left the club, and the remaining members had to pay \$50 more apiece. How many members are in the club?

Do you detect the similarity to the career application in the first text? The two problems are the same, with different numbers. The second is not practical, any more than the first comes up as part of a job.

The reason that this problem—well worth doing by students—appears in more than one text is that it is a superb problem, so superb that it has been appearing in texts for hundreds of years, copied from one author to another. If you want a problem that makes students solve a quadratic equation, here it is.

I keep looking for uses of algebra in jobs, but I keep being disappointed. To be more accurate, I used to keep looking until I became convinced that there were essentially none. For this article I searched again and found a website that promised applications of "college algebra" to the workplace. The first was

> You are a facilities manager for a small town. The town contains approximately 400 miles of road that must be plowed following a significant snowfall. How many plows must be used in order to complete the job in one day if the plows can travel at approximately 7 miles per hour when engaged?

This is another textbook "application" made up, I think, by its writer with no reference to external reality. (It's a big small town that has 400 miles of streets.) The facilities manager knows how many plows there are and can estimate how many more, if any, are needed. The next problem, I think, did arise outside of the head of a textbook writer:

> How much ice cream mix and vanilla flavor will it take to make 1000 gallons of vanilla ice cream at 90% overrun with the vanilla flavor usage rate at 1 oz. per 10 gallon mix? (90% overrun means that enough air is put into the frozen mix to increase its volume by 90%.)

Though dressed up with x's and y's, the solution amounts to calculating that you need $1{,}000/(1 + 0.9) = 526.3$ gallons of mix to puff up into

1,000 gallons of ice cream, so you will need 526.3/10 = 52.6 ounces of flavor.

The employee adding the flavor will not need algebra, nor will he or she need to think through this calculation. There will be a formula, or rule, that gives the result, and that is what happens on the job. Problems that arise on the job will be for the most part problems that have been solved before, so new solutions by workers will not be needed.

I am glad that we do not have to depend on workers' ability to solve algebra problems to get through the day because, as every teacher of mathematics knows, students don't always get problems right. The chair of the department of a Big Ten university once observed, probably after a bad day, that it was possible for a student to graduate with a mathematics major without ever having solved a single problem correctly. Partial credit can go a long way. This was in the 1950s, looked on by many as a golden age of mathematics education.

In one of those international tests of mathematical achievement appeared the problem of finding which of two magazine subscriptions was cheaper: 24 issues with (a) the first four issues free and $3 each for the remainder or (b) the first six issues free and $3.50 each for the remainder. This is not a tough problem, so I leave its solution to you. As easy as it is, only 26% of United States eighth-graders could do it correctly. That percentage was above the international average of 24%. Even the Japanese eighth-graders could manage only 39%. No doubt when the eighth-graders become adults they will be better at solving such problems, but even so I do not want them having to solve problems that when solved incorrectly can do me harm.

Though people know that they do not use algebra every day, or even every month, many seem to think that there are hosts of others who do. Perhaps they have absorbed the textbook writers' insistence on the "real world" uses of algebra, even though the texts actually demonstrate that there are none. Were uses of algebra widespread in the world of work, all textbook writers would have to do is to ask a few people about their last applications of algebra, turn them into problems, and put them in their texts. If 75% of all jobs required algebra, they could get a problem from three of every four people they ask. However, such problems do not appear in the texts. We get instead the endlessly repeated problems about investment clubs losing two members and all of the other chestnuts, about cars going from A to B and farmers fencing fields and so on, that I lack the space to display. The reason that problems drawn from everyday life do not appear in the texts is not that textbook authors lack energy and initiative; it is that they do not exist.

Though they may not use algebra themselves, people are solidly behind having everyone learn algebra. Tom and Ray Magliozzi, the brothers who are hosts of National Public Radio's popular "Car Talk" program, like to pose as vulgarians when they are actually nothing of the kind. On one program, brother Tom made some remarks against teaching geometry and trigonometry in high school. I doubt very much that he was serious. Whether he was serious or not does not affect the content of his remarks or the reaction of listeners. The reaction was unanimous endorsement of mathematics. When mathematics is attacked, people leap to its defense.

In his piece Tom alleged that he had an octagonal fountain in his backyard that he wanted to surround with a border and that he needed to calculate the length of the side of the concentric octagon. After succeeding, using, he said, the Pythagorean theorem, he reflected

> That this was maybe the second time in my life—maybe the first—that I had occasion to use the geometry and trigonometry that I had learned in high school. Furthermore, I had never had occasion to use the higher mathematics that the high school math had prepared me for.
>
> *Never!* Why did I—and millions of other students—spend valuable educational hours learning something that we would never use? Is this education? Learning skills that we will never need?

After some real or pretended populism ("The people who run the education business are money-grubbing, self-serving morons"), he concluded that

> The purpose of learning math, which most of us will never use, is only to prepare us for further math courses—which we will use even less frequently than never.

There were answers, quite a few of them, posted at the "Car Talk" website. All disagreed with Tom's conclusion, which actually has elements of truth. (A reply that started with "I agree" might be thought to be a counterexample, but the irony that followed was at least as heavy as lead.) One response included

> Perhaps you've had only one opportunity to use geometry in your life, but there are a number of occupations in which it's a must. Myself, I'm pleased that my house was designed and built by people who were capable of calculating the correct rise of a roof for proper drainage or the number of cubic feet of concrete needed for a strong foundation.

Here is the common error of supposing that problems once solved must be solved anew every time they are encountered. House builders have handbooks and tables, and use them. Indeed, houses, as well as pyramids and cathedrals, were being built long before algebra was taught in the schools and, in fact, before algebra.

Another common misconception occurs in another response:

> You sure laid a big oblate spheroid shaped one when you went on your tirade against having to learn geometry, trigonometry and other things mathematical.
>
> Who uses this stuff? Geologists, aircraft designers, road builders, building contractors, surgeons and, yes, even radio broadcast technicians (amplitude modulation and frequency modulation are both based on manipulating wave forms described by trig functions—don't get me started on alternating current).
>
> So, Tommy, get a life. The only people who don't use these principles every day are those who can't do and can't teach, and thus are suited only for lives as politicians or talk show hosts.

People seem to think that because something involves mathematics it is necessary to know mathematics to use it. Radio does indeed involve sines and cosines, but the person adjusting the dials needs no trigonometry. Geologists searching for oil do not have to solve differential equations, though differential equations may have been involved in the creation of the tools that geologists use.

I am not saying that mathematics is never required in the workplace. Of course it is, and it has helped to make our technology what it is. However, it is needed very, very seldom, and we do not need to train millions of students in it to keep businesses going. Once, when I was an employee of the Metropolitan Life Insurance Company, I was given an annuity rate to calculate. Back then, insurance companies had rate books, but now and then there was need for a rate not in the book. Using my knowledge of the mathematics of life contingencies, I calculated the rate. When I gave it to my supervisor he said, "No, no, that's not right. You have to do it *this* way." "But," I said, "that's three times as much work." Yes, I was told, but that's the way that we calculate rates. My knowledge of life contingencies got in the way of the proper calculation, done the way it had been done before, which any minimally competent employee could have carried out.

It may be that there could arise, say, a partial differential equation that some company needed to solve, the likes of which it had never seen before. If so, there are plenty of mathematicians available to do the job. They'd work cheap, too.

Jobs do not require algebra. I have expressed this truth many times in talks to any group who would listen, and it was not uncommon for a member of the audience to tell me, after the talk or during it, that I was wrong and that he used algebra or calculus in his job all the time. It always turned out that he used the mathematics because he wanted to, not because he had to.

Even those who are not burdened with the error that algebra is necessary to hold many jobs support the teaching of algebra. Everyone supports the teaching of algebra. The public wants more mathematics taught, to more students. The requirements keep going up, never down.

The reason for this, I am convinced, is that the public knows, or senses, that mathematics develops the power to reason. It shows, better than any other subject, how reason can lead to truth. Of course, other sciences exhibit the power of reason, but there's all that overhead—ferrous and ferric, dynes and ergs—that has to be dealt with. In mathematics, there is nothing standing between the problem and the reasoning.

Economists reason as well, but sometimes two economists reason to two different conclusions. Philosophers reason, but never come to *any* conclusion. In mathematics, problems can be solved using reason, and the solutions can be checked and shown to be correct. Reasoning needs to be learned, and mathematics is the best way to learn it.

People grasp this, perhaps not consciously, and hence want their children to undergo mathematics. Many times people have told me that they liked mathematics (though they call it "math") because it was so definite and it was satisfying to get the right answer. Have you not heard the same thing? They liked being able to reason correctly. They knew that the practice was good for them. No one has ever said to me, "I liked math because it got me a good job."

We no longer have the confidence in our subject that allows us to say that. We justify mathematics on its utility in the world of getting and spending. Our forebears were not so diffident. In 1906 J. D. Fitch wrote

> Our future lawyers, clergy, and statesmen are expected at the University to learn a good deal about curves, and angles, and number and proportions; not because these subjects have the smallest relation to the needs of their lives, but because in the very act of learning them they are likely to acquire that habit of steadfast and accurate thinking, which is indispensable in all the pursuits of life.

I do not know who J. D. Fitch was, but he was correct. Thomas Jefferson said

> Mathematics and natural philosophy are so peculiarly engaging and delightful as would induce everyone to wish an acquaintance with them. Besides this, the faculties of the mind, like the members of a body, are strengthened and improved by exercise. Mathematical reasoning and deductions are, therefore, a fine preparation for investigating the abstruse speculations of the law.

In 1834, the Congressional Committee on Military Affairs reported

> Mathematics is the study which forms the foundation of the course [at West Point]. This is necessary, both to impart to the mind that combined strength and versatility, the peculiar vigor and rapidity of comparison necessary for military action, and to pave the way for progress in the higher military sciences.

Here is testimony from a contemporary student:

> The summer after my freshman year I decided to teach myself algebra. At school next year my grades improved from a 2.6 gpa to a 3.5 gpa. Tests were easier and I was much more efficient when taking them and this held true in all other facets of my life. To sum this up: algebra is not only mathematical principles, it is a philosophy or way of thinking, it trains your mind and makes otherwise

complex and overwhelming tests seem much easier both in school and in life.

Anecdotal evidence to be sure, but then all history is a succession of anecdotes.

That is what mathematics education is for and what it has always been for: to teach reasoning, usually through the medium of silly problems. In the Rhind Papyrus, that Egyptian textbook of mathematics c. 1650 BC, we find

> Give 100 loaves to five men so that the shares are in arithmetic progression and the sum of the two smallest is 1/7 of the three greatest.

The ancient Egyptians were a practical people, but even so this eminently unpractical problem was thought to be worth solving. (The shares are 1 2/3, 10 5/6, 20, 29 1/6, and 38 1/3.) George Chrystal's *Algebra* (1886) has on page 154 more than fifty problems, all with the instruction "Simplify", including

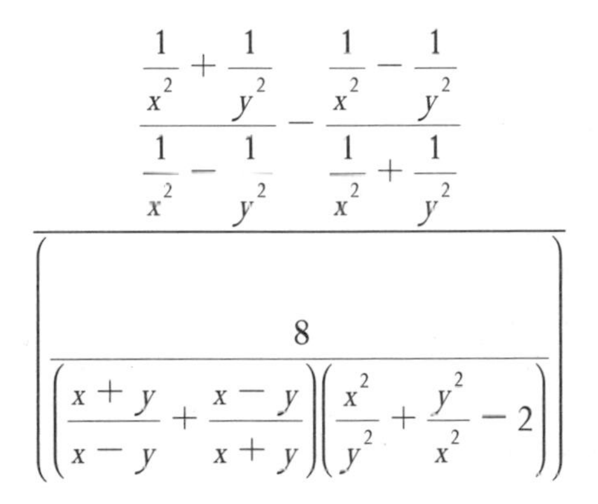

$$\frac{\dfrac{\frac{1}{x^2}+\frac{1}{y^2}}{\frac{1}{x^2}-\frac{1}{y^2}}-\dfrac{\frac{1}{x^2}-\frac{1}{y^2}}{\frac{1}{x^2}+\frac{1}{y^2}}}{\left(\dfrac{8}{\left(\frac{x+y}{x-y}+\frac{x-y}{x+y}\right)\left(\frac{x^2}{y^2}+\frac{y^2}{x^2}-2\right)}\right)}$$

There is no reason given, anywhere in his text, why anyone would want to simplify such things. It was obvious. That is how algebra is learned. As for the reason for learning algebra, that was obvious as well, and it was not for jobs. (The answer to the problem—what fun Chrystal must have had in making it up—is −1.)

I am not so unrealistic as to advocate that textbook writers start to produce texts with titles like *Algebra, a Prelude to Reason*. That would not fly. We do not want to make unwilling students even more unwilling. We cannot go back to texts like Chrystal's. But could we perhaps *tone it down* a little? Can we be a little less insistent that mathematics is essential for earning a living?

What mathematics education is for is not for jobs. It is to teach the race to reason. It does not, heaven knows, always succeed, but it is the best method that we have. It is not the only road to the goal, but there is none better. Furthermore, it is worth teaching. Were I given to hyperbole, I would say that mathematics is the most glorious creation of the human intellect, but I am not given to hyperbole so I will not say that. However, when I am before a bar of judgment, heavenly or otherwise, and asked to justify my life, I will draw myself up proudly and say, "I was one of the stewards of mathematics, and it came to no harm in my care." I will not say, "I helped people get jobs."

A Tisket, a Tasket, an Apollonian Gasket

Dana Mackenzie

In the spring of 2007 I had the good fortune to spend a semester at the Mathematical Sciences Research Institute in Berkeley, an institution of higher learning that takes "higher" to a whole new extreme. Perched precariously on a ridge far above the University of California at Berkeley campus, the building offers postcard-perfect vistas of the San Francisco Bay, 1,200 feet below. That's on the west side. Rather sensibly, the institute assigned me an office on the east side, with a view of nothing much but my computer screen. Otherwise I might not have gotten any work done.

However, there was one flaw in the plan: Someone installed a screen-saver program on the computer. Of course, it had to be mathematical. The program drew an endless assortment of fractals of varying shapes and ingenuity. Every couple minutes the screen would go blank and refresh itself with a completely different fractal. I have to confess that I spent a few idle minutes watching the fractals instead of writing.

One day, a new design popped up on the screen *(see the first figure)*. It was different from all the other fractals. It was made up of simple shapes—circles, in fact—and unlike all the other screen savers, it had numbers! My attention was immediately drawn to the sequence of numbers running along the bottom edge: 1, 4, 9, 16 . . . They were the perfect squares! The sequence was 1 squared, 2 squared, 3 squared, and so on.

Before I became a full-time writer, I used to be a mathematician. Seeing those numbers awakened the math geek in me. What did they mean? And what did they have to do with the fractal on the screen? Quickly, before the screen-saver image vanished into the ether, I sketched it on my notepad, making a resolution to find out someday.

As it turned out, the picture on the screen was a special case of a more general construction. Start with three circles of any size, with each one touching the other two. Draw a new circle that fits snugly into

Figure 1. Numbers in an Apollonian gasket correspond to the curvatures or "bends" of the circles, with larger bends corresponding to smaller circles. The entire gasket is determined by the first four mutually tangent circles; in this case, two circles with bend 1 and two "circles" with bend 0 (and therefore infinite radius). The circles with a bend of zero look, of course, like straight lines. Image courtesy of Alex Kontorovich.

the space between them, and another around the outside enclosing all the circles. Now you have four roughly triangular spaces between the circles. In each of those spaces, draw a new circle that just touches each side. This creates 12 triangular pores; insert a new circle into each one of them, just touching each side. Keep on going forever, or at least until the circles become too small to see. The resulting foamlike structure is called an Apollonian gasket *(see the second figure)*.

Something about the Apollonian gasket makes ordinary, sensible mathematicians get a little bit giddy. It inspired a Nobel laureate to write a poem and publish it in the journal *Nature*. An 18th-century Japanese samurai painted a similar picture on a tablet and hung it in front

---

Figure 2. An Apollonian gasket is built up through successive "generations." For instance, in generation 1 *(top left)*, each of the lighter circles is inscribed in one of the four triangular pores formed by the dark circles. The complete gasket, whimsically named "bugeye" by Katherine Bellafiore Sanden, an undergraduate student of Peter Sarnak at Princeton University, has circles with bends −1 (for the largest circle that encloses the rest), 2, 2, and 3. The list of bends that appears in a given gasket (here, 2, 3, 6, 11, etc.) form a number sequence whose properties Sarnak would like to explain—but, he says, "the necessary mathematics has not been invented yet." Image courtesy of Katherine Bellafiore Sanden.

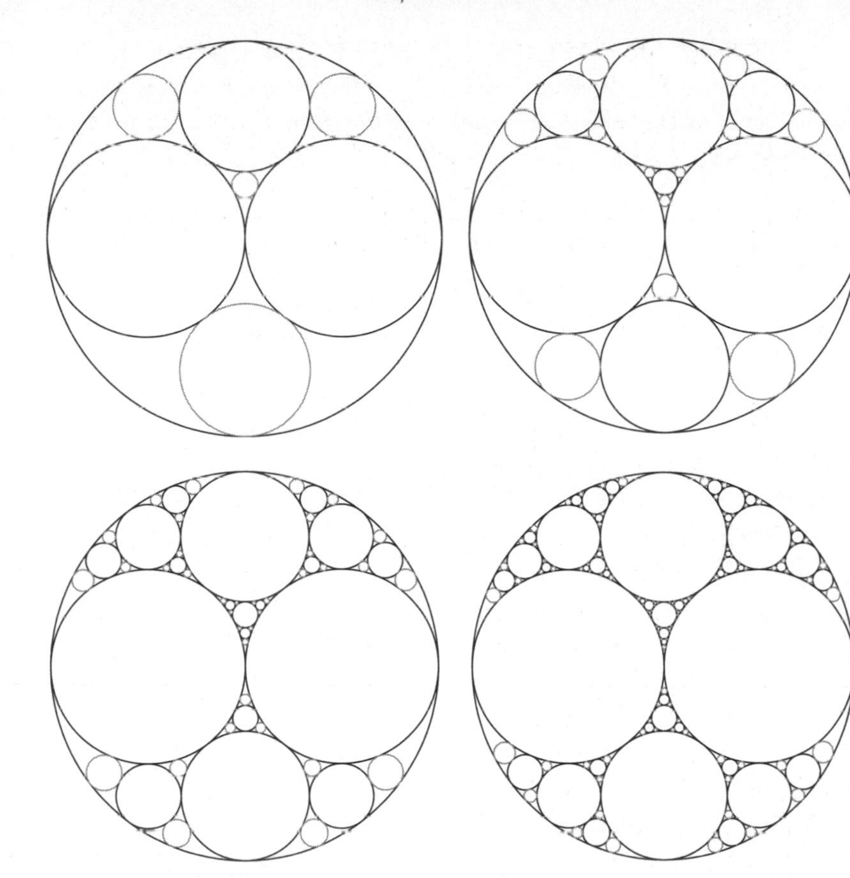

26
14
35
3
6
30
11
23
18
27
15
2

of a Buddhist temple. Researchers at AT&T Labs printed it onto T-shirts. And in a book about fractals with the lovely title *Indra's Pearls*, mathematician David Wright compared the gasket to Dr. Seuss's *The Cat in the Hat*:

> The cat takes off his hat to reveal Little Cat A, who then removes his hat and releases Little Cat B, who then uncovers Little Cat C, and so on. Now imagine there are not one but three cats inside each cat's hat. That gives a good impression of the explosive proliferation of these tiny ideal triangles.

Getting the Bends

Even the first step of drawing an Apollonian gasket is far from straightforward. Given three circles, how do you draw a fourth circle that is exactly tangent to all three?

Apparently the first mathematician to seriously consider this question was Apollonius of Perga, a Greek geometer who lived in the third century B.C. He has been somewhat overshadowed by his predecessor Euclid, in part because most of his books have been lost. However, Apollonius's surviving book *Conic Sections* was the first to systematically study ellipses, hyperbolas, and parabolas—curves that have remained central to mathematics ever since.

One of Apollonius' lost manuscripts was called *Tangencies*. According to later commentators, Apollonius apparently solved the problem of drawing circles that are simultaneously tangent to three lines, or two lines and a circle, or two circles and a line, or three circles. The hardest case of all was the case where the three circles are tangent.

No one knows, of course, what Apollonius' solution was, or whether it was correct. After many of the writings of the ancient Greeks became available again to European scholars of the Renaissance, the unsolved "problem of Apollonius" became a great challenge. In 1643, in a letter to Princess Elizabeth of Bohemia, the French philosopher and mathematician René Descartes correctly stated (but incorrectly proved) a beautiful formula concerning the radii of four mutually touching circles. If the radii are r, s, t, and u, then Descartes's formula looks like this:

$$1/r^2 + 1/s^2 + 1/t^2 + 1/u^2 = 1/2\,(1/r + 1/s + 1/t + 1/u)^2.$$

All of these reciprocals look a little bit extravagant, so the formula is usually simplified by writing it in terms of the *curvatures* or the *bends* of the circles. The curvature is simply defined as the reciprocal of the radius. Thus, if the curvatures are denoted by a, b, c, and d, then Descartes's formula reads as follows:

$$a^2 + b^2 + c^2 + d^2 = (a + b + c + d)^2/2.$$

As the third figure shows, Descartes's formula greatly simplifies the task of finding the *size* of the fourth circle, assuming the sizes of the first three are known. It is much less obvious that the very same equation can be used to compute the *location* of the fourth circle as well, and thus completely solve the drawing problem. This fact was discovered in the late 1990s by Allan Wilks and Colin Mallows of AT&T Labs, and Wilks used it to write a very efficient computer program for plotting Apollonian gaskets. One such plot went on his office door and eventually got made into the aforementioned T-shirt.

Descartes himself could not have discovered this procedure, because it involves treating the coordinates of the circle centers as complex numbers. Imaginary and complex numbers were not widely accepted by mathematicians until a century and a half after Descartes died.

In spite of its relative simplicity, Descartes's formula has never become widely known, even among mathematicians. Thus, it has been rediscovered over and over through the years. In Japan, during the Edo period, a delightful tradition arose of posting beautiful mathematics problems on tablets that were hung in Buddhist or Shinto temples, perhaps as an offering to the gods. One of these "Japanese temple problems," or *sangaku*, is to find the radius of a circle that just touches two circles and a line, which are themselves mutually tangent. This is a restricted version of the Apollonian problem, where one circle has infinite radius (or zero bend). The anonymous author shows that, in this case, $\sqrt{a} + \sqrt{b} = \sqrt{c}$, a sort of demented version of the Pythagorean theorem. This formula, by the way, explains the pattern I saw in the screensaver. If the first two circles have bends 1 and 1, then the circle between them will have bend 4, because $\sqrt{1} + \sqrt{1} = \sqrt{4}$. The next circle will have bend 9, because $\sqrt{1} + \sqrt{4} = \sqrt{9}$. Needless to say, the pattern continues forever. (This also explains what the numbers in the first figure mean. Each circle is labeled with its own bend.)

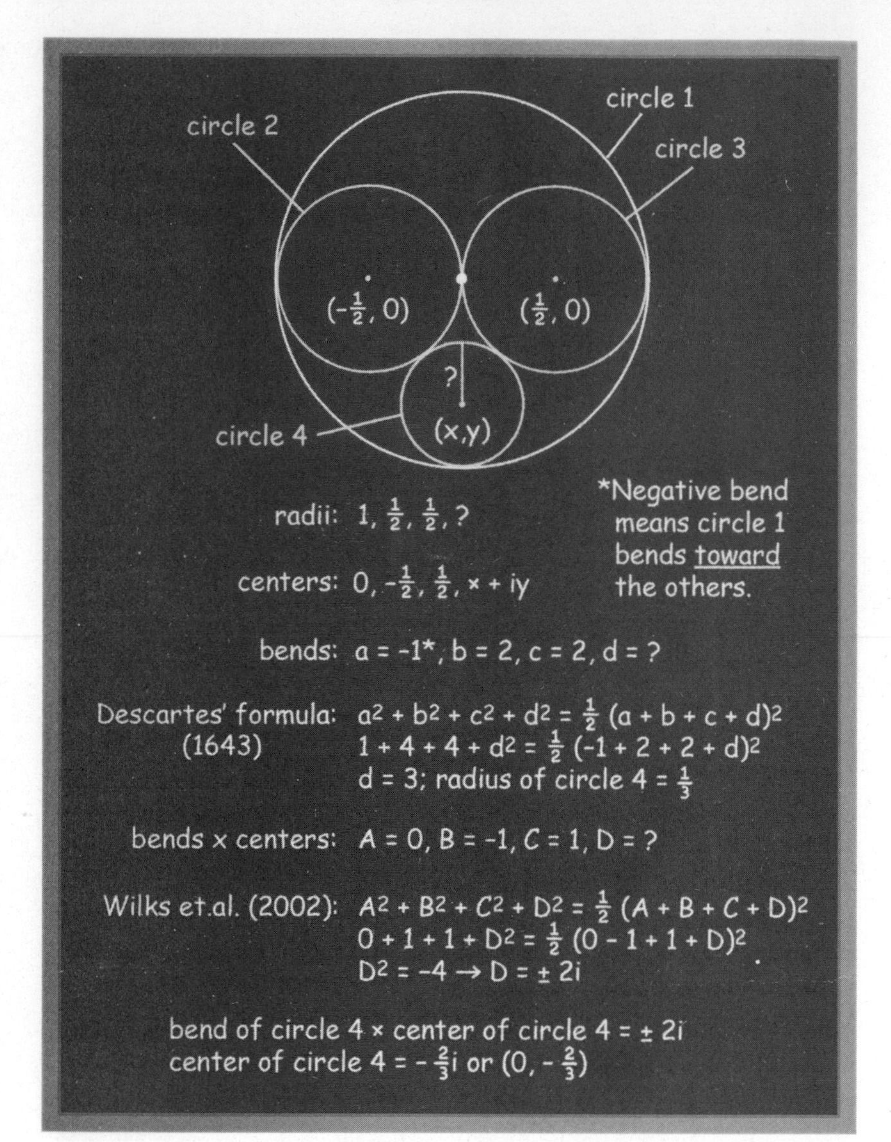

Figure 3. In 1643 René Descartes gave a simple formula relating the radii of any four mutually tangent circles. More than 350 years later, Allan Wilks and Colin Mallows noticed that the same formula relates the coordinates of the centers of the circles (expressed as complex numbers). Here Descartes's formula is used to find the radius and center of the fourth circle in the "bugeye" packing.

Apollonian circles experienced perhaps their most glorious rediscovery in 1936, when the Nobel laureate (in chemistry, not mathematics) Frederick Soddy became mesmerized by their charm. He published in *Nature* a poetic version of Descartes' theorem, which he called "The Kiss Precise":

> Four circles to the kissing come
> The smaller are the benter.
> The bend is just the inverse of
> The distance from the center.
> Though their intrigue left Euclid dumb
> There's now no need for rule of thumb
> Since zero bend's a dead straight line
> And concave bends have minus sign
> The sum of the squares of all four bends
> Is half the square of their sum.

Soddy went on to state a version for three-dimensional spheres (which he was also not the first to discover) in the final stanza of his poem.

Ever since Soddy's prosodic effort, it has become something of a tradition to publish any extension of his theorem in poetic form as well. The following year, Thorold Gosset published an n-dimensional version, also in *Nature*. In 2002, when Wilks, Mallows, and Lagarias published a long article in the *American Mathematical Monthly*, they ended it with a continuation of Soddy's poem entitled "The Complex Kiss Precise":

> Yet more is true: if all four discs
> Are sited in the complex plane,
> Then centers over radii
> Obey the self-same rule again.

(The authors note that the poem is to be pronounced in the Queen's English.)

A Little Bit of Gasketry

To this point I have only written about the very beginning of the gasket-making process—how to inscribe one circle among three given circles. However, the most interesting phenomena show up when you look at the gasket as a whole.

The first thing to notice is the foamlike structure that remains after you cut out all of the discs in the gasket. Clearly the disks themselves take up an area that approaches 100 percent of the area within the outer disk, and so the area of the foam (known as the "residual set") must be zero. On the other hand, the foam also has infinite length. Thus, in fact, it was one of the first known examples of a *fractal*—a curve of dimension between 1 and 2. Even today its dimension (denoted δ) is not known exactly; the best-proven estimate is 1.30568.

The concept of fractional dimension was popularized by Benoît Mandelbrot in his enormously influential book *The Fractal Geometry of Nature*. Although the meaning of dimension 1.30568 is somewhat opaque, this number is related to other properties of the foam that have direct physical meaning. For instance, if you pick any cutoff radius r, how many bubbles in the foam have radius larger than r? The answer, denoted $N(r)$, is roughly proportional to r^{δ}. Or if you pick the n largest bubbles, what is the remaining pore space between those bubbles? The answer is roughly proportional to $n^{1-2/\delta}$.

Physicists are very familiar with this sort of rule, which is called a *power law*. As I read the literature on Apollonian packings, an interesting cultural difference emerged between physicists and mathematicians. In the physics literature, a fractional dimension δ is *de facto* equivalent to a power law r^{δ}. However, mathematicians look at things through a sharper lens, and they realize that there can be additional, slowly increasing or slowly decreasing terms. For instance, $N(r)$ could be proportional to $r^{\delta}\log(r)$ or $r^{\delta}/\log(r)$. For physicists, who study foams empirically (or semiempirically, via computer simulation), the logarithm terms are absolutely undetectable. The discrepancy they introduce will always be swamped by the noise in any simulation. But for mathematicians, who deal in logical rigor, the logarithm terms are where most of the action is. In 2008, mathematicians Alex Kontorovich and Hee Oh of Brown University showed that there are in fact no logarithm terms in $N(r)$. The number of circles of radius greater than r obeys a strict power law, $N(r) \sim Cr^{\delta}$, where C is a constant that depends on the first three circles of the packing. For the "bugeye" packing illustrated in the second figure, C is about 0.201. (The tilde (~) means that this is not an *equation* but an *estimate* that becomes more and more accurate as the radius r decreases to 0.) For mathematicians, this was a major advance. For physicists, the likely reaction would be, "Didn't we know that already?"

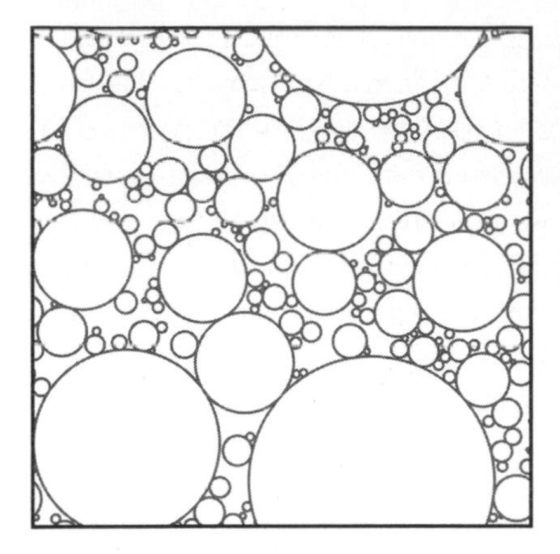

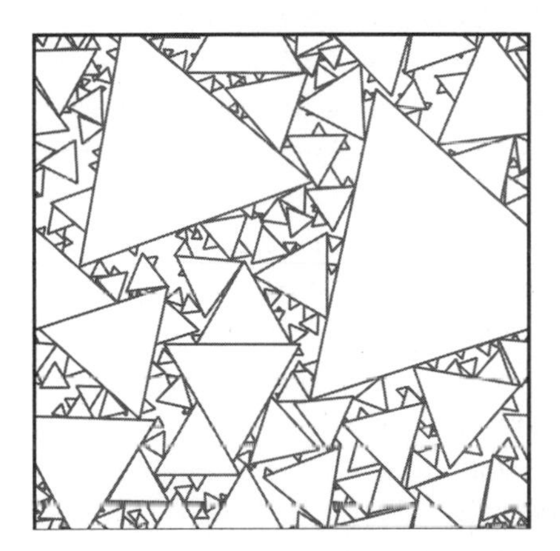

Figure 4. Physicists study random Apollonian packings as a model for foams or powders. In these simulations, new bubbles or grains nucleate in a random place and grow, either with rotation or without, until they encounter another bubble or grain. Different geometries for the bubbles or grains, and different growth rules, lead to different values for the dimension of the residual set—a way of measuring the efficiency of the packing. Image courtesy of Stefan Hutzler and Gary Delaney. First published in Delaney GW, Hutzler S and Aste T (2008), Relation Between Grain Shape and Fractal Properties in Random Apollonian Packing with Grain Rotation http://dx.doi.org/10.1103/PhysRevLett.101.120602, Phys. Rev. Lett., 101, 120602.

Random Packing

For many physical problems, the classical definition of the Apollonian gasket is too restrictive, and a random model may be more appropriate. A bubble may start growing in a randomly chosen location and expand until it hits an existing bubble, and then stop. Or a tree in a forest may grow until its canopy touches another tree, and then stop. In this case, the new circles do not touch three circles at a time, but only one. Computer simulations show that these "random Apollonian packings" still behave like a fractal, but with a different dimension. The empirically observed dimension is 1.56. (This means the residual set is larger, and the packing is less efficient, than in a deterministic Apollonian gasket.) More recently, Stefan Hutzler of Trinity College Dublin, along with Gary Delaney and Tomaso Aste of the University of Canberra, studied the effect of bubbles with different shapes in a random Apollonian packing. They found, for example, that squares become much more efficient packers than circles if they are allowed to rotate as they grow, but surprisingly, triangles become only slightly more efficient. As far as I know, all of these results are begging for a theoretical explanation.

For mathematicians, however, the classical, deterministic Apollonian gasket still offers more than enough challenging problems. Perhaps the most astounding fact about the Apollonian gasket is that if the first four circles have integer bends, then *every other circle* in the packing does too. If you are given the first three circles of an Apollonian gasket, the bend of the fourth is found (as explained above) by solving a quadratic equation. However, every *subsequent* bend can be found by solving a *linear* equation:

$$d + d' = 2(a + b + c)$$

For instance, in the bugeye gasket, the three circles with bends $a = 2$, $b = 3$, and $c = 15$ are mutually tangent to two other circles. One of them, with bend $d = 2$, is already given in the first generation. The other has bend $d' = 38$, as predicted by the formula, $2 + 38 = 2(2 + 3 + 15)$. More importantly, even if we did not know d', we would still be guaranteed that it was an integer, because a, b, c, and d are.

Hidden behind this "baby Descartes equation" is an important fact about Apollonian gaskets: They have a very high degree of symmetry. Circles a, b, and c actually form a sort of curved mirror that reflects

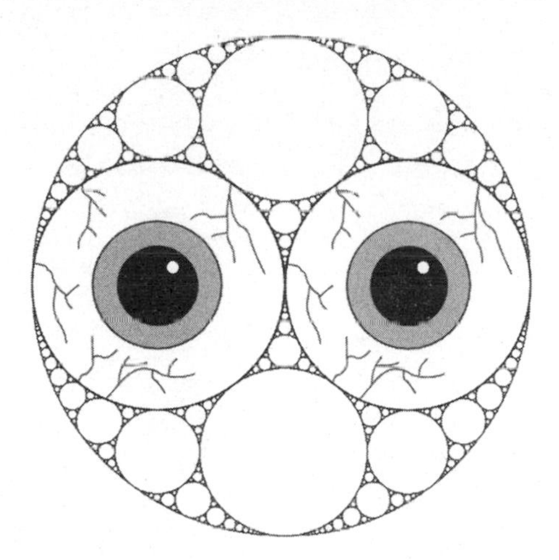

Figure 5. "Bugeye" gasket.

circle d to circle d' and vice versa. Thus the whole gasket is like a kaleidoscopic image of the first four circles, reflected again and again through an infinite collection of curved mirrors.

Kontorovich and Oh exploited this symmetry in an extraordinary and amusing way to prove their estimate of the function $N(r)$. Remember that $N(r)$ simply counts how many circles in the gasket have radius larger than r. Kontorovich and Oh modified the function $N(r)$ by introducing an extra variable of position—roughly equivalent to putting a lightbulb at a point x and asking how many circles illuminated by that lightbulb have radius larger than r. The count will fluctuate, depending on exactly where the bulb is placed. But it fluctuates in a very predictable way. For instance, the count is unchanged if you move the bulb to the location of any of its kaleidoscopic reflections.

This property makes the "lightbulb counting function" a very special kind of function, one which is invariant under the same symmetries as the Apollonian gasket itself. It can be broken down into a spectrum of similarly symmetric functions, just as a sound wave can be decomposed into a fundamental frequency and a series of overtones. From this spectrum, you can in theory find out everything you want to know about the lightbulb counting function, including its value at any particular location of the lightbulb.

For a musical instrument, the fundamental frequency or lowest overtone is the most important one. Similarly, it turned out that the first

Figure 6. A favorite example of Sarnak's is the "coins" gasket, so called because three of the four generating circles are in proportion to the sizes of a quarter, nickel, and dime, respectively. Image courtesy of Alex Kontorovich.

symmetric function was all that Kontorovich and Oh needed to figure out what happens to $N(r)$ as r approaches 0.

In this way, a simple problem in geometry connects up with some of the most fundamental concepts of modern mathematics. Functions that have a kaleidoscopic set of symmetries are rare and wonderful. Kontorovich calls them "the Holy Grail of number theory." Such functions were, for instance, used by Andrew Wiles in his proof of Fermat's last theorem. An interesting new kaleidoscope is enough to keep mathematicians happy for years.

Gaskets Galore

Kontorovich learned about the Apollonian kaleidoscope from his mentor, Peter Sarnak of Princeton University, who learned about it from Lagarias, who learned about it from Wilks and Mallows. For Sarnak, the Apollonian gasket is wonderful because it has neither too few nor too many mirrors. If there were too few, you would not get enough information from the spectral decomposition. If there were too many, then previously known methods, such as the ones Wiles used, would already answer all your questions.

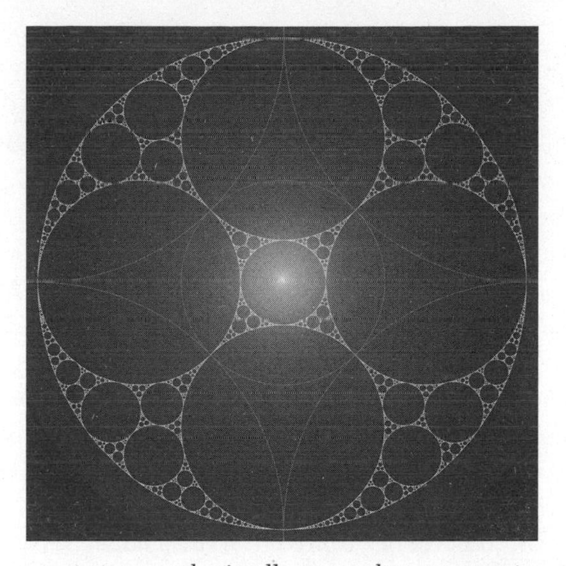

Figure 7. Many variations on the Apollonian gasket construction are possible. In this beautiful example, each pore is occupied by three inscribed circles rather than by one. The faint gray arcs represent five "curved mirrors." Reflections in these curved mirrors—known technically as circle inversions—create a kaleidoscopic effect. Every circle in the gasket is generated by repeated inversions of the first six circles through these curved mirrors. Image courtesy of Jos Leys.

Because Apollonian gaskets fall right in the middle, they generate a host of unsolved number-theoretic problems. For example, which numbers actually appear as bends in a given gasket? These numbers must satisfy certain "congruence restrictions." For example, in the bugeye gasket, the only legal bends have a remainder of 2, 3, 6, or 11 when divided by 12. So far, it seems that every number that satisfies this congruence restriction does indeed appear in the figure somewhere. (The reader may find it amusing to hunt for 2, 3, 6, 11, 14, 15, 18, 23, etc.) "Computation indicates that every number occurs, but we can't prove that even 1 percent of them actually occur!" says Ron Graham of the University of California at San Diego. For other Apollonian gaskets, such as the "coins" gasket in the fifth figure, there are some absentees—numbers that obey the congruence restrictions but don't appear in the gasket. Sarnak believes, however, that the number of absentees is always finite, and beyond a certain point any number that obeys the congruence restrictions does appear somewhere in the gasket. At this point, though, he is far from proving this conjecture—the necessary math just doesn't exist yet.

And even if all the problems concerning the classic Apollonian gaskets were solved, there are still gaskets galore for mathematicians to work on. As mentioned before, they could study random Apollonian gaskets. Another modification is the gasket shown in the last figure, where each pore is filled by three circles instead of one. Mallows and Gerhard Guettler have shown that such gaskets behave similarly to the original Apollonian gaskets—if the first six bends are integers, then all the rest of the bends are as well. Ambitious readers might want to work out the "Descartes formula" and the "baby Descartes formula" for these configurations, and investigate whether there are congruence restrictions on the bends.

Perhaps you, too, will be inspired to write a poem or paint a tablet in honor of Apollonius' ingenious legacy. "For me, what's attractive about Apollonian gaskets is that even my 14-year-old daughter finds them interesting," says Sarnak. "It's truly a god-given problem—or perhaps a Greek-given problem."

The Quest for God's Number

Rik van Grol

The Rubik's cube triggered one of the largest puzzle crazes in the world. The small mechanical puzzle, invented by Ernő Rubik in Hungary, has sold more than 350 million copies. Although it has existed since 1974, the popularity of the cube skyrocketed around 1980, when the cube was introduced outside of Hungary. In the early days, simply solving the puzzle was the main issue, especially because no solution books were available and there was no Internet. But solving the puzzle in the shortest amount of time was also hot news. In the early 1980s the best times were on the order of 24 seconds.

By the end of the 1980s, the craze was starting to ebb, but in certain groups the puzzle remained very much alive, and now the Rubik's cube is making a comeback. Solving the cube the quickest—*speed cubing*—is currently a major activity. The fastest times are under 8 seconds with the average around 11 seconds. The foundation for a fast solving time is a good algorithm, and so the search for efficient algorithms has been an important area of study since the early 1980s. The ultimate goal is to discover the best method of all—the optimal solution algorithm—which has been dubbed *God's algorithm.*

> *God's algorithm is the procedure to bring back Rubik's Cube from any random position to its solved state in the minimum number of steps.*

The maximum of all minimally needed number of steps is referred to as *God's number.* This number can be defined in several ways. The most common is in terms of the number of *face turns* required, but it can also be measured as the number of *quarter turns.* Whereas a quarter turn is either a positive or negative 90° turn, a face turn can be either of these or a 180° turn. A 180° face turn is equal to two quarter turns. Earlier this year, after decades of gradual progress, it was determined that God's number is 20 face turns. Thus, if God's algorithm were used to solve the cube, no starting position would ever require more than 20 face turns.

Apart from determining God's number, another major question has been to find out whether God's algorithm is an elegant sequence of moves that can be "easily" memorized, or if, instead, God's algorithm amounts to a short procedure with giant lookup tables. If it's the latter case, then no one will ever be able to learn how to solve the Rubik's cube in the minimal number of moves (read on to learn why this is true). Still, even if God's algorithm has no practical purpose, it is interesting to know what God's number is.

Playing God (with Small Numbers)

If you start with a solved cube and ask someone to make a few turns, you will (after some practice) be able to return it to the solved state in a minimum number of moves as long as the initial number of scrambling moves is not too large. With fewer than four scrambling moves, it is easy; with four, it becomes tricky. With five, it is simply hard. Some experts can handle six or even seven scrambling moves, but any more and it is essentially impossible to solve the cube in the minimal number of moves.

Generally speaking, most algorithms take between 50 and 100 moves. What if you were to randomly turn the cube 1,000 times? Will it take 1,000 moves to get it back? No, it still takes most algorithms 50 to 100 moves because the algorithms are designed to work from any starting position.

Starting Small: The 2 × 2 × 2 Cube

Unlike the classic 3 × 3 × 3 Rubik's Cube, the 2 × 2 × 2 cube has been completely analyzed. God's number is 11 in face turns and 14 in quarter turns. To find these values, all possible configurations of the 2 × 2 × 2 cube were cataloged, and for each of these configurations, the minimal number of turns needed to reach the solved state was determined. This brute force approach was possible because there are "only" about 3.7 million configurations to study.

To calculate the number of configurations for the 2 × 2 × 2 cube, we start with the observation that the eight corner *cubies* (as they are called) can be permuted in 8! ways. For any such permutation, each corner cubie can be oriented in three ways, leading to 3^8 possible orientations. However, given the orientation of seven corners, the orientation of the eighth is determined by the puzzle mechanism, so the corners really have 3^7 orientations. As the orientation of the whole cube is not fixed in space (any one of the eight corners can be placed in, say, the top-front-right position, and once it is placed there, the entire cube can be rotated so that any one of three faces is on top), the total number of permutations needs to be divided by $8 \times 3 = 24$. Hence, the total number of positions is

$$\frac{8! \times 3^7}{8 \times 3} = 7! \times 3^6 = 3{,}674{,}160.$$

There are only 2,644 positions for which 11 face turns are required to solve the puzzle. Assuming all configurations have the same likelihood of being a starting position, the average number of face turns required to solve the puzzle is 9. Likewise, there are only 276 positions from which 14 quarter turns are required, and on average, 11 quarter turns are required to solve the puzzle.

A Leap in Complexity: The 3 × 3 × 3 Cube

Until recently, God's number for the 3 × 3 × 3 cube was not known. From the late 1970s until now the search area has been limited by two numbers: the lower bound and the upper bound. The lower bound G^{low} is determined by proving that there are positions that require at least G^{low} turns. The upper bound G^{high} is determined by proving that no position requires more than G^{high} turns.

So, how many configurations are there? With 8 corner cubies and 12 edge cubies, there are $8! \times 12! \times 3^8 \times 2^{12}$ different patterns, but not all patterns are possible:

- *With 8 corners there are 8! corner permutations, and with 12 edges there are 12! edge permutations. However, because it is impossible to interchange two edge cubies without also interchanging 2 corner cubies, the total number of permutations should be divided by 2.*
- *Turning of corner cubies (keeping their position, but cycling the colors on their three faces) needs to be done in pairs—only 7 corner cubies can be turned freely.*

- *Flipping of edge cubies (keeping their position, but switching the colors of their two faces) needs to be done in pairs—only 11 edge cubies can be flipped freely.*

Because of the six center pieces, the orientation of the cube is fixed in space, so the number of permutations should not be divided by 24 as with the $2 \times 2 \times 2$ cube. Hence, the number of positions of the $3 \times 3 \times 3$ cube is

$$\frac{8! \times 12!}{2} \times \frac{3^8}{3} \times \frac{2^{12}}{2} = 43{,}252{,}003{,}274{,}489{,}856{,}000 \approx 4.3 \times 10^{19}.$$

This is *astronomically* bigger than 3,674,160 for the $2 \times 2 \times 2$ cube, and it made searching the entire space computationally impossible. For instance, if every one of the 350 million cubes ever sold were put in a new position every second, it would take more than 3,900 years for them to collectively hit every possible position (with no pair of cubes ever sharing a common configuration). Or to put it another way: if a computer were capable of determining the fewest number of moves required to solve the cube for 1,000 different starting positions each second, it would take more than a billion years of computing time to get through every configuration.

Determining God's number by independently improving the upper and lower bounds was a quest that lasted for three decades—but it has finally come to an end. In July 2010 the upper and lower bounds met at the number 20.

Raising the Lower Bound

Using counting arguments, it can be shown that there exist positions requiring at least 18 moves to solve. To see this, one counts the number of distinct positions achievable from the solved state using at most 17 moves. It turns out that this number is smaller than 4.3×10^{19}. This simple argument was made in the late 1970s (see Singmaster's book in the Further Reading section), and the result was not improved upon for many years. Note that this is not a constructive proof; it does not specify a concrete position that requires 18 moves. At some point, it was suggested that the so-called *superflip* would be such a position. The superflip is a state of the cube where all the cubies are in their correct position with the corner cubies oriented correctly, but where each edge cubie is flipped (oriented the wrong way).

It took until 1992 for a solution for the superflip with 20 face turns to be found, by Dik Winter. In 1995 Michael Reid proved that this solution was minimal, and thus a new lower bound for God's number was found. Also in 1995, a solution for the superflip in 24 quarter turns was found by Michael Reid, and it was later proved to be minimal by Jerry Bryan. In 1998 Michael Reid found a new position requiring more than 24 quarter turns to solve. The position, which he calls the *superflip composed with four spots*, requires 26 quarter turns. This put the lower bound at 20 face turns or 26 quarter turns.

Lowering the Upper Bound

Finding an upper bound requires a different kind of reasoning. Perhaps the first concrete value for an upper bound was the 277 moves mentioned by David Singmaster in early 1979. He simply counted the maximum number of moves required by his cube-solving algorithm. Later, Singmaster reported that Elwyn Berlekamp, John Conway, and Richard Guy had come up with a different algorithm that took at most 160 moves. Soon after, Conway's Cambridge Cubists reported that the cube could be restored in at most 94 moves. Again, this reflected the maximum number of moves required by a specific algorithm.

A significant breakthrough was made by Morwen Thistlethwaite. Whereas algorithms up to that point attacked the problem by putting various cubies in place and performing subsequent moves that left them in place, he approached the problem by gradually restricting the types of moves that could be executed. Understanding this method requires a brief introduction to the *cube group*.

As we work with the cube, let's agree to keep the overall orientation of the cube fixed in space. This means that the center cubies on each face will never change. We may then label the faces Left, Right, Front, Back, Up, and Down. The cube group is composed of all possible combinations of successive face turns, where two such combinations are equal if and only if they result in the same cube configuration. We denote the clockwise quarter turns of the faces by L, R, F, B, U, and D, and use concatenation as the group operation. For instance, the product FR denotes a quarter turn of the front face followed by a quarter turn of the right face, while F^2 denotes a half turn of the front face. The group identity, I, is no move at all. So, for example, $F^4 = I$.

Thistlethwaite divided the cube group G_0 into the following nested chain of subgroups:

$$G_0 = \langle L, R, F, B, U, D \rangle$$

$$G_1 = \langle L, R, F, B, U^2, D^2 \rangle$$

$$G_2 = \langle L, R, F^2, B^2, U^2, D^2 \rangle$$

$$G_3 = \langle L^2, R^2, F^2, B^2, U^2, D^2 \rangle$$

$$G_4 = \{I\}$$

Thistlethwaite's algorithm works by first performing a few moves that result in a configuration that no longer requires quarter turns of the U and D faces (although it still requires half turns). From this point on, only moves in the subgroup G_1 are needed. A few additional G_1 moves puts the cube into a position where quarter turns are no longer needed for the F and B faces, and so on. The algorithm uses large lookup tables at each step, and while it is not practical for humans, the successive sets of moves per subgroup are small enough to allow computer analysis. Initially, Thistlethwaite showed that any configuration could be solved in at most 85 moves. In January 1980 he improved his strategy to yield a maximum of 80 moves. Later that same year, he reduced the number to 63, and then again to 52.

After this rush of activity, progress stalled for several years. In 1989 Hans Kloosterman reported an algorithm that required at most 44 moves, which he later improved to 42. In 1992 Herbert Kociemba improved Thistlethwaite's algorithm by reducing it to a two-phase algorithm requiring only the subgroups G_0, G_2, and G_4. Using Kociemba's ideas, Michael Reid announced in 1995 that he had improved the upper bound to 29 face turns.

At about this time, Richard Korf introduced a new approach. Instead of using a fixed algorithm, his strategy simultaneously searched for a solution along three different lines of attack. On average, his algorithm appeared to solve the cube in 18 moves. There was, however, no worst-case analysis, and so the upper bound held still at 29.

Things turned quiet again until 2006, when Silviu Radu initiated a new countdown by reducing the upper bound to 27. The next year, Gene Cooperman brought it down to 26. With the lower bound of 20 face turns in sight, Tomas Rokicki entered the picture, reducing the upper bound to 25 in March 2008. Working with John Welborn, he had it down to 24 by April, then 23 in May, and 22 by August. Finally in July 2010, Rokicki announced an upper bound of 20, the established value of the lower bound and therefore the long-sought-after value of God's number.

Rokicki worked with Kociemba, as well as mathematician Morley Davidson and Google engineer John Dethridge. The team used symmetry arguments to significantly reduce the search space and then managed to partition the space of all configurations that remained into pieces small enough to fit onto a modern computer. They then made use of the enormous computing resources available through Google. Had the entire problem been done on a single good desktop PC, they report it would have taken about 35 years to complete, but by farming out different pieces to a large number of computers, the team was able to complete the calculation in just a few weeks.

It is a remarkable achievement.

Epilogue

Now that the quest for God's number for the classic $3 \times 3 \times 3$ cube has come to an end, it is time to look ahead. Is there an elegant version of God's algorithm—one that a human could implement? And what is God's number for the $4 \times 4 \times 4$, $5 \times 5 \times 5$, $6 \times 6 \times 6$, and $7 \times 7 \times 7$ cubes, all of which are now on the market? Finding God's number for such puzzles is a challenge that should last far into the future. The number of positions for the $7 \times 7 \times 7$ cube reaches 2.0×10^{160}.

Further Reading

Two influential early books are David Singmaster's *Notes on Rubik's Magic Cube* (Enslow Publishers, 1981), which is readable and has many technical details; and Patrick Bossert's *You Can Do the Cube* (Puffin Books, 1981), a how-to guide written when the author was 12 years old that sold more than 1.5 million copies. Another classic, complete with the requisite group theory, is *Inside Rubik's Cube and Beyond*, by Christoph Bandelow (Birkhäuser, 1982). On the web, an excellent how-to guide with several links to other sources can be found at Jaap's Puzzle Page: http://www.jaapsch.net/puzzles/. A brief history of the quest for God's number can be found on Tom Rokicki's site, http://www.cube20.org.

Meta-morphism:
From Graduate Student to Networked Mathematician

Andrew Schultz

While the stereotypical mathematician is a hermit locked alone in his office, the typical mathematician is far from a solitary explorer. A great amount of the mathematics produced today is created collaboratively, spurred into existence during those quintessentially mathematical social interactions: on chalkboards following a seminar talk, on napkins during a coffee break at a conference, on the back of a coaster at a pub. Though it often isn't clear to those wading through graduate programs, one of the key metamathematical skills one should develop while working on a master's or Ph.D. is the ability to participate in this social network. What follows is a rough guide to how you can use graduate school to build the professional relationships that will shape your career.

The Hungry Caterpillar

Stepping into the mathematical social network begins by getting to know your graduate student cohort. It's likely that some of the friendships you form during graduate school will be among the closest in your life, and even those fellow students who aren't your best friends are likely to be professional colleagues long after you've received your degree. It's worth the investment of time and energy to foster these relationships as your first semester begins.

When arriving on campus to start your graduate career, you'll likely convene with the new graduate students in your department and a handful of the faculty for a kind of informal orientation. Ph.D. programs often draw students from a wide variety of backgrounds, so don't be surprised to find people whose professional experience, familial status, or country of origin doesn't match your own. Despite any differences you might notice at first, this group has a common bond with you that you've probably never experienced: shared professional passion and the dedication to pursue an advanced degree over the course of several years. Use this commonality to bridge social or cultural gaps that your classmates might settle into upon arrival. Fortunately your busy class schedule will leave you with plenty of excuses for convening en masse: to tackle lengthy homework assignments, to review topics covered in class, to prepare for qualifying exams. As you work toward your degree, you will rely on the various skills and perspectives that your fellow classmates can offer, so it is in your best interest to meet and spend some time with as many in your incoming class as possible. As you progress through the program, you might be surprised to find your ideal study partner, your favorite office mates, and your mathematical siblings aren't the people you might have guessed when you first arrived.

Although mathematics and the novelty of graduate school are convenient starting points for meeting other incoming graduate students, don't rely on math to be the only tie that binds you: meet people for pizza at the end of a long week; organize a hike at a local nature preserve; or set up an informal, weekly grad student happy hour. Chances are good that graduate students who are further along in the program will be organizing various social events to which you'll be invited, and these will provide you with a good opportunity for meeting people whose experience can be of great benefit to you, both within the program and in your extra-mathematical life. Again, it is to your benefit to meet as many of the graduate students in your department as possible, so band together with a few first-year students and jump into this wider social pool. This larger community will also give you a chance to find people whose nonacademic interests match your own, and your department's graduate student e-mail list can be a boon for collecting people together to join in your favorite sport, play your favorite game, or take in a local theatrical performance. You can also consider broadening your social circle outside the math department by looking for university-wide student groups organized around your particular interests; being a mathematician certainly doesn't oblige you to spend time only with other mathematicians, and the break from an otherwise mathematically centered life could be a welcome respite.

Once you've established yourself among the graduate students, you'll next want to get to know your local faculty. Departments typically have a number of social activities planned through the course of the year, from annual get-togethers like a fall barbecue to weekly afternoon tea breaks. These give you a good opportunity for interacting with faculty members outside of the classroom, and you should take advantage of this opportunity. Chat with your algebra professor about where the course is going, talk to your analysis professor about the Research Experiences for Undergraduates project you worked on, or try to find some nonmathematical interests you share with other faculty. These conversations will give you an indication of which professors you most easily relate to, and this is an important factor to keep in consideration when choosing your thesis adviser. If you don't give yourself the chance to interact with a faculty member who is a candidate for becoming your Ph.D. supervisor, you might find yourself spending an hour each week with someone you can't talk to. Also, when considering a candidate for your supervisor, you will want to take advantage of the connections you've made with older graduate students by asking them about experiences they may have had in working with this person; for obvious reasons, it is particularly helpful if you can get an honest assessment from a current advisee.

Pupal Growth

Once you've gotten through your first year of graduate school, you are likely to have gravitated toward one research group or another within your department, and it's important that you become an active part of this community. Some of this will happen in the traditional classroom, where you already know the rules of engagement, though seminars will play an increasingly important role as you develop as a mathematician. Generally speaking, a seminar talk is a fifty-minute presentation to mathematicians in a specific discipline about recent developments in their field; often, but not always, the speaker at a seminar will be visiting from another mathematics department.

Once you've decided on your dissertation topic, you should start attending local seminars in your research area. Before attending, though, you'll need to adjust your expectations from those you have of a typical class. Seminar speakers have a limited time in which to introduce their topics, discuss connections to larger problems in the area, and then present specific results. Since the target audience is almost always specialists in the field, speakers often don't spend time bringing nonspecialists up to speed. As a graduate student who might have limited background in the discipline, you could very well find that many (if not all) of the seminars you attend are mostly incomprehensible to you. Not only is this okay, but it's the experience of nearly every graduate student who attends a seminar; it's hard to drink from a fire hose. Don't let this discourage you from attending future seminars, and don't turn the seminar into your personal fifty-minute nap session.

Your job when attending seminars is to focus on understanding as much of the talk as you can. Bring along a notepad and write down any questions you think of. Don't expect that your questions will sound as fancy as those being asked by the senior faculty member you're sitting next to; you're just starting in the area, and you're not expected to be making esoteric connections. Instead, bullet-point the big ideas of the talk: What were the basic objects under investigation? What qualitative information did the stated theorems give about these objects? How do the stated theorems depend on or diverge from each other? When the talk is over, you should feel free to participate in the question-and-answer session even if your questions don't sound as sophisticated as others. Afterward you should certainly speak with an experienced faculty member—if at all possible, the seminar speaker—about some of the questions you had. This adds an additional ten minutes to your seminar experience but can put the fifty minutes you've already invested into perspective. What's more, by attending seminars, you'll be training yourself to learn mathematics in an important way: contrary to the foundation-building, semester-long methodology used in teaching known mathematics, this result-focused, hour-long seminar approach is how you're most likely to hear about (and personally disseminate) new developments in your area for the rest of your career. For this reason, it's important to keep attending seminars even if you feel as though you're not understanding all of the talks. Each additional seminar will fill out your understanding of the discipline as a whole, and soon you'll find that a talk you've just attended reminds you of another talk you heard two months before; you'll be weaving your own mathematical tapestry.

The other benefit of attending seminars is that they are occasionally preceded by a seminar lunch or followed by a seminar dinner. Graduate students are always encouraged to attend these informal gatherings, and oftentimes their meals are subsidized. What graduate student doesn't like a cheap meal? These get-togethers are a golden opportunity to interact with faculty members outside of your department (think "future postdoc mentors", "future coauthors", etc.), so you should make a regular habit of attending. Striking up conversations in these settings is usually very easy. Questions like "Where did you attend graduate school?" and "What made you start researching . . . ?" seem obvious, but they are great places to begin. As always, don't feel obliged to stay within the bounds of mathematics when making conversation; after a long day of focusing on work, a nonmathematical topic of conversation could be a welcome change. Without prying or excluding others from the conversation, explore connections that you might have: perhaps the speaker hails from somewhere you've been meaning to visit, or one of your undergraduate professors went to the speaker's graduate school. Remember that these meals are meant to be fun; relax, be yourself, and make a good-faith effort at participating in the conversation.

The Emerging Butterfly

As you progress in your graduate career, you'll likely have the opportunity to speak about mathematics to an audience of mathematicians, either on your own work or on some theory you've been studying for your dissertation. If you are offered such an opportunity, take advantage of it. One doesn't develop the ability to give an interesting mathematics talk without experience, and you'll want to give yourself as many opportunities as possible to hone this critically important craft.

There are a number of excellent guides for how you can give a good mathematics talk (see [1, 2, 3, 4], or talk to someone whose presentation style you admire), but don't forget the interpersonal component of talking to an audience. Do the basics well: make eye contact regularly, gauge the audience's understanding, and make necessary adjustments. The audience will sense and respond to your attitude, so you can help encourage an enthusiastic response by projecting your own interest when describing your results. Along these same lines, avoid self-deprecating humor and resist the urge to downplay the importance of your results because they don't seem as profound as topics you might have heard while attending past seminars. The increased accessibility you detect in your talk comes from the fact that you have spent a lot of time developing the mathematics which bolsters it, and most of your audience won't have the benefit of this prolonged exposure to your topic. In other words: what seems obvious to you is often not immediately obvious to those in attendance. Help the audience follow your talk by providing them with interpretations of the results you present, such as how a certain lemma will be used in developing a later theory, or why a particular result is connected to previous work in the area.

Don't feel that you need to give your first talk in a research seminar filled with faculty. Instead, see if the graduate students in your department have a student-run, general-interest seminar that you can speak in. If no such seminar exists, take the lead and start one. Graduate-student colloquia can be a tremendous opportunity for you and your cohort to sharpen a critical professional skill in a low-stakes, friendly environment, and your department will be stronger for the introduction of such a seminar. Approach your department chair or the director of your graduate program and see if you can get some nominal funding to support the seminar, and use the money to entice student attendance with that siren song of graduate life: free food. Presenting in such a seminar will force you to boil your technical results down into an understandable form, and you'll reap the benefits of seeing how your classmates perform this same reduction. As you go on to present in seminars with faculty attendees or with a more specialized focus, you'll rely on this same skill (though you'll need to adjust the parameters of "understandable" depending on your target audience). Even if you don't plan to keep research as an active part of your professional life after you finish graduate school, this skill remains applicable for the many times you have to talk about mathematics to nonmathematicians: when explaining the importance of a subject to a class of undergraduates, or when justifying some mathematical program to administrators at your college.

After you've had a chance to present work locally, you'll want to take advantage of any opportunities which arise for presenting talks at nearby meetings or at far-flung conferences. There is often support for graduate student travel, either from your department or the conference's organizing body, so don't assume that an empty bank account will prevent you from participating. Occasionally you can also receive support even when you aren't presenting at a given conference, or you might have the opportunity to attend a conference which won't require outside support (if it's at a local university, say). If you are given financial backing, or if the out-of-pocket expenses are manageable, it's always a good idea to attend conferences which cover mathematics of interest to you when you have the opportunity. Don't feel obliged to limit your participation to conferences organized around your specific research area; your mathematical interests are likely varied, and your professional life will be richer by fostering this breadth.

Once you've arrived at a conference, don't forget that there's more to do than simply attend talks (or deliver your own). Conferences represent an opportunity to further your social sphere and make contacts with some of the movers and shakers in your field, or at least a handful of mathematicians who are a bit further along in their mathematical careers than you. This will likely be the first time in a while that you've felt like you truly know no one around, but don't let that be an excuse for making a quick retreat to your room at the end of each talk. Don't be shy about introducing yourself to people during coffee breaks or in the ten-minute pauses between talks or presentations. Again, basic introductions can take you very far, so feel free to start with your name and institution and see where things go. At the beginning of a conference, you can always ask if the person will be presenting later in the week; if you've already seen their talk, you can ask them something you didn't get a chance to ask during the question-and-answer period (you're still writing questions down during talks, right?). Take advantage of this opportunity to establish yourself as an inquisitive, approachable person to a large group of people outside your home institution.

Life as a Pollinator

Life as a professional mathematician requires participation in a social network, and graduate school represents an ideal setting for you to gradually develop the skills and connections which will help you thrive in this web: first with your classmates, then local faculty, and later with the wide mathematical world. Regardless of where you end up after graduate school, continue to take advantage of the opportunities you have to further your own connections, and do your part to help budding mathematicians at your institution join this network: foster their interest in exploring and presenting their own mathematical questions, encourage their attendance in colloquia or seminars (or help organize a student-targeted colloquium), and convince your department to set aside funds so they can attend conferences to meet other mathematicians.

By placing value on the interpersonal aspects of practicing math, we ultimately increase the quality of mathematical content and discourse for those we seek to serve: our institutions, our students, our colleagues, and ourselves.

Acknowledgments

Thanks to Anne Brubaker and the referees of this article for carefully reading through drafts and making helpful suggestions. Thanks to Ravi Vakil for teaching me to get the most out of seminars and conferences, even when I didn't know what the speaker was talking about. Thanks to Project NExT for helping me extend my own mathematical network.

References

[1] C. T. Bennett and F. Sottile, Math talks, *Starting Our Careers: A Collection of Essays and Advice on Professional Development from the Young Mathematicians' Network*, American Mathematical Society, Providence, RI, 1999.

[2] P. Halmos, How to talk mathematics, *AMS Notices* **21** (1974), 155–8.

[3] J. E. McCarthy, How to give a good colloquium, *Canadian Mathematical Society Notes* **31,** no. 5 (1999), 3–4.

[4] J. Swallow, Proving yourself: How to develop an interview lecture, *AMS Notices* **56** (2009), 948–51.

One, Two, Many:
Individuality and Collectivity in Mathematics

Melvyn B. Nathanson

"Fermat's last theorem" is famous because it is old and easily understood, but it is not particularly interesting. Many, perhaps most, mathematicians would agree with this statement, though they might add that it is, nonetheless, important because of the new mathematics created in the attempt to solve the problem. By solving Fermat, Andrew Wiles became one of the world's best known mathematicians, along with John Nash, who achieved fame by being crazy, and Theodore Kaczynski, the Unabomber, by killing people.

Wiles is known not only because of the problem he solved, but also because of how he solved it. He was not part of a corporate team. He did not work over coffee, by mail, or via the Internet with a group of collaborators. Instead, for many years, he worked alone in an attic study and did not talk to anyone about his ideas. This is the classical model of the artist, laboring in obscurity. (Not real obscurity, of course, since Wiles was, after all, a Princeton professor.) What made the solution of Fermat's last theorem so powerful in the public and scientific imagination was the fact that the story comported with the romantic myth: solitary genius, great accomplishment.

This is a compelling myth in science. We have the image of the young Newton, who watched a falling apple and discovered gravity as he sat, alone, in an orchard in Lincolnshire while Cambridge was closed because of an epidemic. We recall Galois, working desperately through the night to write down before his duel the next morning all of the mathematics he had discovered alone. There was Abel, isolated in Norway, his discovery of the unsolvability of the quintic ignored by the mathematical elite. And Einstein, exiled to a Swiss patent office, where he analyzed Brownian motion, explained the photoelectric effect, and discovered relativity. In a speech in 1933, Einstein said that being a lighthouse keeper would be a good occupation for a physicist. These are the kind of stories that give Eric Temple Bell's *Men of Mathematics* its hypnotic power and inspire many young students to do research.

Wiles did not follow the script perfectly. His initial manuscript contained a gap that was eventually filled by Wiles and his former student Richard Taylor. Within epsilon, Wiles solved Fermat in the best possible way. Intense solitary thought produces the best mathematics.

Gel'fand's List

Some of the greatest twentieth-century mathematicians, such as André Weil and Atle Selberg, had few joint papers. Others, like Paul Erdős and I. M. Gel'fand, had many. Erdős was a master collaborator, with hundreds of co-authors. (Full disclosure: I am one of them.) Reviewing Erdős's number theory papers, I find that in his early years, from his first published work in 1929 through 1945, most (60 percent) of his 112 papers were singly authored, and that most of his stunningly original papers in number theory were papers that he wrote by himself.

In 1972–73 I was in Moscow as a postdoc studying with Gel'fand. In a conversation one day he told me there were only ten people in the world who *really* understood representation theory, and he proceeded to name them. It was an interesting list, with some unusual inclusions and some striking exclusions. ("Why is X not on the list," I asked, mentioning the name of a really famous representation theorist. "He's just an engineer," was Gel'fand's disparaging reply.) But the tenth name on the list was not a name, but a description: "Somewhere in China," said Gel'fand, "there is a young student, working alone, who understands representation theory."

Bers Mafia

A traditional form of mathematical collaboration is to join a school. Analogous to the political question, "Who's your rabbi?" (meaning "Who's your boss? Who is the guy whom you support and who helps you in return?"), there is the mathematical question, "Who's your mafia?" The mafia is the group of scholars with whom you share research interests, with whom you socialize, whom you support, and who support you. In the New York area, for example, there is the self-described

"Ahlfors-Bers mafia," beautifully described in a series of articles about Lipman Bers that were published in a memorial issue of the *Notices of the American Mathematical Society* in 1995.

Bers was an impressive and charismatic mathematician at NYU and Columbia who created a community of graduate students, postdocs, and senior scientists who shared common research interests. Being a member of the Bers mafia was valuable both scientifically and professionally. As students of the master, members spoke a common language and pursued common research goals with similar mathematical tools. Members could easily read, understand, and appreciate one another's papers, and their own work fed into and complemented the research of others. Notwithstanding sometimes intense internal group rivalries, members would write recommendations for one another's job applications, review their papers and books, referee their grant proposals, and nominate and promote each other for prizes and invited lectures. Being part of a school made life easy. This is the strength and the weakness of the collective. Members of a mafia, protected and protecting, competing with other mafias, are better situated than those who work alone. Membership guarantees moderate success but makes it unlikely to create really original mathematics.

The Riemann Hypothesis

The American Institute of Mathematics organized its first conference, "In Celebration of the Centenary of the Proof of the Prime Number Theorem: A Symposium on the Riemann Hypothesis," at the University of Washington on August 12–15, 1996. According to its website, "the American Institute of Mathematics, a nonprofit organization, was founded in 1994 by Silicon Valley businessmen John Fry and Steve Sorenson, longtime supporters of mathematical research." The story circulating at the meeting was that the businessmen funding AIM believed that the way to prove the Riemann hypothesis was the corporate model: To solve a problem, put together the right team of "experts" and they will quickly find a solution.

At the AIM meeting, various experts (including Berry, Connes, Goldfeld, Heath-Brown, Iwaniec, Kurokawa, Montgomery, Odlyzko, Sarnak, and Selberg) described ideas for solving the Riemann hypothesis. I asked one of the organizers why the celebrated number theorist Z was not giving a lecture. The answer: Z had been invited, but declined to speak. Z had said that if he had an idea that he thought would solve the Riemann hypothesis, he certainly would not tell anyone because he wanted to solve it alone. This is a simple and basic human desire: Keep the glory for yourself.

Thus, the AIM conference was really a series of lectures on "How *not* to solve the Riemann hypothesis." It was a meeting of distinguished mathematicians describing methods that had failed, and the importance of the lectures was to learn what not to waste time on.

The Polymath Project

The preceding examples are prologue to a discussion of a new, widely publicized Internet-based effort to achieve massive mathematical collaboration. Tim Gowers began this experiment on January 27, 2009, with the post "Is massively collaborative mathematics possible?" on his Weblog http://gowers.wordpress.com. He wrote, "Different people have different characteristics when it comes to research. Some like to throw out ideas, others to criticize them, others to work out details, others to re-explain ideas in a different language, others to formulate different but related problems, others to step back from a big muddle of ideas and fashion some more coherent picture out of them, and so on. A hugely collaborative project would make it possible for people to specialize. . . . In short, if a large group of mathematicians could connect their brains efficiently, they could perhaps solve problems very efficiently as well." This is the fundamental idea, which he restated explicitly as follows: "Suppose one had a forum . . . for the online discussion of a particular problem. . . . The ideal outcome would be a solution of the problem with no single individual having to think all that hard. The hard thought would be done by a sort of super-mathematician whose brain is distributed among bits of the brains of lots of interlinked people."

What makes Gowers's polymath project noteworthy is its promise to produce extraordinary results—new theorems, methods, and ideas—that could not come from the ordinary collaboration of even a large number of first-rate scientists. Polymath succeeds if it produces a superbrain. Otherwise, it's boring.

In appropriately pseudo-scientific form, I would restate the "Gowers hypothesis" as follows: Let qual(w) denote the quality of the mathematical paper w, and let Qual(M) denote the quality of the mathematical papers written by the mathematician M. If w is a paper produced by the massive collaboration of a set $\mathcal{M}$ of mathematicians, then

$$(1) \qquad \mathrm{qual}(w) > \sup\{\mathrm{Qual}(M) : M \in \mathcal{M}\}.$$

A reading of the many published articles and comments on massive collaboration suggests that its enthusiastic proponents believe the following much stronger statement:

$$(2) \qquad \lim_{|\mathcal{M}| \to \infty} (\mathrm{qual}(w) - \sup\{\mathrm{Qual}(M) : M \in \mathcal{M}\}) = \infty.$$

Superficially, at least, this might seem plausible, especially when proposed by one Fields Medalist (Gowers) and enthusiastically supported by another (Terry Tao).

I assert that (1) and (2) are wrong, and that the opposite inequality is true:

$$(3) \qquad \mathrm{qual}(w) < \sup\{\mathrm{Qual}(M) : M \in \mathcal{M}\}.$$

First, some background. Massive mathematical collaboration is one of several recent experiments in scientific social networking. One of the best known is the DARPA Network Challenge. On December 5, 2009, the Defense Advanced Research Projects Agency (DARPA) tethered ten red weather balloons at undisclosed but readily accessible locations across the United States, each balloon visible from a nearby highway, and offered a $40,000 prize to the first individual or team that could correctly give the latitude and longitude of each of the ten balloons. In a press release, DARPA wrote that it had "announced the Network Challenge . . . to explore how broad-scope problems can be tackled using social networking tools. The Challenge explores basic research issues such as mobilization, collaboration, and trust in diverse social networking constructs and could serve to fuel innovation across a wide spectrum of applications."

In less than nine hours, the MIT Red Balloon Challenge Team won the prize. According to the DARPA final project report, "The geolocation of ten balloons in the United States by conventional intelligence methods is considered by many to be intractable; one senior analyst at the National Geospatial Intelligence Agency characterized the problem as impossible. A distributed human sensor approach built around social networks was recognized as a promising, nonconventional method of solving the problem, and the Network Challenge was designed to explore how quickly and effectively social networks could mobilize to solve the geolocation problem. The speed with which the Network Challenge was solved provides a quantitative measure for the effectiveness of emerging new forms of social media in mobilizing teams to solve an important problem."

The DARPA Network Challenge shows that, in certain situations, scientific networking can be extraordinarily effective, but there is a fundamental difference between the DARPA Network Challenge and massive mathematical collaboration. The difference is the difference between stupidity and creativity. The participants in the DARPA Network Challenge had a stupid task to perform: Look for a big red balloon and, if you see one, report it. No intelligence required. Just do it. The widely disbursed members of the MIT team, like a colony of social ants, worked cooperatively and productively for the greater good, but didn't create anything. Mathematics, however, requires intense thought. Individual mathematicians do have "to think all that hard." Individual mathematicians create.

In a recent magazine article ("Massively collaborative mathematics," *Nature*, October 15, 2009), Gowers and Michael Nielsen boasted, "The collaboration achieved far more than Gowers expected, and showcases what we think will be a powerful force in scientific discovery—the collaboration of many minds through the Internet." They are wrong. Massive mathematical collaboration has so far failed to achieve its ambitious goal.

Consider what massive mathematical collaboration has produced, and who produced it. Gowers proposed the problem of finding an elementary proof of the density version of the Hales-Jewitt theorem, which is a fundamental result in combinatorial number theory and Ramsay theory. In a very short time, the blog team came up with a proof, chose a nom de plume ("D.H.J. Polymath"), wrote a paper, uploaded it to arXiv, and submitted it for publication. The paper is: D.H.J. Polymath, "A new proof of the density Hales-Jewett theorem," arXiv:0910.3926.

The abstract describes it clearly: "The Hales-Jewett theorem asserts that for every r and every k there exists n such that every r-colouring of the n-dimensional grid $\{1, \ldots, k\}^n$ contains a combinatorial line. . . . The Hales-Jewett theorem has a density version as well, proved by Furstenberg and Katznelson in 1991 by means of a significant extension of the ergodic techniques that had been pioneered by Furstenberg in his proof of Szemerédi's theorem. In this paper, we give the first elementary proof of the theorem of Furstenberg and Katznelson, and the first to provide a quantitative bound on how large n needs to be."

A second, related paper by D.H.J. Polymath, "A new proof of the density Hales-Jewett theorem," arXiv:0910.3926, has also been posted on arXiv.

These papers are good, but obviously not Fields Medal quality, so Nathanson's inequality (3) is satisfied. A better experiment might be massive collaboration without the participation of mathematicians in the Fields Medal class. This would reduce the upper bound in Gowers' inequality (1), and give it a better chance to hold. It is possible, however, that Internet collaboration can succeed only when controlled by a very small number of extremely smart people. Certainly, the leadership of Gowers and Tao is a strong inducement for a mathematician to play the massive participation game, since, inter alia, it allows one to claim joint authorship with Fields Medalists.

After writing the first paper, Gowers blogged, "Let me say that for me personally this has been one of the most exciting six weeks of my mathematical life. . . . There seemed to be such a lot of interest in the whole idea that I thought that there would be dozens of contributors, but instead the number settled down to a handful, all of whom I knew personally." In other words, this became an ordinary, not massive, collaboration.

This was exactly how it was reported in *Scientific American*. On March 17, 2010, Davide Castelvecchi wrote, "In another way, however, the project was a bit of a disappointment. Just six people—all professional mathematicians and usual suspects in the field—did most of the work. Among them was another Fields Medalist and prolific blogger, Terence Tao of the University of California, Los Angeles."

Human beings are social animals. We enjoy working together, through conversation, letter writing, and e-mail. (More full disclosure: I've written many joint papers. One paper even has five authors. Collaboration can be fun.) But massive collaboration is supposed to achieve much more than ordinary collaboration. Its goal, as Gowers wrote, is the creation of a super-brain, and that won't happen.

Mathematicians, like other scientists, rejoice in unexpected new discoveries and delight when new ideas produce new methods to solve old problems and create new ones. We usually don't care how the breakthroughs are achieved. Still, I prefer one person working alone to two or three working collaboratively, and I find the notion of massive collaboration aesthetically appalling. Better a discovery by an individual than the same discovery by a group.

I would guess that even in the already interactive twentieth century, most of the new ideas in mathematics originated in papers written by a single author. A glance at MathSciNet shows that only two of Tim Gowers's 42 papers have a co-author. (Terry Tao responded to this observation by noting that half of his many papers are collaborative.)

In a contribution to a "New Ideas" issue of *The New York Times Magazine* on December 13, 2009, Jordan Ellenberg described massive mathematical collaboration with journalistic hyperbole: "By now we're used to the idea that gigantic aggregates of human brains—especially when allowed to communicate nearly instantaneously via the Internet—can carry out fantastically difficult cognitive tasks, like writing an encyclopedia or mapping a social network. But some problems we still jealously guard as the province of individual beautiful minds: writing a novel, choosing a spouse, creating a new mathematical theorem. The Polymath experiment suggests this prejudice may need to be rethought. In the near future, we might talk not only about the wisdom of crowds but also of their genius."

It is always good to rethink old prejudices, but sometimes the reevaluation confirms the truth of the original prejudice. Massive collaboration will produce useful results, but it will not meet the standard that Gowers set: No mathematical super-brain will evolve on the Internet and create new theories that will yield brilliant solutions to important unsolved problems. It won't happen. Gowers cited the classification of the finite simple groups as a kind of massive collaboration, and this is a perfect example. It was a useful result. Ignoring the contentious question of whether the proof was or is finally correct, which is an inherent problem of massive collaboration, the work is definitely boring. As far as I know, neither brilliant insights nor new techniques have come out of the proof and been applied to create new areas of mathematics or solve old problems in unrelated fields. It is more engineering than art. Recalling Mark Kac's famous division of mathematical geniuses into two classes, ordinary geniuses and magicians, one can imagine that massive collaboration will produce ordinary work and, possibly, in the future, even work of ordinary genius, but not magic. Work of ordinary genius is not a minor accomplishment, but magic is better.

Reflections on the Decline of Mathematical Tables

MARTIN CAMPBELL-KELLY

For some people it's typewriters. For other people it's mechanical calculating machines that bring a nostalgic tear to the eye. For me it's mathematical tables. The sight—even the smell—of a set of four-figure tables transports me to my distant school and college days. You can still find mathematical tables—their yellowed pages filled with decimal digits and not much else—in secondhand bookstores and occasionally on eBay. I once thought I might like to collect mathematical tables, but then I discovered from the *Index to Mathematical Tables*[1] that many hundreds of tables have been published. Even a selective collection would prove burdensome, if not grounds for divorce. Ironically, the *Index to Mathematical Tables*, a monumental bibliographical endeavor, was published in 1962, just as tables were going out of business.

Mathematical tables were excruciatingly tedious to calculate. Take for example logarithmic tables, which ruled the calculating roost for about 300 years. Logarithms were invented by the great Scottish philosopher John Napier around 1614, and a practical table of logarithms to base 10, the *Arithmetica Logarithmica*, was calculated by the English mathematician Henry Briggs and published in 1624. While the logarithm was an invention of genius, computing them to 14 decimal places was assuredly a labor of Hercules. It is said that when Napier and Briggs first met "almost one quarter of an hour was spent, each beholding the other with admiration, before one word was spoken."

Logarithms were so laborious to calculate that subsequent tables were not recomputed but were compiled from the existing canons. The raw logarithms would be reduced to four, five, six, or seven decimal places and conveniently arranged and printed using the best typography of the day. It is said that the most accurate table of logarithms ever produced was Charles Babbage's seven-figure tables. Babbage was somewhat a connoisseur of tables and owned approximately 300 volumes. Like his predecessors, he did not recompute the logarithms but copied them from existing tables. Where there was a discrepancy in his sources he would recompute the offending entry. It was almost a literary exercise—an anthology of the best logarithms, so to speak.

But Babbage did produce some astronomical and actuarial tables from scratch. Babbage used computers for the calculations—not machines but human drudges, because "computer" was then an occupation, not a machine. It was while he was checking his astronomical tables that he made the oft-quoted remark "I wish to God these calculations had been executed by steam." Shortly after, in 1824, he secured funding from the British government to build a "difference engine" for calculating tables (powered by hand, not steam). Babbage's engine was a magnificent failure. He overdesigned, mismanaged, and nothing materialized except a prototype and some plans for a full-scale machine. Babbage's difference engine was finally completed in 1991 at the London Science Museum, in time to celebrate the bicentenary of Babbage's birth. Amazingly, it worked beautifully.

Although a few difference engines along Babbage's lines were built in the second half of the nineteenth century, they tended to be fickle and expensive and there was no real market for them. So, for the first half of the twentieth century, most tables were produced by human computers, sometimes using a mechanical desk calculator or a punched card machine, but as often as not just plain pencil and paper.

The biggest human computing organization in the world was established in New York in 1938 and is described by David Grier in his remarkable book *When Computers Were Human*.[2] The so-called Mathematical Tables Project was a Depression-era make-work project for Roosevelt's Works Progress Administration. It was intended to employ out-of-work clerks and bookkeepers in the useful work of making mathematical tables. Most of the project's human computers used nothing more than colored pencils and preprinted calculating sheets. At its peak in World War II, the project employed 450 computers producing tables for the war effort. Table making was then a flourishing business. In 1943, a new journal—*Mathematical Tables and Other Aids to Computation*—recorded the practical and theoretical advances being made.

Another massive wartime table-making project was established by the Moore School of Electrical Engineering, University of Pennsylvania, and the Ballistics Research Laboratory at Aberdeen Proving Ground, Maryland. The war had created an unprecedented demand for "firing tables" for newly developed artillery and for existing weapons to be deployed in new theaters of war. A firing table provided the elevation and azimuth to enable a gun to be aimed with accuracy for a given target distance. A team of 200 female computers, equipped with mechanical calculating machines, helped to produce the tables. But this was not nearly enough computing power. Each firing table contained about 3,000 entries, and each entry took about one or two person-days to compute. Two Moore School academics, John Mauchly and Presper Eckert, proposed the construction of an electronic computer to perform these ballistics calculations—this was the ENIAC, the Electronic Numerical Integrator and Computer. Eckert and Mauchly got the go-ahead to build the ENIAC in spring 1943, although it was not completed until late 1945 and so was too late to help in the war effort. While ENIAC was not the first, it was certainly the most famous early electronic digital computer.

Before the ENIAC was finished, however, it was realized that although it could calculate ballistics tables to perfection, it was otherwise rather limited in the kind of computations it could perform. In the summer of 1944 the Princeton mathematician John von Neumann learned of the ENIAC, and worked with the Moore School team to design a better machine. Dubbed the EDVAC, for Electronic Discrete Variable Automatic Computer, it was a design of such versatility that its architecture has underpinned computer design ever since. It became known as the stored-program computer.

In the 1950s the new stored-program computers started to proliferate. Suddenly table making was made easy. Among other computational tasks, computers began to grind out tables as never before. We can see from the *An Index of Mathematical Tables* that in the 1950s more tables were published than in any previous decade. But of course, if you had access to a computer you no longer had much need for mathematical tables. For example, trigonometrical and hyperbolic functions could be computed using a simple subroutine to compute values on the fly in the course of a program.

At first it was not entirely clear whether computers would stimulate the production and consumption of tables or eliminate them entirely. An aging generation of table makers clung to the former hope. But the writing was on the wall. By the end of the 1950s, most big academic and research institutions had a computer, and it was clear that by the end of the 1960s they all would. In the late 1940s *Mathematical Tables and Other Aids to Computation* had become the journal of record of the Association for Computing Machinery, but it seemed like the tail was wagging the dog, and in 1954 ACM launched its own *Journal of the ACM*. In 1959 *Mathematical Tables and Other Aids to Computation* bent to the wind and renamed itself the *Mathematics of Computation*. Over the next two decades the publication of new tables slowed to a trickle.

Four-figure tables still found a place in school and undergraduate studies, but the advent of the electronic calculators in the 1970s finally put an end to them. I recall an occasion in 1990 when I saw in a thrift store a two-foot-high stack of brand-new four-figure tables. On inspection it turned out they were printed in 1978. I suppose they had been stored for a decade in the publisher's warehouse and had now been remaindered—with one last stop in a thrift store before finally being pulped. I was powerless to prevent their fate, though I rescued one for my bookshelves for the princely sum of 10 pence.

It is one of the great ironies of computing that in the 1950s what all go-ahead table makers wanted was a digital computer, but within a decade the computer had made both them and their tables obsolete. As a wise man once cautioned, "Be careful what you wish for."

References

1. Fletcher, A., Miller, J.C.P., Rosenhead, L., and Comrie, L. J. *An Index to Mathematical Tables.* Addison-Wesley, 1962.
2. Grier, D. A. *When Computers Were Human*, Princeton University Press, 2005.

Under-Represented Then Over-Represented: A Memoir of Jews in American Mathematics

Reuben Hersh

Over-Represented

When I studied at the Courant Institute of NYU from 1957 to 1962, its Jewish (specifically, Ashkenazi) flavor was impossible to miss. Of course, it was in large part the creation of Richard Courant, who came to New York in 1934 as a Jewish refugee expelled by Adolf Hitler from his post as the leader of the great and famous mathematical school at Göttingen in Germany. Two of NYU's most important professors, Kurt Friedrichs and Fritz John, had been Courant's students at Göttingen [**13**]. Lipman Bers was also a refugee. Of the younger members of the brilliant faculty, Peter Lax (my mentor and adviser) and his good friend Louis Nirenberg, world leaders in their specialty of partial differential equations, were themselves graduates of NYU (and Jewish). More on the applied side were Joe Keller and Harold Grad (also Jewish, also NYU graduates). Jack Schwartz, a New York Jew, had recently come down from New Haven to join the Courant faculty. Martin Davis had been one of Jack's fellow undergraduates at City College. Morris Kline (Jewish, of course) had actually preceded Courant as an NYU professor. In addition there were Anneli Lax, Warren Hirsch, Jerry Berkowitz, Lazar Bromberg, and Max Goldstein. (What about Wilhelm Magnus, a German like Courant, Friedrichs, and John? Only later on did I come to understand that he was not Jewish, but Catholic and a staunch anti-Nazi. He left Germany, not as a refugee from Hitler, but as a postwar immigrant—sponsored and invited by Courant.) And my first boss at Courant, the professor for whom I was a homework grader, had a Jewish last name (Morawetz) and an Irish first name (Cathleen!) I soon came to understand that her father was the well-known Irish mathematician John Lighton Synge, and her husband the well-known Jewish chemist Herbert Morawetz. The only real anomaly here was Jim Stoker—a highly respected geometer and applied mathematician, but apparently a—a what? A WASP (white Anglo-Saxon Protestant)! Yet not such an anomaly. In the mid-thirties Stoker went to the Federal Institute of Technology in Zürich to get a Ph.D. in mechanics. One of the first courses he took was on geometry with Heinz Hopf. He fell in love with the subject and the teacher, and took a Ph.D. in math instead. When he'd finished, Hopf wrote to Courant about this *junger Amerikaner* who would fit in very well with Courant's plans. I never did get to know Stoker, somehow. I did meet, talk with, and take courses from all the others I have mentioned. Yes, you could say Jews were over-represented at Courant in my time.

But after all, the department had virtually been created by a Jewish refugee, at an institution with a student body that Constance Reid describes as "composed largely of the sons and daughters of working-class Jewish immigrants." I later read, in her biography of Courant [**11**], that in deciding to settle at NYU, Courant had been counting on New York's large supply of smart youngsters, what he referred to as "a reservoir of talent," to fill the ranks of NYU's graduate program, and indeed, at least to some degree, of its mathematics faculty.

So the "over-representation" there seemed perfectly natural.

From NYU, after five years as a grad student, I was lucky enough to spend two years as an instructor in Palo Alto, California, at Stanford University, a great opportunity for me to develop mathematically and to meet mathematicians. First of all, Ralph Phillips, the collaborator of my mentor Peter Lax. Not a New Yorker—a "real American," you might say, if you think New York is not exactly the real America. But still—Jewish! My supervisor in my teaching duties was Paul Berg—like myself, a Jewish NYU product. Three young hotshots were Don Ornstein, Paul Cohen, and Bob Osserman. My fellow instructors included Si Hellerstein, Steve Shatz, Lenny Sarason, and Rohit Parikh. Rohit is from India, and I don't think he is Jewish. The chairman at Stanford was David Gilbarg. Yes, Jewish. Gilbarg's collaborator in nonlinear elliptic partial differential equations was Bob Finn. Sounds pretty non-Jewish. Remember Mark Twain and Huckleberry Finn? But when I got to know Bob Finn, it turned out that "Finn" was a shortening of "Finkelstein." So what? My own surname, Hersh, was invented by my father, Hersh Fish

Laznowski, when he became a U.S. citizen some time around 1921, and decided to become Americanized as Philip Hersh.

In my two years at Stanford, I met some of the department's very famous faculty of European origin. Most conspicuous, at least to me, was Lars Hörmander, a tall blond Swede who was notorious for being mathematically perfect. Certainly not Jewish. (I didn't know that one of his mentors and advisers back in Lund had been the Hungarian Jew Marcel Riesz.) Anyway, the other famous Europeans at Stanford were Stefan Bergman from Poland, George Polya and Gabor Szego from Hungary, Charles Loewner from Czechoslovakia, and Hans Samelson from Germany. There was also Menahem Max Schiffer from Israel and Sam Karlin in statistics. All brilliant; all Jews.

Well, why not? It has been said, more than once, that by driving the Jewish mathematicians and physicists from Europe to America, Hitler gave the U.S. a present more valuable than anything else you can think of. Stanford grabbed more than its share.

When my two years at Stanford were up, I took employment at the University of New Mexico in Albuquerque, where I have pretty much remained ever since. The chairman here was Julius Blum—another refugee. He had escaped from Berlin, and earned his Ph.D., in statistics, at Berkeley. He was close friends with another statistician in the UNM department—Judah Rosenblatt. Judah's statistics Ph.D. was from Columbia. I was quickly informed of Judah's other big distinction—his grandfather was none other than Yossele Rosenblatt, the most famous American cantor, whose recordings of Judaic musical liturgy were beloved by many, especially by the Orthodox. Along with Julius and Judah, the department was dominated by Bernie Epstein, author of a popular graduate text on PDE's, and Ignace Kolodner, another refugee, who also was a Courant Ph.D.! This was far from New York or California, in the semidesert of tricultural (Hispanic, Native American, and Anglo-Cowboy) New Mexico.

Jewish over-representation!

Over-Representation?

Why? How come so many Jews? A natural question, but one seldom asked in conversation, and never asked in print. Too ticklish, too much chance to be misunderstood, or give offense, or get in trouble one way or the other. You don't want to seem anti-Jewish, you don't want to seem too Jewish, you don't want to seem hung up on the Jewish question; much better to just act like you don't notice something a little unexpected, calling for explanation. Strangely enough, even in historical articles about the immigration of European refugee mathematicians to the U.S. in the 1930s, the words "Jew" and "Jewish" are usually avoided. Now Ioan James, a leading homotopy-theorist who held the famous Savilian Chair at Oxford once held by G. H. Hardy, and since retiring turned to mathematical biography, has published a collection of brief biographies concentrating on Jewish mathematicians and physicists [**4**]. James' book, coming out in 2009, 45 years after I first arrived in New Mexico from Palo Alto, is almost the first place I have seen Jewish over-representation acknowledged as an obvious fact of some interest.

What exactly is "over-representation"? Well, I haven't attempted a head count. For one thing, I don't want to get into the question, "Who is a Jew?" Carl Gustav Jacob Jacobi, Gotthold Eisenstein, and Leopold Kronecker are always listed as the first important Jewish mathematicians, yet all three were Christians—that is to say, they all three underwent conversion or baptism. A rabbi would say, "Jewish means son or daughter of a Jewish mother." Of Courant's two protégés at the Courant Institute, Kurt Friedrichs was "Jewish" only by marriage to a Jewish wife. Fritz John was "half-Jewish" (on his father's side). In the simplistic view of the general population, they were both somehow "Jewish refugees." Hermann Weyl, Hilbert's greatest pupil, was also "Jewish" only by virtue of his marriage to a Jewish woman. Cathleen Morawetz, of course, is really Irish, regardless of her Czech Jewish husband and married name. The famous topologist Mary Ellen Rudin at the University of Wisconsin is another daughter of gentiles with a Jewish name by marriage. (She was born and raised in rural Texas, and turned into a mathematician by Robert Lee Moore himself.) And what about the famous "refugee" algebraist, Emil Artin? He had no "Jewish blood," but he was married to the half-Jewish Natascha. One of the oft-repeated stories of Nazi idiocy is of the offer Artin received from Helmut Hasse (Courant's successor as head of mathematics at Göttingen). To convince Artin not to leave Göttingen for the U.S. in 1934, Hasse actually offered to have Artin's quarter-Jewish children declared officially "Aryan"! (Of course, no such offer could be made to their half-Jewish mother.) (By the way, Hasse, a German nationalist who comfortably served under the Nazis, was actually, secretly, contaminated by the blood of a Jewish forebear;

in fact, he was distantly related to the composer Felix Mendelssohn. Hasse was very proud of that before the Nazis, but tried to hide it after they came into power. Carl Ludwig Siegel, a colleague and a staunch anti-Nazi, always referred to him as Herr Hasse-Mendelssohn.)

Of course, it's not just America. As a grad student of Peter Lax, I became aware of the "Hungarian miracle," meaning, the amazing number of first-class Hungarian mathematicians in the 1920s and '30s. The list would start with the brothers Marcel and Frigyes Riesz, the collaborators Gabor Szego and George Polya, the "titled" John von Neumann and Theodore von Karman, and Lipot Fejer, Rozsa Peter, Michael Fekete, Paul Erdős, Paul Turan, Alfred Renyi, Arthur Erdelyi, Cornelius Lanczos. No one mentions the very strange fact that every single one of them was Jewish! (Nowadays, of course, not all great Hungarian mathematicians are Jews.)

Or look at Italy, whose Jews are Sephardic, not Ashkenazim [**1**]. We Ashkenazi don't recognize Italian names as Jewish. When Ascher (later to be Oscar) Zariski left the Ukraine, he went to Rome to study algebraic geometry with Guido Castelnuovo, Federigo Enriques, and Francesco Severi. Two out of the three were Jews. (Severi, the non-Jew, would later disgrace himself as a collaborator with Mussolini's fascism.) And there are more Italian Jews in mathematics: Giulio Ascoli, Vito Volterra, Guido Fubini, Luigi Cremona, both Corrrado and Beniamino Segre, and Salvatore Pincherle. (I am told that in Italy, a surname that is also a place name is an indicator of Jewishness.) Not to mention Beppo Levi and Tullio Levi-Civita.

So counting American Jewish mathematicians is a hopeless task for several reasons, of which the lack of a definition of "Jewish" is only one. Nevertheless, if we don't insist on numerical precision, I take it as obvious and uncontroversial that the proportion of Jews among American mathematicians has been, in recent decades, much greater than the proportion of Jews in the U.S. population as a whole. That is what I mean by "over-representation."

What is not usually mentioned is the remarkable contrast with the situation earlier—before World War II.

Under-Represented

There were really only four prominent Jewish mathematicians in the U.S. before World War II—James Joseph Sylvester, Norbert Wiener,

Solomon Lefschetz, and Salomon Bochner. Is that not strange? Huge crowds of Jews in the '50s and after—almost none in the '30s and earlier (apart from refugees arriving after 1934.)

Sylvester, of course, doesn't really count. He was English, not American. As a victim of English anti-Jewish discrimination, he came to the U.S. twice. He came at the very beginning of his career to the University of Virginia, where he was victimized as a Jew, a foreigner, and a suspected opponent of slavery, and forced to flee in fear of his life after an altercation with a "student" [**8**]. Then, much later, as a famous algebraist, he was recruited to create the first real math department in the U.S., at Johns Hopkins. After a few fruitful years there, he went home to receive, finally, his rightful position at Oxford, as Savilian professor.

Norbert Wiener was MIT's first, and for a long time almost the only, American Jewish mathematician. From his autobiographies [**14, 15**] we learn the astounding fact that he didn't learn he was Jewish until he was grown up, and then found out that he was possibly descended from the great and famous Jewish physician of the golden age of Arabic rule in Spain, Moses Maimonides. Norbert's father Leo was a Harvard professor of languages, and Norbert felt strongly that his own mathematical attainments entitled him to a position in the mathematics department at Harvard. However, that department never hired a Jew in a regular faculty position, until Oscar Zariski was hired in 1947. Why not Wiener? Well, there may have been more than one reason. Wiener was certainly a great mathematician, but he was also insecure, neurotic, alternately pretentious and apologetic, near-sighted, and rotund. Wiener himself was sure that the obstacle to his getting an offer from Harvard was the notorious, unconcealed anti-Semitic bias of the dominant, most influential Harvard mathematician, the great and famous George David Birkhoff. To be fair, Birkhoff was not equally and uniformly hostile to every single Jew. For example, he sponsored an invitation to the young Polish Jewish prodigy, Stanislaus Ulam, to become a member of the Harvard Fellows, in 1939 when Ulam was lucky enough to be visiting the U.S. at the time of Hitler's attack on Poland. Ulam was a banker's son, and had fine upper-class Polish manners. And ultimately, after the war, when Zariski received the long overdue offer from Harvard that let him escape from the heavy burden of teaching at Johns Hopkins, he was surprised and pleased to learn that Birkhoff had supported his joining the department. In fact, Birkhoff didn't mind having some Jewish students.

There is a solidly authenticated report of a phone conversation between Birkhoff and the chair of the math department at the University of Rochester, where a Jewish refugee had not received the job offer that Birkhoff felt Rochester should have made. It seems Birkhoff assumed, rightly or wrongly, that Rochester's failure to come through with the expected offer was an expression of anti-Semitic bias, for he was heard shouting over the phone at the Rochester math chairman, "Who do you think you are, Harvard?"

James [4] reports a conversation between Birkhoff and an officer of the Rockefeller Foundation (p. 260) "who noted afterwards, 'B. speaks long and earnestly concerning the Jewish question and the importation of Jewish scholars. He has no theoretical prejudice against the race and on the contrary every wish to be absolutely fair and sympathetic. He does however think that we must be more realistic than we are at present concerning the dangers in the situation and he is privately and entirely confidentially more or less sympathetic with the difficulties of Germany. He does not approve of their methods, but he is inclined to agree that the results were necessary.'"

Well, there's only one Harvard. What about the other elite Ivy Leaguers?

Yale? In 1947 Nathan Jacobson, the leading algebraist, was the first Jew to make it into Yale's math department. He wrote about some of his experiences as such, in notes to his collected papers [3]. There's a book by Oren [9] about how the barrier against Jews on the Yale faculty was gradually broken down.

Princeton? That looks better, for the great topologist Solomon Lefschetz joined that department as early as 1924. Lefschetz has a remarkable story. His family were Russian merchants who moved to Paris. In France he studied engineering rather than mathematics, because as a foreigner he had no chance for an academic appointment in that country. He came to the U.S. in 1905 to get some practical experience as an engineer. While he was working for Westinghouse in Pittsburgh, a terrible accident occurred. Both of his hands were destroyed! But instead of yielding to despair, he changed careers. He did graduate work at Clark University in Worcester, Massachusetts, earned a doctorate in algebraic geometry, and became a professor in Kansas. There, in total mathematical isolation, he made seminal discoveries in algebraic topology that attracted attention at Princeton. The American topologist

James Waddell Alexander, in the math department at Princeton, got Lefschetz a visiting appointment, and then a regular position. In time, Lefschetz became chairman at Princeton, and is given credit for the leadership that made it one of the foremost mathematics departments in the U.S. and in the world. All this is well known, and on the record. What I have not seen in print, but learned directly from Lefschetz's student Abe Hillman, is the fact that, as one would expect, the appointment of the foreign Jew Solomon Lefschetz to the faculty at Princeton was far from easy. The administration of Princeton University resisted bitterly. But Alexander was not only a leading topologist; he was a member of a first-rank family in Princeton, socially and financially [**5**]. He was a socialist, an active supporter of Norman Thomas's campaigns for the presidency. (He was also a prominent mountaineer; in fact, he preferred to enter his office by climbing the outside of Fine Hall and then coming in through the window.) His great-great-grandfather Archibald Alexander was the first professor and Principal of Princeton Theological Seminary, from 1812 until 1851. Several members of the family were president or vice-president of the Equitable Life Insurance Company. Alexander's father was a well-known artist, whose circle of friends in Paris and America included Claude Debussy, Henry James, Stephane Mallarme, Auguste Rodin, and John Singer Sargent. Because of his social and financial connections, he was able to bring pressure beyond what a mere mathematics department could exert, and succeed in making Solomon Lefschetz a Princeton professor.

Like Norbert Wiener, Solomon Lefschetz did not fit in perfectly with the WASP-y academia of the 1920s Ivy League. Gian-Carlo Rota has painted an unforgettable picture of him, in his memoir [**12**]. Far from being timid or retiring because of his severe physical handicap, Lefschetz was a roaring lion, fearless and intimidating in all mathematical or academic controversies. He became president of the American Mathematical Society in 1935, but not without opposition from G. D. Birkhoff, who wrote in a private letter to the secretary of the AMS, "I have a feeling that Lefschetz will be likely to be less pleasant even than he had been, in that from now on he will try to work strongly and positively for his own race. They are exceedingly confident of their own power and influence in the good old USA." Birkhoff was deluded. Far from favoring Jews, Lefschetz as a Princeton professor usually refused to accept Jewish students, for he thought they probably would not be able to get academic jobs. Perhaps he thought that having a Lefschetz as their adviser would only make it harder for them. Lefschetz did hire the Jewish refugee Salomon Bochner—a major coup in his campaign to raise his department to world class. But the two great mathematicians soon clashed. My friend Martin Davis recalls that at their beer parties the grad students made sure Solomon and Salomon did not overlap, for they did not speak to each other.

I am also indebted to Abe Hillman for some word-of-mouth history of Columbia University's math department. Like NYU, it is located in New York City, a major center of Jewish population. But unlike NYU, it doesn't seek to identify with the city; rather, it seeks to be viewed as in the same class as the other elite Ivy League schools, Princeton, Yale, and Harvard. Indeed, since its New York location might render it susceptible to a large Jewish participation, it has an even stronger motivation to preserve its non-Jewish image. Nevertheless, there was a Jewish mathematician at Columbia as early as 1900. Edward Kasner was the first Jewish appointee, and his appointment is credited to the efforts of his mentor Cassius Jackson Keyser, a leading and influential member of the Columbia math department. J. F. Ritt was appointed in 1921, the second Jewish member of the department, and the adviser of my friend Abe Hillman before he switched from Columbia to Princeton. Hillman told me that as a student Ritt had transferred from City College to George Washington University in his senior year, because he believed that a degree from City had some degree of Jewishness associated to it. He always signed himself as J. F. Ritt, not Joseph Fels, as another measure of self-protection.

A third Jewish mathematician of note was associated with Columbia. Jesse Douglas, a student of Kasner, was one of the very first winners of the Fields Medal, along with Lars Ahlfors in 1936, for his solution of the Plateau problem, to construct a minimal surface bounded by an arbitrary space curve. Douglas's name is almost forgotten today. He is a rather tragic figure, one of several important mathematicians gravely handicapped by what are now called bipolar, and used to be called manic-depressive, symptoms. He had a junior position at MIT, which he lost as a result of inability to perform consistently in the classroom. Although a Columbia graduate, and a member of the National Academy of Sciences, he never was offered a regular position at Columbia. According to Hillman, Ritt was opposed to hiring Douglas at Columbia, for two reasons: because it might attract unfavorable attention to have three Jews in the math department there, and also because Douglas was the student of Ritt's rival, Kasner, the other Jew in the math department. Douglas was forced to support himself by holding three different part-time teaching jobs in three different colleges in the New York area. Hillman was able to help him by recruiting support from Herbert Robbins, who was at Columbia, but in the statistics department (not the math department), and was willing to do Douglas a good turn. With Robbins' support, Douglas did get a full-time job at City College.

Here at the University of New Mexico, there was a brief interaction with the "Jewish problem" in the 1930s, now totally forgotten even though it was written up by Carroll Newsom [**7**] in his autobiography. Newsom was chairman of the math department at UNM in the 30s, before he went on to higher things as president of NYU and then head of the publishing giant Prentice-Hall. Newsom writes that he was asked to help in the effort to find jobs for Jewish refugees from Europe, and he decided to do so. In fact, he hired Arthur Rosenthal, who was a well-known Austrian Jewish mathematician, co-author of the major treatise *Set Functions*, with Hans Hahn [2]. Newsom writes that he was subject to serious attack by New Mexicans who objected to giving a job to a foreigner and a Jew. Newsom did not give in to this pressure, and in fact took pride in his own courage in hiring Rosenthal. Rosenthal left New Mexico after a few years, and went to Purdue.

So in elite U.S. math departments in the '20s and '30s of the last century there was no over-representation of Jews, but rather under-representation. Ralph Phillips, who got his degree in 1939, had the advantage or disadvantage that his name does not sound Jewish. He applied for jobs in a number of departments, and received invitations for interviews. But then, when his prospective employers met him and learned that he was Jewish, their interest in hiring him evaporated. MIT was one great university that canceled its interest in Phillips when they found out he was a Jew.

In his article about these experiences [**10**], Phillips mentions Birkhoff's malign influence. But Birkhoff was not without defenders. Saunders MacLane, who collaborated with G. D. Birkhoff's son Garrett in their well-known algebra textbook, wrote a response defending Birkhoff from Phillips [**6**]. He did not deny that Birkhoff was a bit of an anti-Semite, but he argued that it was unfair to single out Birkhoff, for in those days it was normal or common to be anti-Semitic—"everybody" did it. MacLane's defense seems defective, however, for Cassius Keyser and James Alexander prove that not everyone was anti-Semitic, even in the '20s and '30s.

Today

What this story makes plain is that there was a great transformation in American society, with respect to Jews, as a consequence of World War II. The U.S. was attacked by Hitler's ally, Japan, and so became committed to all-out war against Hitler and Nazism. But Hitler and Nazism meant first of all, and above all, extreme and unlimited hatred of Jews. So anti-Semitism became un-American. Hitler was America's enemy, and Hitler was the supreme anti-Semite. So it became untenable to exclude Jews from academia, or from Wall Street, or from the cabinet of the U.S. President.

Once the barriers were down, it turned out that a lot of Jewish students were interested in math, and before long under-representation became over-representation.

It is pretty clear that there used to be major cultural differences between the community of American Jews and the mainstream, non-Jewish, American community. Jews were bookish, studious; they were used to arguing and reading. Their tradition of business and commerce was associated with calculation and arithmetic. All this, it is easy to believe, makes it natural that a disproportionate number are attracted to math. Some people even think that there is something Talmudic about mathematics!

What is certain is that the Jewish domination of American mathematics has passed its peak. One need only look at the names of the winners of scientific talent contests in recent years. No longer are most of the names Jewish. Instead, most of the names are Chinese, or Vietnamese, or Japanese, or Korean, or Indian. There are some Jewish names, but most of these turn out to be the names of children who have come to the U.S. from Bulgaria or Romania. American-born Jews are a diminishing presence in American mathematics.

Why? Easily explained, although again I only have strong impressions and anecdotal evidence. To put it in brief, we have become assimilated and Americanized. Unlike our grandparents' generation, we are just as likely to play golf and drink cocktails as the gentiles (don't say *goyim*). Our bright youngsters choose law school or business school, not science. We get divorced; even vote Republican. Jews have been accepted in America, and so (allowing for lots of exceptions) we have become, more and more, just like other Americans.

Acknowledgments

Thanks to Jerry Alexanderson, Chandler Davis, Martin Davis, Bonnie Gold, Jerry Goldstein, Richard J. Griego, Michael Henle, Peter Lax, Elena Marchisotto, Warren Page, Peter Ross, and Steve Rosencrans for helpful comments and suggestions.

References

1. A. Guerraggio and P. Nastasi, *Italian Mathematics Between the Two World Wars*, Birkhauser Verlag, 2005.
2. H. Hahn and A. Rosenthal, *Set Functions*, University of New Mexico Press, 1948.
3. N. Jacobson, "A personal history and commentary," in *Collected Mathematical Papers*, Birkhauser, 1989.
4. I. James, *Driven to Innovate: A Century of Jewish Mathematicians and Physicists*, Peter Lang, Oxford, 2009.
5. ———, *Remarkable Mathematicians*, Cambridge University Press, Cambridge, U.K., 2002.
6. S. MacLane, "Jobs in the 1930s and the views of George D. Birkhoff," *Math. Intelligencer* **16** (1994) 9–10. doi:10.1007/BF03024350.
7. C. V. Newsom, *Problems Are for Solving: An Autobiography*, Dorrance, Bryn Mawr, Pa., 1983.
8. J. J. O'Connor and E. F. Robertson, *James Joseph Sylvester*, MacTutor website; available at http://www–history.mcs.st–andrews.ac.uk/Biographies/Sylvester.html.
9. D. A. Oren, *Joining the Club*, Yale University Press, 2001.
10. R. Phillips, "Reminiscences about the 1930s," *Math. Intelligencer*, **16** (1994) 6–8. doi:10.1007/BF03024349.
11. C. Reid, *Hilbert-Courant*, Springer-Verlag, New York, 1986.
12. G.-C. Rota, *Indiscrete Thoughts*, Birkhauser, 1997.
13. R. Siegmund-Schultze, *Mathematicians Fleeing from Nazi Germany: Individual Fates and Global Impact*, Princeton University Press, Princeton N.J. 2009.
14. N. Wiener, *Ex-Prodigy, My Childhood and Youth*, MIT Press, Cambridge, Mass., 1953.
15. ———, *I Am a Mathematician*, MIT Press, Cambridge, Mass., 1956.

Did Over-Reliance on Mathematical Models for Risk Assessment Create the Financial Crisis?

David J. Hand

The German chemist Baron Justus von Liebig said, "We are too much accustomed to attribute to a single cause that which is the product of several" (von Liebig, 1872). It seems to me that this is perfectly reasonable. So that our limited human brains can cope with the awesome complexity of the natural world, and the corresponding complexity of the artificial globalized society and economy we have constructed, we naturally seek to reduce them to simple components. And then we try to identify the most important of these components and focus our attention on this.

An illustration of this single cause attribution in the current economic crisis is given by the case of Joe Cassano. Cassano was head of the financial products division of AIG and was responsible for underwriting several hundred billions of dollars worth of debt in the form of credit default swaps. Jackie Speier, a member of the U.S. House of Representatives, is quoted as saying that Cassano is "almost single-handedly . . . responsible for bringing AIG down and by reference the economy of [the US]".

But the fact is that, as Justus von Liebig noted, such single cause attribution is incorrect. The truth is almost always more complex than this. In the present economic crisis, a variety of contributory factors came together, to act synergistically, although many of these have been singled out for special attention. A number of books and reports (e.g., Turner, 2009) have now appeared which attempt to disentangle the various threads which intertwined to precipitate the crisis, but the meeting at which this paper was presented did not attempt to tease these threads apart. Rather, it focused on just one of them: the role of mathematics and mathematical models in the crisis. From that perspective, what is important is that the various events produced an environment stimulating financial innovation and a huge growth in securitized credit instruments and derivatives, with the result being a dramatically increased mathematical complexity in the tools for modeling and pricing financial risk.

The particular question I was asked to address is the one I have adopted as my title. Note that it uses the word "over-reliance." The Turner report is equally careful in its choice of words. Neither asks if the mathematics was incorrect, or even if it was inappropriate, but they merely raise the question of whether people were unwise in the extent to which they relied on it. Noting that, I nevertheless think it will clarify things if we put the question in a broader context. So, we can ask several rather distinct questions.

First, was the mathematics wrong? Whether wrong or not, we can also ask, was the mathematics unrealistic? And if it was unrealistic we can go on to ask if it was based on false premises, or was inappropriate in some way.

We can also ask if any limitations of the mathematics were not communicated properly. And then, if so, we have to note that communication is a two-way process, and ask why the failure in communication occurred. Was it that someone didn't raise appropriate warnings? Or was it that warnings were sounded, but others didn't listen? And then, if people didn't listen, why not? Was it that they couldn't understand the warnings being given, or were they unwilling to listen?

While I don't intend to discuss each of these questions in detail, I will make some comments on some of them. So, first, was the mathematics wrong?

The short answer to this is "no." Mathematics is concerned with deducing the consequences of a given set of initial premises. While it is possible to make mistakes in the deductive process, in the present case, since so many people were using and had checked the deductions, it is essentially inconceivable that the mathematics was wrong.

But perhaps that's a very purist view of mathematics. To a lay person, if a piece of mathematics ends up drawing incorrect conclusions it may well be a quibble as to whether the mathematics per se is wrong or whether the problems lie elsewhere. To take an example from my own work, I have recently demonstrated a fundamental flaw in a particular index used very widely for measuring the performance of risk score cards in the personal banking sector (Hand, 2009). While the deductive mathematics underlying this measure is fine, what is not so fine are the assumptions on which the mathematics is built. Buried deep in these assumptions is one which means that different risk scorecards are evaluated using different measures. To a lay person, who does not distinguish between the underlying assumptions and the deductive process, the consequence is simply that one cannot apply this mathematically derived measure: that the mathematics is "wrong."

Although, in this example, there is nothing incorrect with the mathematics per se, the mathematics is somehow missing the point. "Missing the point" is a consequence of basing the analysis on false premises. The notion that sophisticated securitized credit instruments improve financial stability appears to have been a fundamental premise, a core belief, but it appears to have little or no supporting empirical evidence. Similarly, there appears to have been an assumption that natural selection in the financial markets would mean that innovations, and in particular mathematical innovations, would be beneficial, since those that did not work would be selected out. If something failed, people would not use it. In fact, however, anyone familiar with evolutionary processes will know that evolution can lead in unpredictable directions, especially in complex environments. In any case, it is important to be clear what we mean here. By "beneficial" we mean beneficial to the economy, and the people in it, in some sense. But evolution is blind. Evolution does not tend towards a particular prespecified objective.

A familiar premise of the mathematical models was that assets could be sold rapidly and easily if necessary. But, as we saw some years ago, with the collapse of Long-Term Capital Management, this is not true if everyone tries to do it simultaneously. We have an unmeasured liquidity risk, so that one of the premises of the mathematical models fails.

At a lower level, there are also criticisms of things such as assumed distributional forms, and independence of events and players. Normality is well-known not to apply here. In fact, as statisticians know very well, normal distributions do not occur in nature. A 1989 paper in *Psychological Bulletin* (Micceri, 1989) has the provocative title "The Unicorn, the Normal Curve, and Other Improbable Creatures." Of course, statisticians are aware that certain kinds of statistical techniques are robust to non-normality—such as distributions of the means of samples. But they also know this is not the case for measures describing the size of the tails of distributions, which is often of primary concern in risk evaluation.

The eminent statistician George Box said that "all models are wrong, but some are useful" (Box, 1979). In particular, this means that one should always have a healthy skepticism about models. Furthermore this skepticism should be greater for some usages: consideration of tail areas of distributions should engender more skepticism than results based on the central limit theorem. To put it bluntly, one should avoid the hubris of assuming that one's models are correct. One should always allow for model uncertainty when deciding what capital reserves to keep, what risks to take. While the model might tell us how risky a venture appears to be, and how much it seems that we need to keep in reserve to allow for that eventuality, we should always ask, "But what if the model is wrong?"

The late Leo Breiman had something to say about this. He said, "When a model is fit to data to draw quantitative conclusions . . . the conclusions are about the model's mechanism, and not about nature's mechanism . . . It follows that if the model is a poor emulation of nature, the conclusions may be wrong" (Breiman, 2001).

My own view here is that we are really trying to model human behavior. And humans can be unpredictable. They can even be perverse.

To drive home the dangers and the difficulties of building models which adequately reflect human behavior, I want to describe a very simple example from my own research. Much of my work is concerned with building statistical models for the retail banking sector: that sector concerned with credit cards, personal loans, car finance, individual current accounts, mortgages, and so on. The last, of course, is particularly relevant since the subprime crisis was certainly one of the precipitating factors. The illustration, however, concerns how people behave when using credit cards in petrol stations, and, in particular, the shape of the distribution of amount spent per transaction, as shown in the card transaction data. Full details of this analysis are given in Hand and Blunt (2001).

One might predict that the shape of the distribution would be roughly normal, but with a slight right skew since the values can only be positive. And, broadly speaking, this is correct. However, what is unlikely to be predicted is a number of very pronounced spikes in the distribution. The data set is very large, so any spikes represent real underlying behavioral aspects—they are not random fluctuations.

Investigation soon reveals an explanation for some of these spikes: they arise as a consequence of marketing initiatives by the petrol companies, incentivizing customers to spend more. Others, however, do not have so ready an explanation.

In particular, close examination of the histogram of transaction amounts shows pronounced spikes at values ending in £5 and £10. It seems that many people, even though paying by credit card, choose to spend round numbers of pounds. Supporting evidence for this explanation is given by the existence of slight tails to the right, but not to the left, of these values: it is as if people were aiming at £20 (for example) and occasionally overshooting by accident.

Continuing the analysis, even closer examination shows that there are also spikes, albeit not so pronounced, at other whole numbers of pounds. For example, the histogram cells at £13 and £16 are substantially higher than the cells of widths a penny or two either side of these values. People clearly prefer to spend whole numbers of pounds.

Hand and Blunt (2001) explore the data in more depth. The deliberate choice of round numbers does not stop at whole numbers of pounds. There are also spikes at 50p, and also at 25p and 75p, and also at other values ending in 5p and 10p, all progressively lower.

The point of this little example is to show that data describing human behavior, even apparently simple behavior, can conceal unimagined complexities. When dealing with human beings we are not merely dealing with intrinsic randomness—as when we study the intrinsic randomness in quantum physics. When we study human beings we certainly have intrinsic randomness, but we also have other things to contend with—human motivations, intransigence, greed, and so on. Electrons may have their uncertainties, but they are not greedy.

That brings me to another of the questions mentioned above. Was it in fact *inappropriate* mathematics?

Models such as pricing models are fine, in isolation, and at a low level. But difficulties can arise when they are put together, and embedded in a larger system. In such a situation a larger scale model is needed, a model of the entire complex system, and not of merely a tiny part of it. An econometric model, in fact.

I suppose a very concrete example of this sort of limitation is in automated trading systems which all react the same way to given market conditions—as we discovered in 1987. The correlation between their behavior induces a massive swing in one direction or another—a run on a stock or a drying up of liquidity. The famous Prisoner, facing his Dilemma, would doubtless have something to say here.

I think a key issue in all of this is *communication*. If there was an over-reliance on mathematical models it was the reliance placed on them by the higher echelons of management. Perhaps the phrase "naive belief" might be better than over-reliance.

A few weeks ago the *Financial Times* contacted me, asking me to comment on a criticism of the mathematicians who had developed the models, namely the suggestion that it was the mathematicians' fault, because they had been unable to communicate the risks to senior management. Apparently, their fault was that they had developed models which senior management could not understand. This seems an extraordinary position to me. It suggests that, despite not understanding the models and their implications, the managers were happy to act on the basis of the recommendations deriving from them. As I wrote in an article solicited by the London Mathematical Society, if I jumped into the cockpit of a Boeing 747, and crashed it because I didn't know how to fly it, you would hardly blame Joe Sutter, the 747 production chief.

It appears that Joe Cassano was put in charge of the AIG financial markets operation after it was up and running, and that he inherited the mathematical drivers without properly understanding them. If this is true, one might blame him for agreeing to run something he did not understand, and also blame those who appointed him for not grasping what was needed to run such an operation. The only defense I can think of, and a poor one at that, is that the mathematical models were developed after the senior bankers had begun their careers and were already in senior posts. Since the half-life of material learned at university is five years (a ballpark figure: clearly it varies between disciplines), this at least explains, even though it does not justify, why they didn't understand what they were doing.

With this breakdown in communication in mind, one might ask, were managers in fact given warnings on which they failed to act? Such a scenario is not at all far-fetched. Harry Markopolos had been trying to raise concerns about Bernard Madoff since 1999. In 2005 he sent a report to the SEC, and the SEC made a cursory examination of Madoff and declared his activities legitimate.

More generally, if a given strategy appears to be earning large sums of money for your bank, and in particular large sums of money for you, then you might not be inclined to look too rigorously at criticisms of it. This is, perhaps, a natural human trait: along with greed, an unwillingness to believe bad news, and a tendency to follow the herd.

There have been many financial crises in the past. One has to ask whether the causes of this one are different in kind. It is certainly true that mathematical financial innovations played a role. It is also true that, had people been more aggressively warned about their limitations, arising from the premises on which they were based, and had they listened and taken those warnings on board, then things might well have turned out differently. But in order for that to happen, many other things would also have had to have been done differently. It seems to me that the bottom line is that, unless there is an incentive for people to behave differently, then greed will trump other factors. Fraud illustrates this, but there are also other structural incentive issues. For example:

- the fact of hedge fund managers taking a percentage of profit in good years, but not of loss in bad;
- the fact that if you tried to raise concerns you might be ignored;
- the possibility of the risk rating agencies benefiting if they gave good ratings;
- the fact that employees of regulatory authorities might subsequently want to work for the banks on huge salaries, and so would often be keen to maintain good relations;
- and the risk that moral hazard is built into the system, with a perception that people didn't need to worry about the premises because if things went wrong they would be baled out.

At bottom, can I really argue that one cannot blame the mathematicians? They built and applied the tools, so surely they should share the responsibility? This is, of course, well-worn ethical ground, having been covered in other contexts, such as the relation between nuclear physicists and the atomic bomb. My own view is that it is nonsensical to say that everyone has to share the blame. In a mugging, who bears the responsibility: the man who wields the knife, the owner of the cutlery factory which made it, the receptionist at the entrance of the cutlery factory?

I think it would be a terribly retrograde step if we experienced a backlash against quantitative tools. Analogous quantitative tools have revolutionized so many other aspects of banking, of customer relations, and of life in general, and have had an immense impact for the good. But the mathematical tools have to be used in a proper context. Putting the quants in a back room, instructing them to work their mathematical magic, and then blindly applying the results to the outside world without considering the wider implications is a recipe for disaster. And we are now consuming the product of that recipe.

Acknowledgments

This is a modified version of a paper presented at a meeting of the Foundation for Science and Technology on 10 June 2009.

References

Box G.E.P (1979) *Robustness in the Strategy of Scientific Model Building*. Technical Report, Madison Mathematics Research Center, Wisconsin University.

Breiman L. (2001) Statistical modeling: the two cultures. *Statistical Science*, **16**, 199–215.

Hand D.J. (2009) Measuring classifier performance: a coherent alternative to the area under the ROC curve. *Machine Learning*, **77**, 103–23.

Hand D.J. and Blunt G. (2001) Prospecting for gems in credit card data. *IMA Journal of Management Mathematics*, **12**, 173–200.

Micceri T. (1989) The Unicorn, the normal curve, and other improbable creatures. *Psychological Bulletin*, **105**, 156–66.

Turner A. (2009) *The Turner Review: a regulatory response to the global banking crisis*. Pub. ref. 003289, Financial Services Authority.

von Liebig J. (1872) In a personal communication to Emile Duclaux, quoted in Dubos, R. J. (1950) *Louis Pasteur, Free Lance of Science*, Little, Brown and Co., Boston.

Fill in the Blanks:

Using Math to Turn Lo-Res Datasets into Hi-Res Samples

Jordan Ellenberg

In the early spring of 2009, a team of doctors at the Lucile Packard Children's Hospital at Stanford University lifted a 2-year-old into an MRI scanner. The boy, whom I'll call Bryce, looked tiny and forlorn inside the cavernous metal device. The stuffed monkey dangling from the entrance to the scanner did little to cheer up the scene. Bryce couldn't see it, in any case; he was under general anesthesia, with a tube snaking from his throat to a ventilator beside the scanner. Ten months earlier, Bryce had received a portion of a donor's liver to replace his own failing organ. For a while, he did well. But his latest lab tests were alarming. Something was going wrong—there was a chance that one or both of the liver's bile ducts were blocked.

Shreyas Vasanawala, a pediatric radiologist at Packard, didn't know for sure what was wrong and hoped the MRI would reveal the answer. Vasanawala needed a phenomenally hi-res scan, but if he was going to get it, his young patient would have to remain perfectly still. If Bryce took a single breath, the image would be blurred. That meant deepening the anesthesia enough to stop respiration. It would take a full two minutes for a standard MRI to capture the image, but if the anesthesiologists shut down Bryce's breathing for that long, his glitchy liver would be the least of his problems.

However, Vasanawala and one of his colleagues, an electrical engineer named Michael Lustig, were going to use a new and much faster scanning method. Their MRI machine used an experimental algorithm called compressed sensing—a technique that may be the hottest topic in applied math today. In the future, it could transform the way that we look for distant galaxies. For now, it means that Vasanawala and Lustig needed only 40 seconds to gather enough data to produce a crystal-clear image of Bryce's liver.

Compressed sensing was discovered by chance. In February 2004, Emmanuel Candès was messing around on his computer, looking at an image called the Shepp-Logan Phantom. The image—a standard picture used by computer scientists and engineers to test imaging algorithms—resembles a *Close Encounters* alien doing a quizzical eyebrow lift. Candès, then a professor at Caltech, now at Stanford, was experimenting with a badly corrupted version of the phantom meant to simulate the noisy, fuzzy images you get when an MRI isn't given enough time to complete a scan. Candès thought a mathematical technique called l_1 minimization might help clean up the streaks a bit. He pressed a key and the algorithm went to work.

Candès expected the phantom on his screen to get slightly cleaner. But then suddenly he saw it sharply defined and perfect in every detail—rendered, as though by magic, from the incomplete data. Weird, he thought. Impossible, in fact. "It was as if you gave me the first three digits of a 10-digit bank account number—and then I was able to guess the next seven," he says. He tried rerunning the experiment on different kinds of phantom images; they resolved perfectly every time.

Candès, with the assistance of postdoc Justin Romberg, came up with what he considered to be a sketchy and incomplete theory for what he saw on his computer. He then presented it on a blackboard to a colleague at UCLA named Terry Tao. Candès came away from the conversation thinking that Tao was skeptical—the improvement in image clarity was close to impossible, after all. But the next evening, Tao sent a set of notes to Candès about the blackboard session. It was the basis of their first paper together, "Near Optimal Signal Recovery from Random Projections: Universal Encoding Strategies?" And over the next two years, they would write several more. That was the beginning of compressed sensing, or CS, the paradigm-busting field in mathematics that's reshaping the way people work with large data sets. Only six years old, CS has already inspired more than a thousand papers and pulled in millions of dollars in federal grants. In 2006, Candès' work on the topic was rewarded with the $500,000 Waterman Award, the highest honor bestowed by the National Science Foundation. It's not hard to see why. Imagine MRI machines that

Figure 1. How Math Gets the Grain Out

Compressed sensing is a mathematical tool that creates hi-res data sets from lo-res samples. It can be used to resurrect old musical recordings, find enemy radio signals, and generate MRIs much more quickly. Here's how it would work with a photograph.

1 Undersample

A camera or other device captures only a small, randomly chosen fraction of the pixels that normally comprise a particular image. This saves time and space.

2 Fill in the dots

An algorithm called l_1 minimization starts by arbitrarily picking one of the effectively infinite number of ways to fill in all the missing pixels.

3 Add shapes

The algorithm then begins to modify the picture in stages by laying colored shapes over the randomly selected image. The goal is to seek what's called *sparsity*, a measure of image simplicity.

4 Add smaller shapes

The algorithm inserts the smallest number of shapes, of the simplest kind, that match the original pixels. If it sees four adjacent green pixels, it may add a green rectangle there.

5 Achieve clarity

Iteration after iteration, the algorithm adds smaller and smaller shapes, always seeking sparsity. Eventually it creates an image that will almost certainly be a near-perfect facsimile of a hi-res one.

take seconds to produce images that used to take up to an hour, military software that is vastly better at intercepting an adversary's communications, and sensors that can analyze distant interstellar radio waves. Suddenly, data becomes easier to gather, manipulate, and interpret.

Compressed sensing works something like this: You've got a picture—of a kidney, of the president, doesn't matter. The picture is made of 1 million pixels. In traditional imaging, that's a million measurements you have to make. In compressed sensing, you measure only a small fraction—say, 100,000 pixels randomly selected from various parts of the image. From that starting point there is a gigantic, effectively infinite number of ways the remaining 900,000 pixels could be filled in.

The key to finding the single correct representation is a notion called sparsity, a mathematical way of describing an image's complexity, or lack thereof. A picture made up of a few simple, understandable elements—like solid blocks of color or wiggly lines—is sparse; a screenful of random, chaotic dots is not. It turns out that out of all the bazillion possible reconstructions, the simplest, or sparsest, image is almost always the right one or very close to it.

But how can you do all the number crunching that is required to find the sparsest image quickly? It would take way too long to analyze all the possible versions of the image. Candès and Tao, however, knew that the sparsest image is the one created with the fewest number of building blocks. And they knew they could use l_1 minimization to find it and find it quickly.

To do that, the algorithm takes the incomplete image and starts trying to fill in the blank spaces with large blocks of color. If it sees a cluster of green pixels near one another, for instance, it might plunk down a big green rectangle that fills the space between them. If it sees a cluster of yellow pixels, it puts down a large yellow rectangle. In areas where different colors are interspersed, it puts down smaller and smaller rectangles or other shapes that fill the space between each color. It keeps doing that over and over. Eventually it ends up with an image made of the smallest possible combination of building blocks and whose 1 million pixels have all been filled in with colors.

That image isn't absolutely guaranteed to be the sparsest one or the exact image you were trying to reconstruct, but Candès and Tao have shown mathematically that the chance of its being wrong is infinitesimally small. It might still take a few hours of laptop time, but waiting an extra hour for the computer is preferable to shutting down a toddler's lungs for an extra minute.

Compressed sensing has already had a spectacular scientific impact. That's because every interesting signal is sparse—if you can just figure out the right way to define it. For example, the sound of a piano chord is the combination of a small set of pure notes, maybe five at the most. Of all the possible frequencies that might be playing, only a handful are active; the rest of the landscape is silent. So you can use CS to reconstruct music from an old undersampled recording that is missing information about the sound waves formed at certain frequencies. Just take the material you have and use l_1 minimization to fill in the empty spaces in the sparsest way. The result is almost certain to sound just like the original music.

With his architect glasses and slightly poufy haircut, Candès has the air of a hip geek. The 39-year-old Frenchman is soft-spoken but uncompromising when he believes that something isn't up to his standards. "No, no, it is nonsense," he says when I bring up the work of a CS specialist whose view on a technical point differs—very slightly, it seems to me—from his own. "No, no, no, no. It is nonsense and it is nonsense and it is wrong."

Candès can envision a long list of applications based on what he and his colleagues have accomplished. He sees, for example, a future in which the technique is used in more than MRI machines. Digital cameras, he explains, gather huge amounts of information and then compress the images. But compression, at least if CS is available, is a gigantic waste. If your camera is going to record a vast amount of data only to throw away 90 percent of it when you compress, why not just save battery power and memory and record 90 percent less data in the first place? For digital snapshots of your kids, battery waste may not matter much; you just plug in and recharge. "But when the battery is orbiting Jupiter," Candès says, "it's a different story." Ditto if you want your camera to snap a photo with a trillion pixels instead of a few million.

The ability to gather meaningful data from tiny samples of information is also enticing to the military: Enemy communications, for instance, can hop from frequency to frequency. No existing hardware is fast enough to scan the full range. But the adversary's signal, wherever it is, is sparse—built up from simple signals in some relatively tiny but unknown portion of the frequency band. That means CS could be used to distinguish enemy chatter on a random band from crackle. Not surprisingly, DARPA, the Defense Department's research arm, is funding CS research.

Compressed sensing isn't useful just for solving today's technological problems; the technique will help us in the future as we struggle with how to treat the vast amounts of information we have in storage. The world produces untold petabytes of data every day—data that we'd like to see packed away securely, efficiently, and retrievably. At present, most of our audiovisual info is stored in sophisticated compression formats. If, or when, the format becomes obsolete, you've got a painful conversion project on your hands. But in the CS future, Candès believes, we'll record just 20 percent of the pixels in certain images, like expensive-to-capture infrared shots of astronomical phenomena. Because we're recording so much less data to begin with, there will be no need to compress. And instead of steadily improving compression algorithms, we'll have steadily improving decompression algorithms that reconstruct the original image more and more faithfully from the stored data.

That's the future. Today, CS is already rewriting the way we capture medical information. A team at the University of Wisconsin, with participation from GE Healthcare, is combining CS with technologies called HYPR and VIPR to speed up certain kinds of magnetic resonance scans, in some cases by a factor of several thousand. (I'm on the university's faculty but have no connection to this particular research.) GE Healthcare is also experimenting with a novel protocol that promises to use CS to vastly improve observations of the metabolic dynamics of cancer patients. Meanwhile, the CS-enabled MRI machines at Packard can record images up to three times as quickly as conventional scanners.

And that was just enough for 2-year-old Bryce. Vasanawala, in the control room, gave the signal; the anesthesiologist delivered a slug of sedative to the boy and turned off his ventilator. His breathing immediately stopped. Vasanawala started the scan while the anesthesiologist monitored Bryce's heart rate and blood oxygenation level. Forty seconds later, the scan was done and Bryce had suffered no appreciable oxygen loss. Later that day, the CS algorithm was able to produce a sharp image from the brief scan, good enough for Vasanawala to see the blockages in both bile ducts. An interventional radiologist snaked a wire into each duct, gently clearing the blockages and installing tiny tubes that allowed the bile to drain properly. And with that—a bit of math and a bit of medicine—Bryce's lab test results headed back to normal.

The Great Principles of Computing

Peter J. Denning

Computing is integral to science—not just as a tool for analyzing data, but as an agent of thought and discovery.

It has not always been this way. Computing is a relatively young discipline. It started as an academic field of study in the 1930s with a cluster of remarkable papers by Kurt Gödel, Alonzo Church, Emil Post, and Alan Turing. The papers laid the mathematical foundations that would answer the question "what is computation?" and discussed schemes for its implementation. These men saw the importance of automatic computation and sought its precise mathematical foundation. The various schemes they each proposed for implementing computation were quickly found to be equivalent, as a computation in any one could be realized in any other. It is all the more remarkable that their models all led to the same conclusion that certain functions of practical interest—such as whether a computational algorithm (a method of evaluating a function) will ever come to completion instead of being stuck in an infinite loop—cannot be answered computationally.

At the time that these papers were written, the terms "computation" and "computers" were already in common use, but with different connotations from today. Computation was taken to mean the mechanical steps followed to evaluate mathematical functions; computers were people who did computations. In recognition of the social changes they were ushering in, the designers of the first digital computer projects all named their systems with acronyms ending in "-AC", meaning automatic computer—resulting in names such as ENIAC, UNIVAC, and EDSAC.

At the start of World War II, the militaries of the United States and the United Kingdom became interested in applying computation to the calculation of ballistic and navigation tables and to the cracking of ciphers. They commissioned projects to design and build electronic digital computers. Only one of the projects was completed before the war was over. That was the top-secret project at Bletchley Park in England, which cracked the German Enigma cipher using methods designed by Alan Turing.

Many people involved in those projects went on to start computer companies in the early 1950s. Universities began offering programs of study in the new field in the late 1950s. The field and the industry have grown steadily into a modern behemoth whose Internet data centers are said to consume almost three percent of the world's electricity.

During its youth, computing was an enigma to the established fields of science and engineering. At first, computing looked like only the applied technology of math, electrical engineering, or science, depending on the observer. However, over the years, computing provided a seemingly unending stream of new insights, and it defied many early predictions by resisting absorption back into the fields of its roots. By 1980 computing had mastered algorithms, data structures, numerical methods, programming languages, operating systems, networks, databases, graphics, artificial intelligence, and software engineering. Its great technological achievements—the chip, the personal computer, and the Internet—brought it into many lives. These advances stimulated more new subfields, including network science, Web science, mobile computing, enterprise computing, cooperative work, cyberspace protection, user-interface design, and information visualization. The resulting commercial applications have spawned new research challenges in social networks, endlessly evolving computation, music, video, digital photography, vision, massive multiplayer online games, user-generated content, and much more.

The name of the field has changed several times to keep up with the flux. In the 1940s it was called *automatic computation* and in the 1950s, *information processing*. In the 1960s, as it moved into academia, it acquired the name *computer science* in the U.S. and *informatics* in Europe. By the 1980s computing comprised a complex of related fields, including computer science, informatics, computational science, computer engineering, software engineering, information systems, and information technology. By 1990 the term *computing* had become the standard for referring to this core group of disciplines.

Computing's Paradigm

Traditional scientists frequently questioned the name *computer science*. They could easily see an engineering paradigm (design and implementation of systems) and a mathematics paradigm (proofs of theorems), but they could not see much of a science paradigm (experimental verification of hypotheses). Moreover, they understood science as a way of dealing with the natural world, and computers looked suspiciously artificial.

The founders of the field came from all three paradigms. Some thought computing was a branch of applied mathematics, some a branch of electrical engineering, and some a branch of computational-oriented science. During its first four decades, the field focused primarily on engineering: The challenges of building reliable computers, networks, and complex software were daunting and occupied almost everyone's attention. By the 1980s these challenges largely had been met, and computing was spreading rapidly into all fields, with the help of networks, supercomputers, and personal computers. During the 1980s computers became powerful enough that science visionaries could see how to use them to tackle the hardest questions—the "grand challenge" problems in science and engineering. The resulting "computational science" movement involved scientists from all countries and culminated in the U.S. Congress's adoption of the High-Performance Computing and Communications (HPCC) Act of 1991 to support research on a host of large problems.

Today, there is an agreement that computing *exemplifies* science and engineering, and that neither science nor engineering *characterizes* computing. Then what does? What is computing's paradigm?

The leaders of the field struggled with this paradigm question from the beginning. Along the way, there were three waves of attempts to unify views. Allen Newell, Alan Perlis, and Herb Simon led the first one in 1967. They argued that computing was unique among all the sciences in its study of information processes. Simon, a Nobel laureate in economics, went so far as to call computing a science of the artificial. A catchphrase of this wave was "computing is the study of phenomena surrounding computers."

The second wave focused on programming, the art of designing algorithms that produce information processes. In the early 1970s, computing pioneers Edsger Dijkstra and Donald Knuth took strong stands favoring algorithm analysis as the unifying theme. A catchphrase of this wave was "computer science equals programming." In recent times, this view has foundered because the field has expanded well beyond programming, whereas the public understanding of a programmer has narrowed to just those who write code.

The third wave came as a result of the Computer Science and Engineering Research Study (COSERS), led by Bruce Arden in the late 1970s. Its catchphrase was "computing is the automation of information processes." Although its final report successfully exposed the science in computing and explained many esoteric aspects to the layperson, its central view did not catch on.

An important aspect of all three definitions was the positioning of the computer as the object of attention. The computational-science movement of the 1980s began to step away from that notion, adopting the view that computing is not only a tool for science, but also a new method of thought and discovery in science. The process of dissociating from the computer as the focal point came to completion in the late 1990s when leaders in the field of biology—epitomized by Nobel laureate David Baltimore and echoing cognitive scientist Douglas Hofstadter—said that biology had become an information science and DNA translation is a natural information process. Many computer scientists have joined biologists in research to understand the nature of DNA information processes and to discover what algorithms might govern them.

Take a moment to savor this distinction that biology makes. First, some information processes are natural. Second, we do not know whether all natural information processes are produced by algorithms. The second statement challenges the traditional view that algorithms (and programming) are at the heart of computing. Information processes may be more fundamental than algorithms.

Scientists in other fields have come to similar conclusions. They include physicists working with quantum computation and quantum cryptography, chemists working with materials, economists working with economic systems, and social scientists working with networks. They have all said that they have discovered information processes in their disciplines' deep structures. Stephen Wolfram, a physicist and creator of the software program *Mathematica*, went further, arguing that information processes underlie every natural process in the universe.

All this leads us to the modern catchphrase: "Computing is the study of information processes, natural and artificial." The computer is a tool in these studies but is not the object of study. As Dijkstra once said, "Computing is no more about computers than astronomy is about telescopes."

The term *computational thinking* has become popular to refer to the mode of thought that accompanies design and discovery done with computation. This term was originally called *algorithmic thinking* in the 1960s by Newell, Perlis, and Simon, and was widely used in the 1980s as part of the rationale for computational science. To think computationally is to interpret a problem as an information process and then seek to discover an algorithmic solution. It is a very powerful paradigm that has led to several Nobel Prizes.

Great Principles of Computing

The maturing of our interpretation of computing has given us a new view of the content of the field. Until the 1990s, most computing scientists would have said that it is about algorithms, data structures, numerical methods, programming languages, operating systems, networks, databases, graphics, artificial intelligence, and software engineering. This definition is a technological interpretation of the field. A scientific interpretation would emphasize the fundamental principles that empower and constrain the technologies.

My colleagues and I have developed the Great Principles of Computing framework to accomplish this goal. These principles fall into seven categories: computation, communication, coordination, recollection, automation, evaluation, and design *(see the first table for examples)*.

Each category is a perspective on computing, a window into the knowledge space of computing. The categories are not mutually exclusive. For example, the Internet can be seen as a communication system, a coordination system, or a storage system. We have found that most computing technologies use principles from all seven categories. Each category has its own weight in the mixture, but they are all there.

In addition to the principles, which are relatively static, we need to take account of the dynamics of interactions between computing and other fields. Scientific phenomena can affect one another in two ways: implementation and influence. A combination of existing things implements a phenomenon by generating its behaviors. Thus, digital hardware physically implements computation; artificial intelligence implements aspects of human thought; a compiler implements a high-level language with machine code; hydrogen and oxygen implement water; complex combinations of amino acids implement life.

Influence occurs when two phenomena interact with each other. Atoms arise from the interactions among the forces generated by protons,

Table 1.
The great principles of computing framework

Category	*Focus*	*Examples*
Computation	What can and cannot be computed	Classifying complexity of problems in terms of the number of computational steps to achieve a solution
Communication	Reliably moving information between locations	Information measured as entropy, compression of files, error-correcting codes, cryptography
Coordination	Effectively using many autonomous computers	Protocols that eliminate conditions that cause indeterminate results
Recollection	Representing, storing, and retrieving information from media	All storage systems are hierarchical, but no storage system can offer equal access time to all objects. All computations favor subsets of their data objects in any time interval
Automation	Discovering algorithms for information processes	Most heuristic algorithms can be formulated as searches over enormous data spaces. Many human cognitive processes can be modeled as information processes
Evaluation	Predicting performance of complex systems	Most computational systems can be modeled as networks of servers whose fast solutions yield close approximations of real throughput and response time

(*continued*)

TABLE 1.
(continued)

Category	*Focus*	*Examples*
Design	Structuring software systems for reliability and dependability	Complex systems can be decomposed into interacting modules and virtual machines. Modules can be stratified corresponding to their time scales of events that manipulate objects

Source: http://greatprinciples.org

neutrons, and electrons. Galaxies interact via gravitational waves. Humans interact with speech, touch, and computers. And interactions exist across domains as well as within domains. For example, computation influences physical action (electronic controls), life processes (DNA translation), and social processes (games with outputs). The second table illustrates interactions between computing and each of the physical, life, and social sciences, as well as within computing itself. There can be no question about the pervasiveness of computing in all fields of science.

What Are Information Processes?

There is a potential difficulty with defining computation in terms of information. Information seems to have no settled definition. Claude Shannon, the father of information theory, in 1948 defined information as the sequence of yes-or-no questions one must ask to decide what message was sent by a source. He purposely skirted the issue of the meaning of bit patterns, which seems to be important to defining information. In sifting through many published definitions, Paolo Rocchi in 2010 concluded that definitions of information necessarily involve an objective component—signs and their referents, or in other words, symbols and what they stand for—and a subjective component—meanings.

How can we base a scientific definition of information on something with such an essential subjective component?

Biologists have a similar problem with "life." Life scientist Robert Hazen notes that biologists have no precise definition of life, but they do have a list of seven criteria for when an entity is living. The observable affects of life, such as chemistry, energy, and reproduction, are sufficient to ground the science of biology. In the same way, we can ground a science of information on the observable affects (signs and referents) without having a precise definition of meaning.

A representation is a pattern of symbols that stands for something. The association between a representation and what it stands for can be recorded as a link in a table or database, or as a memory in people's brains. There are two important aspects of representations: *syntax* and *stuff*. Syntax is the rules for constructing patterns; it allows us to distinguish patterns that stand for something from patterns that do not. Stuff is the measurable physical states of the world that hold representations, usually in media or signals. Put these two together and we can build machines that can detect when a valid pattern is present.

A representation that stands for a method of evaluating a function is called an algorithm. A representation that stands for values is called data. When implemented by a machine, an algorithm controls the transformation of an input data representation to an output data representation. The algorithm representation controls the transformation of data representations. The distinction between the algorithm and the data representations is pretty weak; the executable code generated by a compiler looks like data to the compiler and like an algorithm to the person running the code.

Even this simple notion of representation has deep consequences. For example, as Gregory Chaitin has shown, there is no algorithm for finding the shortest possible representation of something.

Some scientists leave open the question of whether an observed information process is actually controlled by an algorithm. DNA translation can thus be called an information process; if someone discovers a controlling algorithm, it could be also called a computation.

Some mathematicians define computation as separate from implementation. They treat computations as logical orderings of strings in abstract languages, and are able to determine the logical limits of computation.

However, to answer questions about the running time of observable computations, they have to introduce costs—the time or energy of storing, retrieving, or converting representations. Many real-world problems require exponential-time computations as a consequence of these implementable representations. My colleagues and I still prefer to deal with implementable representations because they are the basis of a scientific approach to computation.

These notions of representations are sufficient to give us the definitions we need for computing. An information process is a sequence of representations. (In the physical world, it is a continuously evolving, changing representation.) A computation is an information process in which the transitions from one element of the sequence to the next are controlled by a representation. (In the physical world, we would say that each infinitesimal time and space step is controlled by a representation.)

Where Computing Stands

Computing as a field has come to exemplify good science as well as engineering. The science is essential to the advancement of the field because many systems are so complex that experimental methods are the only way to make discoveries and understand limits. Computing is now seen as a broad field that studies information processes, natural and artificial.

This definition is wide enough to accommodate three issues that have nagged computing scientists for many years: Continuous information processes (such as signals in communication systems or analog computers), interactive processes (such as ongoing Web services), and natural processes (such as DNA translation) all seemed like computation but did not fit the traditional algorithmic definitions.

The great-principles framework reveals a rich set of rules on which all computation is based. These principles interact with the domains of the physical, life, and social sciences, as well as with computing technology itself.

Computing is not a subset of other sciences. None of those domains is fundamentally concerned with the nature of information processes and their transformations. Yet this knowledge is now essential in all the other domains of science. Computer scientist Paul Rosenbloom of the University of Southern California in 2009 argued that computing is a new great domain of science. He is on to something.

TABLE 2.

How computing interacts with other domains of science

	Physical	*Social*	*Life*	*Computing*
Computing implemented by	mechanical, optical, electronic, quantum, and chemical computing	mechanical robots, human cognition, and games with inputs and outputs	genomic, neural, immunological, DNA translation, evolutionary computing	compilers, operating systems, emulation, abstractions, procedures, architectures, and languages
Computing implements	modeling, simulation, databases, data systems, quantum cryptography	artificial intelligence, cognitive modeling, autonomic systems	artificial life, biomimetics, systems biology	
Computing influenced by	sensors, scanners, computer vision, optical character recognition, localization	learning, programming, user modeling, authorization, and speech understanding	eye, gesture, expression, and movement tracking; biosensors	networking, security, parallel computing, distributed systems, grids
Computing influences	locomotion, fabrication, manipulation, open-loop control	screens, printers, graphics, speech generation, network science	bioeffectors, haptics, sensory immersion	
Bidirectional influence	robots, closed-loop control	Human–computer interaction, games	Brain–computer interfaces	

Note: Computing interacts in many ways with the other domains of science. Computing implements a phenomenon by generating its behaviors. Computing influences the other domains by interacting with them.

Bibliography

Arden, B. W., ed. 1983. *What Can Be Automated: Computer Science and Engineering Research Study (COSERS)*. Cambridge, MA: The MIT Press.

Bacon, D., and W. van Dam. 2010. Recent progress in quantum algorithms. *Communications of the ACM* 53:84–93.

Baltimore, D. 2001. How Biology Became an Information Science. In *The Invisible Future*, P. Denning, ed. New York: McGraw-Hill.

Chaitin, G. 2006. *Meta Math! The Quest for Omega*. New York: Vintage Press.

Denning, P. 2003. Great Principles of Computing. *Communications of the ACM* 46:15–20.

Denning, P. 2007. Computing is a natural science. *Communications of the ACM* 50:15—18.

Denning, P., and C. Martell. Great Principles of Computing Website. http://greatprinciples.org.

Denning, P., and P. Freeman. 2009. Computing's paradigm. *Communications of the ACM* 52: 28–30.

Hazen, R. 2007. *Genesis: The Scientific Quest for Life's Origins*. Washington, D.C.: Joseph Henry Press.

Hofstadter, D. 1985. *Metamagical Themas: Questing for the Essence of Mind and Pattern*. New York: Basic Books.

Newell, A., A. J. Perlis, and H. A. Simon. 1967. Computer science. *Science* 157:1373–4.

Rocchi, P. 2010. *Logic of Analog and Digital Machines*. Hauppauge, N.Y.: Nova Publishers.

Rosenbloom, P. S. 2004. A new framework for computer science and engineering. *IEEE Computer* 31–6.

Shannon, C., and W. Weaver. 1949. *The Mathematical Theory of Communication*. Champaign, Ill.: University of Illinois Press. Available at http://cm.bell-labs.com/cm/ms/what/shannonday/paper.html.

Simon, H. 1969. *The Sciences of the Artificial*. Cambridge, Mass.: MIT Press.

Wolfram, S. 2002. *A New Kind of Science*. Champaign, Ill.: Wolfram Media.

Computer Generation of Ribbed Sculptures

James Hamlin and Carlo H. Séquin

The 28-foot-tall *Solstice* sculpture by Charles Perry (Figure 1), located in downtown Tampa, Florida, is a prime example of the "ribbed sculptures" to be discussed here. Ribbed sculptures offer a translucent, "airy" presence in indoor as well as outdoor settings. Because of the substantial open space between the ribs, they do not cast harsh shadows or block views completely. Moreover, they are reasonably cost effective to be constructed at a large scale—much less expensive than large free-form bronze sculptures, investment cast from many individual molds.

These ribbed sculptures may trace their roots to the pioneering work of some constructivist artists as well as to the mathematical string models of conic and bilinear surfaces that one can find in science museums. The ribbed approach to defining a shape in space is particularly valid and economical at an architectural scale, as it makes use of Naum Gabo's vision that space could be represented without having to employ a lot of mass. Gabo demonstrated this principle with more than two dozen versions of his *Linear Construction* made of nylon monofilament strung over transparent plastic frames. Serious artists often implement many versions of some worthwhile concept, trying to find a perfect combination of the many variables that define a particular instance. With the availability of interactive computer graphics tools, we now have the possibility to do much of this exploration and fine-tuning with virtual models, if we succeed in capturing our conceptual idea in the form of a computer program with an appropriate set of adjustable control parameters. Such virtual evaluation may be just a convenience when designing table-top sculptures; it becomes a crucial tool in the design of architectural sculpture.

In this paper we discuss our efforts to capture a variety of ribbed sculptures by Charles Perry [4] in this manner. By extracting an implicit framework that underlies most of his ribbed sculptures, we define a

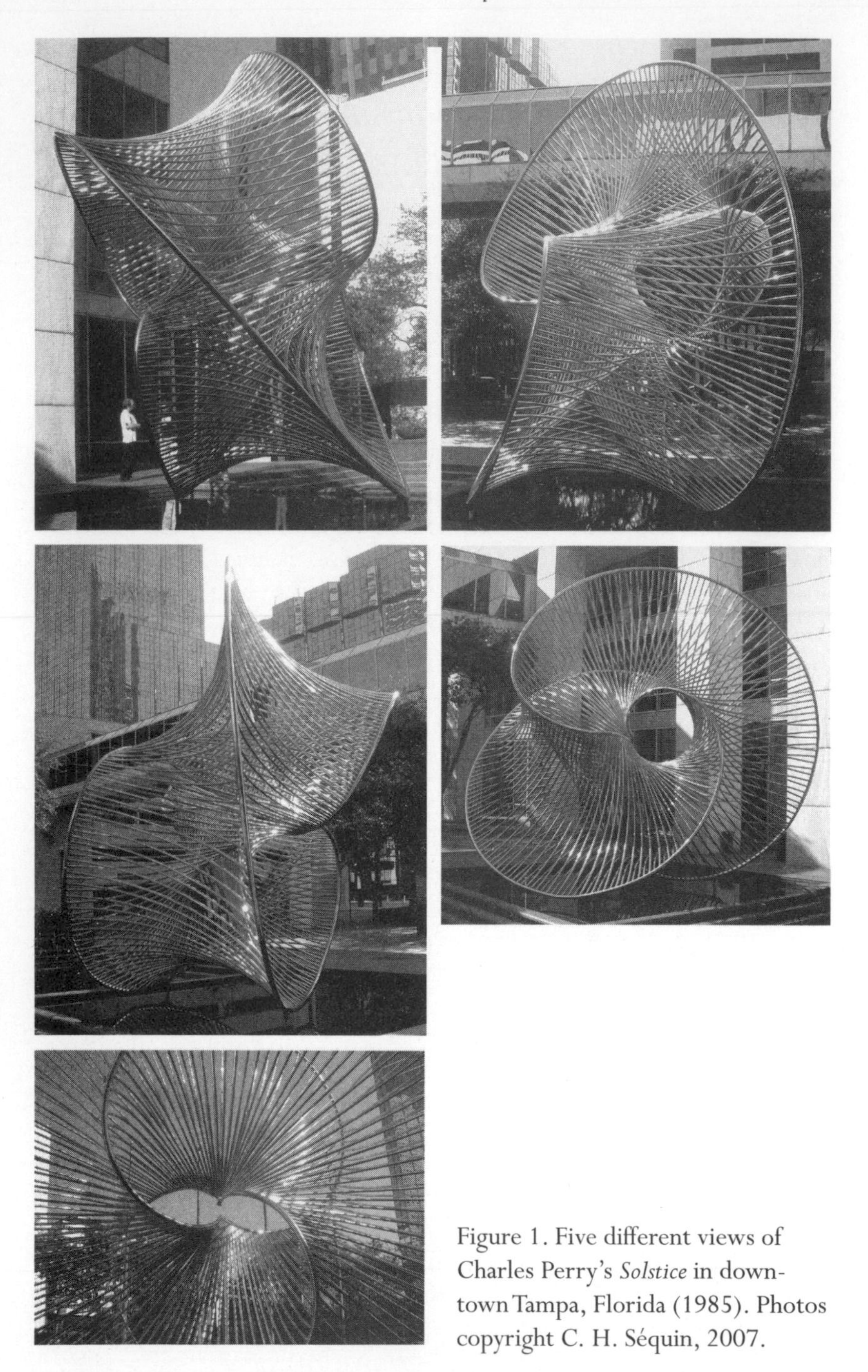

Figure 1. Five different views of Charles Perry's *Solstice* in downtown Tampa, Florida (1985). Photos copyright C. H. Séquin, 2007.

broad approach that enables a wide variety of new ribbed sculptures and mathematical visualization models. We will start our discussion with an analysis of *Solstice*, because it has a particularly compelling underlying representation that lends itself to an elegant parameterization.

As one walks around Perry's *Solstice* sculpture, one is amazed at the richness of diverse views that present themselves from different viewing directions (Figure 1). From some angles the sculpture looks like intertwined organic forms; other views inspire an association of a roller coaster on steroids. But from a few privileged vantage points, a wonderful symmetry is revealed, and the pattern becomes surprisingly regular. At this point an inquisitive mind just wants to know what is going on and whether there might be a simple generative principle that lies at the root of this elegant masterpiece. In the case of *Solstice*, this analytic task was made easy, since Charles Perry was quite forthcoming with explanations of how he planned this sculpture and how he went about constructing it. In a personal communication he wrote:

> The perimeter of *Solstice* is created by placing an equilateral triangle on a ring, so the centroid of the triangle connects with the ring. The triangle is rotated by two-thirds twists as it rotates around the ring. The figure produced by the three vertices of the triangle is a two-thirds twist torus Möbius.
>
> Intuition told me of the right diameter of the tube for the edge of the torus. I made a 12 inch model and worked from that some how. I found that there are four equal quarters going around the torus. I then made a full-scale mockup of 1/4th of the edge. It looked like a section of a roller coaster in my clean new studio. It was in three dimensions. I took this template to a tube-bender; they had a skilled old man who could bend the tube in compound curves to match the template. This was done in sections. In the studio I matched and welded these pieces. I had to cut off the excess ends. Now I had four equal tubes, probably about fifteen feet long. I then determined where the cross tubes would be closest to each other. This part was done by referencing the 1-foot model.
>
> Now, these holes for the cross tubes had to be a variable distance from each other and had to rotate around the edge tube as they progressed. Masking tape, Magic Marker and a center punch for each hole was the method. Certainly I had to measure the total length of each quarter edge and divide this into the number of

holes. All the cross tubes are equal. There are more than 600 cross tubes, and thus more than 1200 roto-broached holes, each at a different angle. The four equal edge pieces and the 600 tubes were shipped separately to Tampa.—I don't even know how I registered the four edge pieces when it was assembled in Tampa.

Perry's description and images of *Solstice* from different angles (Figure 1) allow an easy construction of a generating paradigm that can be captured in a computer program. Some of the crucial parameters in this generative program can then be made into variables, and by setting these variables to new (somewhat constrained) values, novel sculpture designs of "the same kind" can be generated. Before any of these modified designs are sent to the machine shop, one has to make an artistic judgment, whether the new forms have enough aesthetic merit to warrant an actual construction. If the answer is affirmative, then many more details will have to be worked out about how exactly to bend the many ribs into their specified shape and how to connect them to the supporting tubular rails. Considerable engineering effort goes into working out those details.

In this paper we are mainly concerned with the primary design aspect of variations of *Solstice* and of other ribbed sculptures by Charles Perry [4]. We have generalized this paradigm and captured it in several small computer programs that allow us to design a wide variety of such sculptures. Our new designs are presented in virtual form by means of computer graphics renderings. Most of the construction details and engineering issues are ignored at this stage. As the reader will see, even the geometrical design phase offers several intriguing puzzles and programming challenges.

Geometrical Emulation of Solstice

Based on the information obtained directly from Charles Perry, we know that the thicker tubular "guide rail" lies on the surface of a torus. In the case of *Solstice*, it forms a (3,2) torus knot, i.e., the guide rail runs three times around the big loop of the torus and passes twice through its tunnel before it closes again onto itself (Figure 2a).

The thinner "ribs" attached to this guide rail are not simply the edges of a rotating equilateral triangle; straight edges would look rather stiff. The ribs are planar curves, three of which form an approximate "hyperbolic"

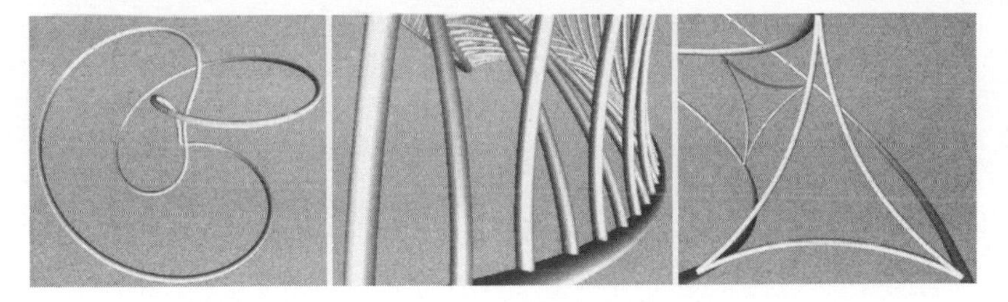

Figure 2. Graphical illustration of the key design parameters in the Solstice program: (a) the torus knot formed by the guide rail, (b) a rib-end offset resulting in "evenly" spaced ribs, (c) a negative bulge of the individual ribs.

triangle composed of three inward-bending concave circular arcs. Furthermore, the ribs do not form closed, three-sided, planar loops, but rather form a spiral staircase. The two ribs that would end in the same triangle vertex have been offset along the guide rail by half the distance between subsequent triangles, so that the ribs seem to land on the guide rails individually, with apparently uniform spacing (Figure 2b). Of course, on the inside of the torus the spacing is much denser than on the outside, since the truly relevant spacing parameter is the "equatorial" angle around the torus. Strictly speaking, this forces the geometry of the many ribs to vary ever so slightly. However, the geometrical deviations are small, and they are within the tolerances to which the actual tubular ribs can be bent. Thanks to their curvature, the ribs can thus readily be fit into the toroidal guide rail at assembly time.

To capture this constructive paradigm, we wrote a program module to generate a guide rail in the form of a sweep surface along an arbitrary (p,q) torus knot. The parametric representation of this sweep line, lying on a torus surface with big radius R and small radius r, is:

$$x = \left(R + r\cos\left(\frac{q\phi}{p}\right)\right)\cos\phi$$

$$y = \left(R + r\cos\left(\frac{q\phi}{p}\right)\right)\sin\phi$$

$$z = r\sin\left(\frac{q\phi}{p}\right)$$

where $0 \le \phi \le 2p\pi$.

This loose frame is then populated with a parameterized set of ribs, which themselves are circular arcs. More explicitly, the guide rail is specified by the integer constants p and q of the torus knot, by the radii, R and r, of the major and minor circles that define the torus on which this knot is embedded, and by the diameter of the guide rail tube itself.

Next we specify the total number of ribs and the offset of the two rib endpoints from an exact cross-sectional plane (cutting a minor circle from the tube of the torus). This offset is measured as an angle along the circular sweep path that defines the major loop of the torus. An offset of zero keeps the ribs entirely in the cross-sectional plane, and an offset of 1 degree would move one of the endpoints of the rib forward by that amount in the major sweep direction. Actually, in our programs this offset is typically defined as a fraction of the total sweep angle of the torus knot guide rail. This makes it easier to interleave properly the rib endings on the guide rail. For a total number of 500 ribs an offset of 0.1% would evenly stagger all rib endings with a minimal amount of helical twist.

The individual ribs themselves are circular arcs between their two endpoints on the guide rail. The amount of arching of each rib can be specified as the turning angle that this circular arch segment is bending through. Alternatively, the amount of bulging can be characterized by the maximal distance of the arc from the chord connecting the two rib endpoints. Different versions of our programs have used different approaches. In either case, the amount of bulging is normalized so that a unit of "1" leads to ribs that would hug the torus surface (at least for small values of the offset parameter); this normalization varies with the variable p. A "bulge" of zero always results in a straight rib, and a negative bulge value indicates that the arc is curving in the inward direction (Figure 2c).

By tuning all the above mentioned parameters carefully, a rather faithful emulation of Perry's *Solstice* sculpture is obtained. In the emulation shown in Figure 3 the program parameters listed in Table 1 were used. (As far as the number of ribs is concerned, we guess that Perry's memory was off by a factor of 2.)

Solstice *Variations*

With the basic generating paradigm captured in our program, it is now easy to make variations of this sculpture. In a first example we reduce the amount of twisting in the overall toroid to obtain a shape more

TABLE 1.
Program parameters used in the emulation of Perry's *Solstice* sculpture

Parameter	*Quantity*
p	3
q	2
R	6.5
r	6.0
Guide-rail diameter	0.15
Rib diameter	0.1
Number of ribs	300
Rib offset	1.8°
Rib bulge	−0.5

closely related to Helaman Ferguson's bronze sculpture called *Umbilic Torus NC* [1], in which the total twist of the triangular cross section is only 120° (Figure 4a). This can be readily achieved by changing the parameter q to 1, yielding a guide rail in the form of a (3,1) torus knot—which is actually not knotted at all.

Alternatively we may choose to increase the twist in the toroidal sweep structure. Avoiding the case (3,3), where the guide rail would break up into 3 separate loops, the next connected candidate is the (3,4) torus knot. This however looks too twisty for our taste; instead we explore the case of the (4,3) torus knot. This results in a quadrilateral cross section that makes a 3/4 turn while traveling once around the toroidal sweep (Figure 4b). This configuration certainly has the potential for another large-scale ribbed sculpture.

Another experiment is to simply switch the values of p and q of the original *Solstice*, thereby generating a (2,3) torus knot, while leaving the rib specifications unchanged. Even though topologically the structure of the knot has not been changed by this switch, the result now looks quite different (Figure 4c). There are two main reasons: First, the geometry of the guide rail now has a totally different structure, even though it describes the same mathematical knot; second, the behavior of the ribs has changed dramatically. There are now only two passes of the guide rail through every minor circle of the torus, and this is not sufficient to

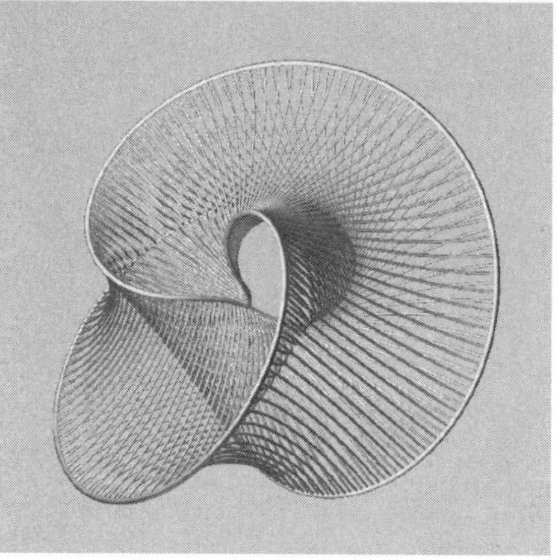

Figure 3. Emulation of *Solstice*: (a) Perry's sculpture in Tampa (Photo copyright C. H. Séquin, 2007), (b) computer emulation shown from the same angle.

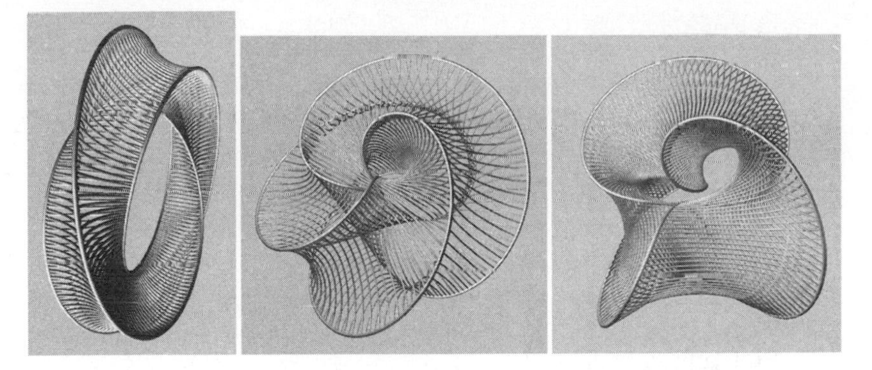

Figure 4. Solstice variations: (a) emulating the (3,1) torus knot of Helaman Ferguson; (b) (4,3) torus knot variation; (c) (2,3) torus knot variation.

form a rib triangle. Also, retaining the parameterized rib endpoint values of the original *Solstice* leads to an effective angular offset of about 120°. This happens because the rib endpoints are specified by a parameter that relates to the whole length of the guide rail. While in the original *Solstice* one third of the length of the rail completed one sweep around the torus, in this new variant with only two major loops in the guide rail, it brings the second rib endpoint only about 2/3 around the torus. Thus the new configuration of the ribs yields the look of a puffed-up cushion and adds more "volume" to the appearance of the sculpture.

Other Ribbed Sculptures by Charles Perry

Charles O. Perry has created several other *Ribbed Sculptures* [4] that can be modeled with this paradigm. His sculptures inspired us to extend and generalize the generating paradigm of populating one or more guide rails with a set of closely spaced ribs. First we made a few modified versions of the *Solstice* emulation program to capture the geometries of some other Perry sculptures. Keeping the programs separate kept them lightweight and easy-to-modify and is preferable to a monolithic heavyweight program in this early phase of exploration.

Ribbed Mace (1998), located in Falls Church, VA (Figure 5a), is probably the simplest of Perry's ribbed structures, yet it is definitely eye-catching. It uses two separate semicircular guide rails in planes that stand at right

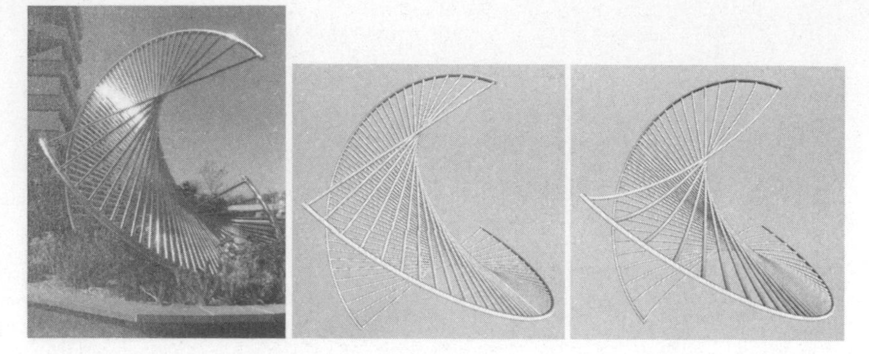

Figure 5. *Ribbed Mace*: (a) Perry's sculpture (Photo copyright C. Perry, 1998), (b) emulation (differing rib density), (c) variation with curved ribs.

Figure 6. *Harmony*: (a) Perry's sculpture (Photo copyright C. Perry, 1990), (b) an emulation with higher rib density.

angles to each other. Its 49 straight ribs form an elegant ruled surface that connects the two guide rails. In our first emulation we have slightly increased the number of ribs and placed them evenly spaced onto the guide rails (Figure 5b). Figure 5c shows a variation with curved ribs, which seems to give the sculpture a more lightweight, winglike look.

Harmony (1990), located in Hartford, CT (Figure 6a), also uses two separate semicircular guide rails. However, it comprises four distinct

Figure 7. *Early Mace*: (a) Perry sculpture (Photo copyright C. Perry, 1971), (b) emulation, (c)(d) rib variations.

ribbed surfaces, each composed of 18 ribs of varying curvature. Figure 6b shows our emulation of this sculpture, but with 21 ribs in each of the four sets.

Early Mace (1971), located in Atlanta, GA (Figure 7a), uses ribs in the shape of circular arcs connecting two separate guide rails. In this case the rails are two pairs of (almost) great semicircles on an invisible sphere, held together by two small semicircles at both ends; this gives the mace shape some thickness. The ribs form inwards-bending quarter arcs. Figure 7b shows an emulation of this sculpture, while Figures 7c and 7d

Figure 8. *Eclipse*: (a) Perry's sculpture (Photo copyright C. Perry, 1973), (b) and (c) modified "nest" modules.

demonstrate what happens when the ribs are first straightened and finally bent outward to follow the surface of the sphere, respectively.

Eclipse (1973), located in San Francisco, CA (Figure 8a), is a much more complicated ribbed structure. It is a construction with dodecahedral symmetry, where each face of the famous Platonic solid with 12 pentagons is replaced with a tapered and twisted "nest" of curved pentagonal rings. Again, the geometry of these rings, as well as their scale and relative offset from one ring to the next one, can easily be captured in parameterized form and exposed to experimentation. In a graduate course on "Computer-Aided Design and Rapid Prototyping" at U.C. Berkeley [8] some students played with this design and created new versions of the "nest" geometry (Figures 8b and 8c).

We have also captured the core features of Perry's *Eclipse* in an emulation program using ribbed surfaces. We allow the user to vary the number of ribs, the degree of twist in each "nest," and the recursive scaling of the extruded pentagons that forms each nest. Figure 9 demonstrates

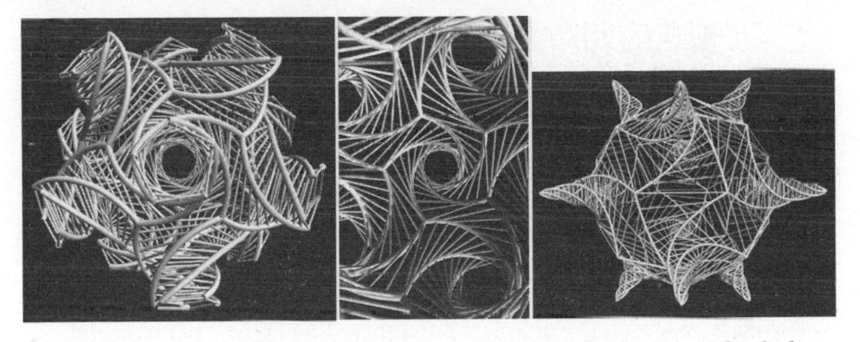

Figure 9. Variations on Perry's *Eclipse*: (a) 12 cylindrical nests around a dodecahedron, (b) inside view of such a structure, (c) highly twisted, pointy nests.

the variety of forms that one can create with these simple parameters. Figure 9a shows a ribbed dodecahedron with cylindrical nests with a moderate twist. Figure 9b shows a view from the inside of such a structure. Figure 9c is another variant with highly twisted conical nests, which scale down the extruded pentagons as they move outwards and thereby form twelve pointy spikes.

Generalized Ribbed Surfaces

In order to create a general ribbed surface [2, 3], we need to define either one or two parameterized guide rail curves and the rail cross sections that will be swept along the rail curve(s). For the ribs we have to specify the total number of ribs, their cross-sectional shape, and two parameter intervals for the locations of their endpoints beginning on guide rail Gb and ending on guide rail Ge. For example, if there are two separate guide rails, Gb and Ge, for the rib endpoints, the beginnings might be spread over the interval [0, 0.5] on rail Gb, while the endpoints span the interval [0, 1.0] on Ge (Figure 10a).

We must also provide an application-dependent set of geometric functions that define the sweep curves for the individual ribs spaced out along the rails. We have employed several alternative coordinate systems associated with the endpoints of the ribs, which allow us to specify the rib shapes in the most convenient manner for a particular sculpture family [2]. For instance, when the ribs are represented as cubic Hermite curves, the ribs are specified by their two endpoint positions and their two end tangent vectors. The endpoints are simply points along the

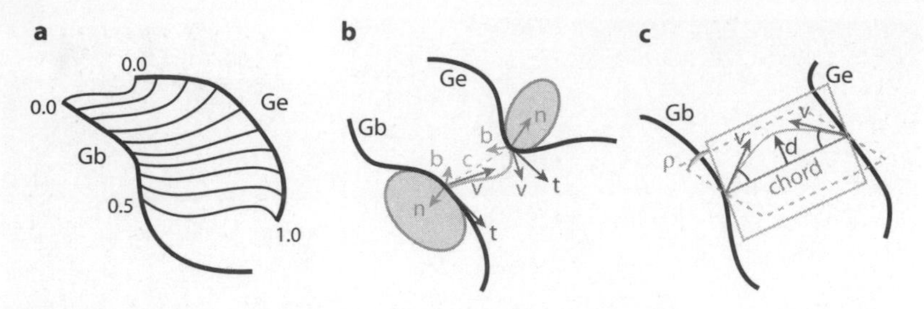

Figure 10. Different rib specification schemes: (a) basic concept: sequence of ribs spanning two rail, (b) rib specification based on Frenet frames at end points, (c) planar ribs as a modification of the chord between end points.

guide rail curves, evaluated at the proper parameter values of the guide-rail sweep for that rib. The end tangents might point in any direction, and different coordinate systems have different advantages for specifying these tangent vectors, as will be discussed below. The orientation angles of the end tangents in these rib-local coordinate systems may vary as a function of the sweep parameter s. In the simplest case, the rib end-condition parameters are specified for the start and for the end of the guide-rail sweep and are linearly interpolated for all ribs in between. More complicated functions could easily be introduced.

The most general approach simply uses the Frenet frame of the rail curve as the coordinate system to define a rib end tangent vector. The rib endpoints are placed at the origin of this Frenet frame, and the rib endpoint tangents v are defined in terms of the rail curve tangent (t), normal (n), and binormal (b) (Figure 10b).

Often we are more concerned with the overall shape of each rib itself, rather than the way it lands on the guide rail. In this case we specify the geometry of the rib endings in a coordinate system that is more intimately tied to the bulk of the rib. The chord that connects the two endpoints of the rib sliding along the rails forms the dominant axis of such a coordinate system. A plane that passes through this dominant axis can be specified by a single angle parameter ρ. The plane defining the zero angle depends on the shape of the guide rails; as a default we may try to keep it as close as possible to perpendicular to the rails at both rib endpoints.

In many cases, it is also desirable to keep the ribs planar and symmetric. In this case all that needs to be specified is the orientation angle ρ of the rib plane around the dominant coordinate axis and the tangent an-

Figure 11. Mathematical models: (a) Hyperboloid, (b) hemi-cube, (c) hemi-dodecahedron.

gles τ that the rib ends form with the chord. Alternatively, the amount of bending of the rib in the given plane can be specified as an offset distance d of the rib midpoint from the chord midpoint (Figure 10c).

The most suitable parameterization of the rib end conditions is application-specific and may reflect both aesthetic and pragmatic requirements. For example, if planar ribs approximating elliptical arcs are the preferred solution, the second coordinate system described above might be used (Figure 10c), perhaps with a further abstraction to specify the rib shape in terms of elliptic eccentricity. Other ways of defining the geometry of the individual ribs may use a blend of the two systems described here [2].

Ribbed Surfaces and Mathematical Visualization

"String art" models of mathematical surfaces depicting hyperbolic paraboloids and other ruled surfaces (Figure 11a) are found in many science museums. Because of their "transparency" such ribbed surfaces are particularly useful when depicting complex, possibly self-intersecting, geometrical objects. Figures 11b and c show ribbed models of a hemi-cube and of a hemi-dodecahedron. These are nonorientable, generalized polyhedra (or 2D cell-complexes) that can be obtained by taking surface elements of the cube or of the dodecahedron and identifying antipodal points, edges, and faces. Thus the 3-dimensional visualization of the hemi-cube has just three bilinearly warped "squares" (Figure 11b), and the hemi-dodecahedron is composed of six warped pentagons

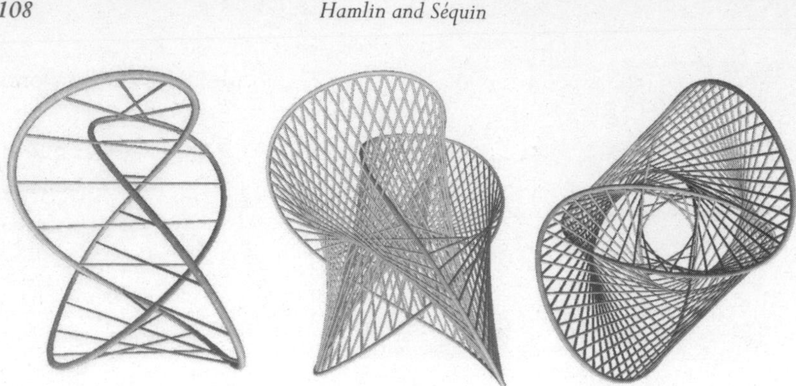

Figure 12. *Figure-8 Knot*: (a) Guide rail with just a few ribs with an offset of 50%, (b) and (c) the resulting structure with many more ribs with an offset of 25%, seen from the side and from the top.

(differently colored in Figure 11c), forming a tetrahedral structure with six angled edges. The latter is the basic building block of the 4-dimensional 57-Cell [9]. Slight modifications of one of the programs described above can readily be used to generate such mathematical visualization models and to build them on a rapid-prototyping machine (Figure 11b).

Designing Virtual Ribbed Sculptures

Using the generalized paradigm of ribbed surfaces, we have created some parameterized sculpture models of our own design. First, we present the design of one of these sculptures step-by-step to show how one may use the discussed broad concept to arrive at a pleasing sculptural form with a level of complexity that makes an uninitiated viewer wonder how this structure may have been conceived.

Figure-8 Knot

Knotted curves are ideally suited as guide rails. Thus we begin by defining a guide rail in the shape of a symmetrical presentation of the Figure-8 knot (Figure 12a). Since we are working with a closed curve, rib placement is reduced to specifying a single "offset" parameter that specifies how far apart (in a fractional sense) the two endpoints of each rib are. Starting from the already interesting topology of the selected knot, we now introduce additional structured complexity by adding one or more ribbed surfaces.

Inserting ribs that span a significant fraction of the knotted rail curve can produce rather confusing results. Thus, to begin with, we use a rather sparse set of thin ribs, so that the structure of the rail curve remains dominant. Figure 12a shows what happens when we introduce a few skinny ribs that span exactly 50% of the Figure-8 rail curve. Because of the inherent symmetry of the chosen knot representation, all ribs remain parallel to the ground plane and thus result in an easily comprehensible but fairly dull and unattractive "ladder."

As we change the rib offset to about 25% of the total rail length, a more interesting pattern emerges, creating a nicely structured surface with ribs that point in many different directions (Figure 12b). Figure 12c shows this same structure from the top.

Up to this stage of the design process, the intervals between the rib endpoints along the guide rail have been left to chance, and they may actually collide (Figure 13a, top). We now fine-tune these rib endings, aiming for evenly interleaved ribs as in Perry's *Solstice* (Figure 13a, bottom). To obtain a precise staggering, we want our offset to be a multiple of $(50/n)$%, where n is the number of ribs. With a few extra lines of code in the sculpture design program, this special constraint can be maintained automatically.

So far in this example we have used only straight ribs. Now that we have achieved a satisfactory configuration of the rib endpoints, we can start to fine-tune the rib shapes. In this example we use a cubic Hermite representation for the rib curves and manipulate the endpoint tangent vectors in the Frenet frame of the rail curve at each endpoint. For example, if the end tangent vectors of the ribs were kept aligned with the local normal vector of the rail curve, the ribs would take off from the apex of one of the "lobes" of the Figure-8 knot in the direction towards the center of that lobe. A slider controls by how much this tangent vector deviates from the osculating plane of the guide curve. Our refined sculpture design is shown in Figures 13b and c.

DNA Arch

In a second experimental design we use hemi-elliptical ribs spanning double helix guide rails that are bent into a parabolic arch. The two guide

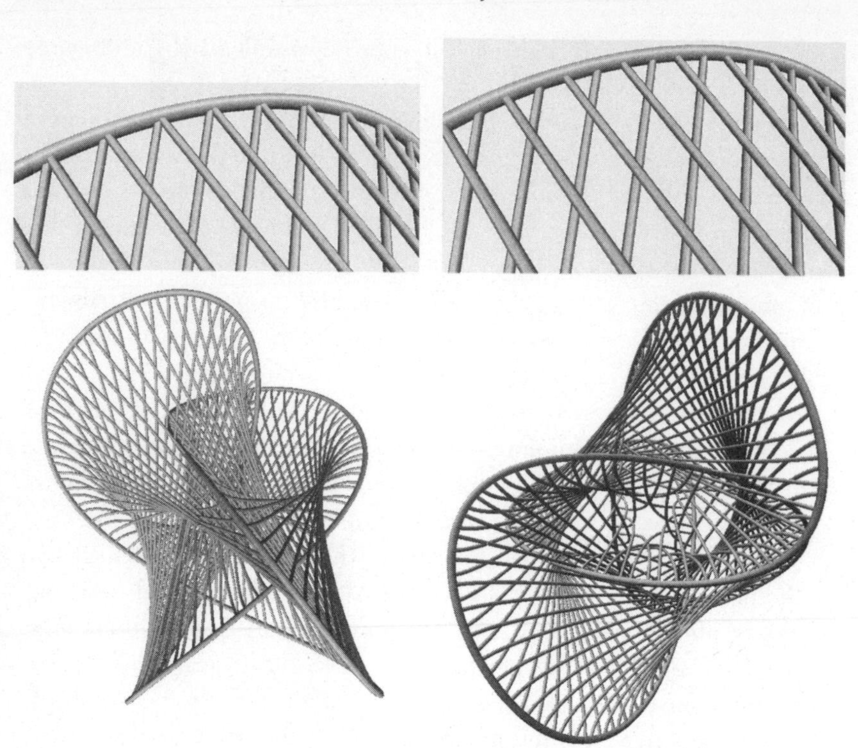

Figure 13. Adjusting the rib endpoint spacing: (a) and (b) coinciding versus evenly spaced endpoints, (c) and (d) refined *Figure-8 Knot* sculpture with curved ribs, seen from the side and from the top.

rails forming the double helix twist around one another by a variable number of times, which can be set via a slider. The radius of the intertwined helices is specified separately at the endpoints and at the apex of the arch and varies smoothly between these values. The double helix is curved to follow a parabolic path, which is specified with a few parameters that define the overall arch structure: its height and the width of its base. The rib end conditions are specified using a more general coordinate system based on the chord vector, the guide rail tangent vector, and their cross product (Section 5); the ribs take off in the direction of this cross product. These end conditions are specified at the two ends of the arch so that the ribs take off in the upward direction, away from the ground plane. These end-condition values are then interpolated for each rib along the whole arch. Figure 14 shows two variations of such sculptures output by this program.

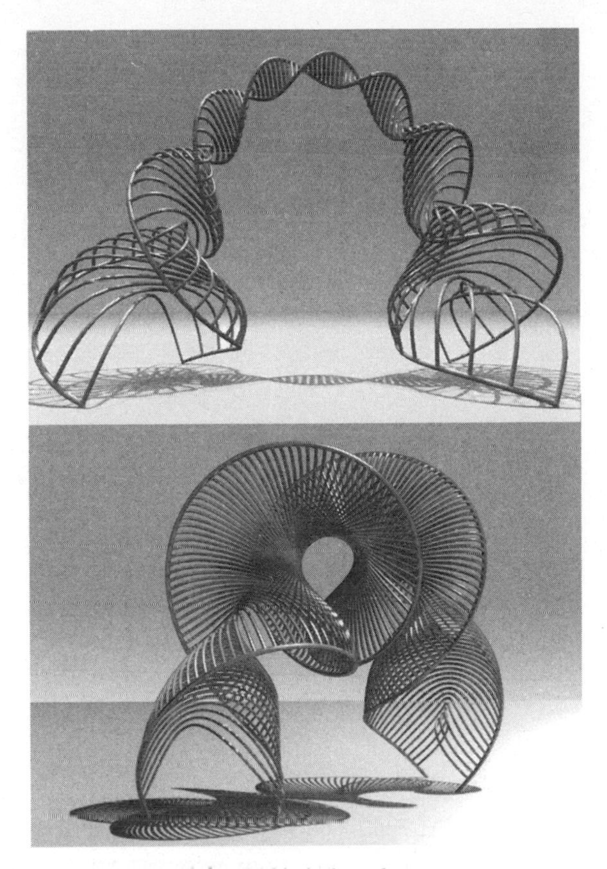

Figure 14. Two variations of the *DNA-Arch* sculpture.

Rendering

Finally, we briefly comment on the rendering of the images of our sculpture models presented in this paper. The geometry of our sculptures is defined in relatively small, special-purpose programs that have been written explicitly to capture a particular type of ribbed surface. Some of them were stand-alone C++ programs; others were special modules added to our SLIDE design environment [10]. None of these have any sophisticated rendering capabilities; they are just sufficient to make visible the emerging geometry during the interactive design phase. Once the geometry has been well defined, one would sometimes like to make a more realistic looking rendering of a proposed sculpture,

Figure 15. Virtual rendering of a sculpture immersed in the central glade of the Berkeley campus.

ideally immersed in the environment for which a particular sculpture may have been designed.

For these higher-quality renderings we used the open-source RenderMan implementation Pixie [5]. Our programs support output in RenderMan's RIB format, and this output is then augmented with environment geometry and images. For example, in Figure 15, we used photographs taken on the UC Berkeley campus to produce reflection maps and a believable backdrop for a rendering of our *Solstice* emulation.

Discussion and Conclusions

With enough effort, any natural or man-made form can be modeled with computer graphics techniques as a one-of-a-kind effort. Our efforts described in this paper were focused on capturing a whole family of similar shapes with a single parameterized representation. The ensuing programs should not only reproduce the inspirational sculptures from which we started with reasonable accuracy, but should then open a rich playing field for further experimentation with the extracted paradigm. Ideally these programs provide a judicious user with a powerful tool with which additional aesthetically pleasing forms can be created, that still belong in the same shape family.

We believe we succeeded quite nicely in defining a few basic geometrical operators that jointly are able to faithfully reproduce some of Perry's ribbed sculptures, and which can generate other attractive sculptural forms. However, the existence of such a tool does not automatically turn every user into an accomplished artist—just as the emergence of photographic cameras did not prompt every user to turn out nothing but successful "light-paintings." Once again we noticed, what had been observed previously with generator programs such as *Sculpture Generator I* [6] or with the SLIDE environment tailored to make a variety of *Viae Globi* sculptures [7]. Novice users typically are tempted to embrace too much complexity and produce what one anonymous reviewer aptly characterized in the following way: "If they begin with baroque they segue inexorably to rococo and oblivion." Indeed we have observed again this tendency among casual users to push some of the discrete variables of the program towards higher numbers, producing results that look more like a pipe-maze in a power plant than an inspiring sculpture.

In the work presented here this extraction and generalization has taken place at two distinct levels. First there is the analysis of Perry's *Solstice* as one member of a much larger family of "ribbed" (p,q) torus knots, leading to the possibility of experimenting with different values for p and q, but also altering the rib shapes and the offset of their endpoints along the torus-knot guide rail. Second there is the broader generalization of the notion of ribbed surfaces, composed of any set of slowly varying rib curves supported by one or more arbitrary guide rails. The combination of these two levels of abstraction leads to a fertile domain of intriguing geometrical shapes, which may make informative mathematical models as well as exciting monumental tubular sculptures.

Acknowledgments

We are indebted to Charles Perry for giving us permission to use the pictures of some of his ribbed sculptures and for letting us reveal the underlying logic of his *Solstice* sculpture by capturing its generating paradigm in a computer program, which then allowed us to create sibling designs to his masterpiece.

References

[1] H. Ferguson, *Umbilic Torus NC*. (1988). In *Helaman Ferguson—Mathematics in Stone and Bronze*, Meridian Creative Group, Erie, Pa., pp 6–7, (1994); see also: http://www.helasculpt.com/gallery/umbilictorusNC27inch/ (accessed April 10, 2010).

[2] J. F. Hamlin, *Applications of Ribbed Surfaces*. Master's Thesis, U.C. Berkeley (2009): http://www.eecs.berkeley.edu/Pubs/TechRpts/2009/EECS-2009-123.html (online library, accessed April 10, 2010).

[3] J. F. Hamlin and C. H. Séquin, *Ribbed Surfaces for Art, Architecture, and Visualization*. Computer-Aided Design and Applications, Vol. 6, No. 6, pp 749–58, (2009).

[4] C. Perry, *Ribbed Sculptures*. (1966–1998): http://www.charlesperry.com/Ribbed.html (accessed April 10, 2010).

[5] *Pixie rendering software*: http://www.renderpixie.com/ (accessed April 10, 2010).

[6] C. H. Séquin, *Interactive Generation of Scherk–Collins Sculpture*. Symposium on Interactive 3D Graphics, Providence, R.I., pp 163–6; ACM, New York (1997).

[7] C. H. Séquin, *Viae Globi—Pathways on a Sphere*. Proc. Mathematics and Design Conference, Deakin University, Geelong, Australia, July 3–5, 2001, pp 366–74; see also: http://www.cs.berkeley.edu/~sequin/PAPERS/MD2001_ViaeGlobi.pdf (accessed April 10, 2010).

[8] C. H. Séquin, *Rapid Prototyping of CAD Models*. CS Graduate course 294-10, Fall 2003: http://www.cs.berkeley.edu/~sequin/CS294/ (accessed April 10, 2010).

[9] C. H. Séquin and J. F. Hamlin, *The Regular 4-Dimensional 57-Cell*. Presented at SIGGRAPH'07, Sketches and Applications, San Diego, Aug. 4–9, (2007); see also: http://www.cs.berkeley.edu/~sequin/PAPERS/2007_SIGGRAPH_57Cell.pdf (accessed April 10, 2010).

[10] J. Smith, *SLIDE design environment*. (2003): http://www.cs.berkeley.edu/~ug/slide/ (accessed March 10, 2010).

Lorenz System Offers Manifold Possibilities for Art

Barry A. Cipra

The Lorenz attractor has been a favorite of mathematical lepidopterists ever since chaos theory took off in the 1970s. First described by MIT meteorologist Edward Lorenz in 1963, the delicate butterfly wings that unfurl from an unassuming cocoon of simple equations have graced the pages of innumerable papers in dynamical systems. But for Bernd Krauskopf and Hinke Osinga, there's an equally attractive geometric counterpart to Lorenz's eponymous point set: a smoothly convoluted surface known as the Lorenz *manifold*.

Krauskopf and Osinga, longstanding collaborators in the Department of Engineering Mathematics at the University of Bristol in the U.K., presented recent results of their studies of the Lorenz manifold in a minisymposium devoted to the legacy of Edward Lorenz at SIAM's dynamical systems conference in Snowbird last May. Their work has focused in part on visualization of the complicated surface, and one of the results of those efforts is literally tangible. In a collaboration with Krauskopf and Osinga, sculptor Benjamin Storch created a precise realization in stainless steel of a portion of the Lorenz manifold.

The Lorenz system is given by a set of three ordinary differential equations or, more precisely, by a parameterized family of ODEs:

$$\begin{aligned} x' &= \sigma(y - x) \\ y' &= \rho x - y - xz \\ z' &= xy - \beta z, \end{aligned}$$

where the parameters σ, β, and ρ are positive real numbers. The classic Lorenz equations—and the ones used for the sculpture—use the values $\sigma = 10$, $\beta = 8/3$, and $\rho = 28$. This endows the system with fixed points at $(0,0,0)$, $(6\sqrt{2}, 6\sqrt{2}, 27)$, and $(-6\sqrt{2}, -6\sqrt{2}, 27)$. The latter two are, roughly speaking, the eyes of the butterfly wings.

According to legend, Lorenz devised these equations as a toy weather model (picking parameter values that would produce results mimicking unstable convection patterns in the atmosphere) and had a computer (a Royal McBee LGP-30) churn out numerical solutions. At one point he restarted the computation using intermediate values from the computation's output, only to discover that seemingly insignificant roundoff—the machine computed to six digits, but reported only three—caused the restarted computation to quickly diverge from its previous output. Lorenz's report of this sensitivity to initial conditions was one of the slow-burning embers that ultimately erupted in the blaze of chaos theory that swept across physics in the 1980s.

Though emblematic of chaos, the Lorenz system was not truly known to be chaotic until 2002, when Warwick Tucker, now at the University of Uppsala in Sweden, proved that the attractor is indeed "strange," the mathematical term of art for an attractor that displays sensitivity to initial conditions. (Actually, even "attractor" is a term of art. There is a subtle but significant difference between an attractor and a mere attracting set. The rigorous study of dynamical systems can be a definitional headache.)

Cryptically speaking, the Lorenz attractor is what you see when you watch particles swept along by the equations' dynamics; the Lorenz manifold is what you *don't* see. Less cryptically, the manifold consists of trajectories that are sucked toward the fixed point (0,0,0). The linearized system there has two negative eigenvalues, which means that these trajectories constitute a smooth two-dimensional surface.

Two trajectories are easy to spot precisely: the positive and negative z-axis. But everything else on the Lorenz manifold can only be computed numerically; there are no explicit formulas that describe the surface. Krauskopf and Osinga have developed methods for doing the requisite computations. They have also proved theorems that guarantee the accuracy of the resulting approximations.

Their basic idea for computing the Lorenz manifold is to start with a small circle around the origin, in the plane of the manifold, and then "grow" the circle outward in a radially uniform fashion (i.e., in some sense ignoring the dynamical equations). The result at each stage is a closed curve numerically approximated by a polygon. The next closed curve is produced by the dynamics in a rather subtle way, through suitable boundary-value problems that specify the vertices of the new polygon. When computed correctly, each new closed curve (an approximate geodesic level set) consists of points a fixed (geodesic) distance from those of the previous curve. The polygonal approximation of the new closed curve is refined, as needed, by increasing the number of vertices.

It might seem to be "truer" to the dynamics to simply transport a closed curve by following the backward flow for an instant (say Δt) to obtain a new closed curve. But any arbitrarily chosen initial small circle very quickly deforms badly—and thus fails to give a useful representation of the surface. The geometric approach by Krauskopf and Osinga deals with this problem and generates a nice mesh representation of the surface.

As the computed closed curves expand away from the initial circle, they begin to twist and turn like a smoke ring caught in a gentle—and symmetric—vortex. (The manifold respects the equations' 180-degree rotational symmetry about the z-axis.) In essence, the Lorenz manifold organizes the trajectories that pass near the origin as they spiral toward the butterfly wings of the attractor, flitting from one to the other.

For the sculpture, Krauskopf and Osinga computed the Lorenz manifold to a geodesic distance of 140.75. An 8-cm-wide ribbon of stainless steel, polished on one side and brushed on the other, the sculpture shows the outermost 20-unit-wide band of the manifold.

Storch fashioned the sculpture, which is titled "Manifold," by welding together three pairs of laser-cut sheet metal pieces (see Figure 1). The pieces in each pair are identical, in accord with the rotational symmetry of the manifold. The bottom two barely bend out of the plane spanned by the two stable eigenvectors (and were cut from a thicker sheet of steel to help support the overall weight of the sculpture), but the others required careful hammering of the flat metal to produce the manifold's graceful curvature.

Comparison of photographs of the sculpture with computer-generated images, Krauskopf and Osinga point out, reveals the extent of Storch's skill: The surfaces match up almost perfectly. Indeed, the mathematicians enjoy showing a short video of the manifold slowly turning, then polling viewers on whether they are watching an actual video or a computer animation. (For the record, this reporter guessed right.)

The stainless steel piece is not the researchers' first foray into a physical rendering of the Lorenz manifold. In 2003, Osinga created a crocheted model of the entire surface out to geodesic distance 110.75.

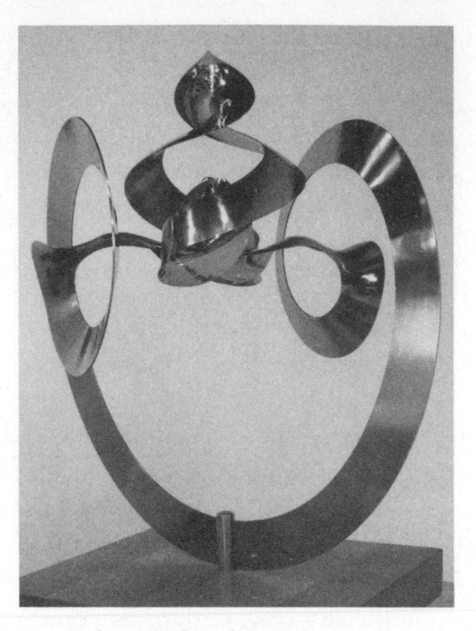

Figure 1. *Manifold*, a stainless steel sculpture by Benjamin Storch. Created in collaboration with Bernd Krauskopf and Hinke Osinga, the sculpture hints at the fascinating dynamics implicit in the equations of the Lorenz system. Reprinted with permission of Hinke Osinga.

Osinga's crocheted piece (Figure 2) adheres closely to the numerical algorithm for computing the manifold: Where the algorithm adds a mesh point, the crochet instructions add a stitch. Attached to a rod representing the z-axis and stiffened with garden wire surrounding its rim and two special trajectories to the origin, the 25,511-stitch piece gives a sense of the Lorenz manifold's complex shape.

In their more purely mathematical work, Krauskopf and Osinga have been studying bifurcations in the Lorenz system as the ρ-parameter varies (keeping σ and β fixed at 10 and 8/3). The Lorenz manifold and its one-dimensional counterpart, the unstable manifold for the origin (i.e., the trajectory that heads to the origin if you reverse time), interact in complicated ways with the stable and unstable manifolds associated with the other fixed points and certain periodic orbits of the system. In particular, the Lorenz manifold and the unstable manifolds of the equations' other two fixed points (which are both two-dimensional surfaces) intersect as an infinite collection of "heteroclinic orbits."

Figure 2. Model of the Lorenz manifold crocheted by Osinga (2003). Reprinted with permission of Hinke Osinga.

A typical heteroclinic orbit, once it leaves the vicinity of its initial fixed point, darts back and forth, loosely looping along the Lorenz attractor, until it finally heads toward the origin. The darting and looping can be described symbolically as a list of *r*'s and *l*'s, each symbol specifying a loop around the "right" or "left" fixed point (see Figure 3). As described in a paper with Eusebius Doedel of Concordia University in Montreal, Krauskopf and Osinga have computed these orbits with up to nine loops, a total of 512 cases, and investigated their dependence on ρ.

In brief, the heteroclinic orbits all stem from a bifurcation known as a "homoclinic explosion point" at $\rho \approx 13.9265$. What's more, each heteroclinic orbit persists as the ρ-parameter increases, up to a "fold," beyond which the continuation doubles back in ρ, with the orbit ending at another homoclinic explosion point. The researchers have mapped out the combinatorial structure that determines which heteroclinic orbits end at which homoclinic explosions.

Much remains to be learned about the Lorenz system, such as how the Lorenz manifold responds to changes in the other two parameters, σ and β. Almost half a century after the Royal McBee performed its

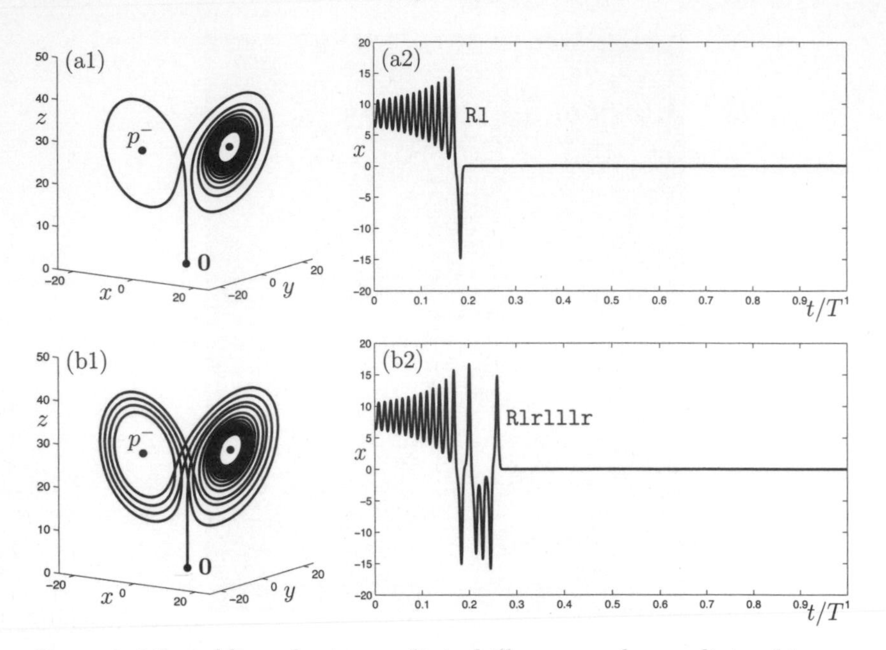

Figure 3. Like soldiers obeying a sadistic drill sergeant, heteroclinic orbits connecting the origin to the Lorenz system's other two fixed points lurch from side to side after emerging from a spiral. Their forced march can be described by a finite sequence of *r*'s and *l*'s ("rights" and "lefts"). In the two examples shown here, the orbits begin with a spiral, symbolically denoted "*R*," around the fixed point with positive *x*-value. Reprinted with permission from Eusebius J. Doedel, Bernd Krauskopf and Hinke M. Osinga, Global bifurcations of the Lorenz manifold, *Nonlinearity*, Vol. 19, No. 12, (2006) pages 2947–2972. Published by IOP Science.

tantalizing computations, the Lorenz attractor and its associated dynamics continue to attract mathematicians, like moths to a flame.

For Further Reading

- B. Krauskopf, H. M. Osinga, and B. Storch, *The sculpture* Manifold*: A band from a surface, a surface from a band*, in Reza Sarhangi and Carlo Séquin, eds., Proceedings of Bridges Leeuwarden: Mathematical Connections in Art, Music, and Science, Leeuwarden, 2008, 9–14.
- H. M. Osinga and B. Krauskopf, *Crocheting the Lorenz manifold*, The Mathematical Intelligencer, 26:4 (2004), 25–37.
- E. J. Doedel, B. Krauskopf, and H. M. Osinga, *Global bifurcations of the Lorenz manifold*, Nonlinearity, 19:12 (2006), 2947–72; with multimedia supplement at http://www.iop.org/EJ/mmedia/0951-7715/19/12/013/.

The Mathematical Side of M. C. Escher

DORIS SCHATTSCHNEIDER

While the mathematical side of Dutch graphic artist M. C. Escher (1898–1972) is often acknowledged, few of his admirers are aware of the mathematical depth of his work. Probably not since the Renaissance has an artist engaged in mathematics to the extent that Escher did, with the sole purpose of understanding mathematical ideas in order to employ them in his art. Escher consulted mathematical publications and interacted with mathematicians. He used mathematics (especially geometry) in creating many of his drawings and prints. Several of his prints celebrate mathematical forms. Many prints provide visual metaphors for abstract mathematical concepts; in particular, Escher was obsessed with the depiction of infinity. His work has sparked investigations by scientists and mathematicians. But most surprising of all, for several years Escher carried out his own mathematical research, some of which anticipated later discoveries by mathematicians.

And yet with all this, Escher steadfastly denied any ability to understand or do mathematics. His son George explains:

> Father had difficulty comprehending that the working of his mind was akin to that of a mathematician. He greatly enjoyed the interest in his work by mathematicians and scientists, who readily understood him as he spoke, in his pictures, a common language. Unfortunately, the specialized language of mathematics hid from him the fact that mathematicians were struggling with the same concepts as he was. Scientists, mathematicians and M. C. Escher approach some of their work in similar fashion. They select by intuition and experience a likely-looking set of rules which defines permissible events inside an abstract world. Then they proceed to explore in detail the consequences of applying these rules. If well chosen, the rules lead to exciting discoveries, theoretical developments and much rewarding work. [18, p. 4]

In Escher's mind, mathematics was what he encountered in schoolwork—symbols, formulas, and textbook problems to solve using prescribed techniques. It didn't occur to him that formulating his own questions and trying to answer them in his own way was doing mathematics.

Until 1937

M. C. Escher grew up in Arnhem, Holland, the youngest in a family of five boys. His father was a civil engineer and his four older brothers all became scientists. The home atmosphere may have instilled in him some habits of scientific inquiry, including the patient, methodical approach that would characterize his later work. Also, the young boys were given regular lessons in woodworking techniques that would later become very useful to Escher in making woodcuts.

His school life may have been less useful than his home life. Recalling his school years, Escher once confessed, "I was an extremely poor pupil in arithmetic and algebra, and I still have great difficulty with the abstractions of figures and letters. I was slightly better at solid geometry because it appealed to my imagination, but even in that subject I never excelled at school" [1, p. 15]. He did well in drawing, however, and his high school art teacher encouraged him to make linocuts.

In 1919 Escher entered the Haarlem School for Architecture and Decorative Arts intending to study architecture, but with the advice of his drawing and graphic arts teacher, Samuel Jessurun de Mesquita, and the consent of his parents, soon switched to a program in graphic arts. Among his prints executed while in Haarlem are three that show plane-filling; two of these are based on filling rhombuses, and one has a rectangle filled with eight different elegant heads, four upside down, each repeated four times [53, pp. 7–8]. Plane-filling would soon become an obsession.

Upon finishing his studies at the Haarlem School in 1922, he traveled for most of a year throughout Italy and Spain, filling a portfolio with sketches of landscapes and details of buildings, as well as meticulous drawings of plants and tiny creatures in nature. During this odyssey, he visited the Alhambra in Granada, Spain, and there marveled at the wealth of decoration in majolica tiles, sketching a section that especially attracted him "for its great complexity and geometric artistry" [1, pp. 24, 41]. This first encounter with the tilings in the Alhambra likely increased his interest in making his own tilings. In any case, during the mid-1920s, he produced a few periodic "mosaics" with a single shape, some of them hand-printed on silk [53, p. 11]. Unlike the Moorish tiles that always had geometric shapes, Escher's tile shapes (which he called "motifs") had to be recognizable (in outline) as creatures, even if only of the imagination. These early attempts show that he understood (intuitively, at least) how to utilize basic congruence-preserving transformations—translations, half-turns (180° rotations), reflections, and glide-reflections—to produce his tilings.

Escher married in 1924, and the couple settled in Rome, where two sons were born. Until 1935 he continued to make frequent sketching trips, most in southern Italy, returning to his Rome studio to compose his sketches for woodcuts and lithographs. In 1935, with the growing rise of fascism in Italy and his sons in ill health, Escher felt it best to move his family from Italy to Switzerland. In 1936 he undertook a long sea journey, and during the trip he spent three days at the Alhambra, joined by his wife Jetta. There they made careful color sketches of many of the majolica tilings. This second Alhambra visit, coupled with his move from the scenery of Italy, marked an enormous change in his work: landscapes would be replaced by "mindscapes."[1] No longer would his sketches and prints be inspired by what he found in mountainous villages, nature, and architecture. Now the ideas would be found only in the recesses of his mind.

Escher later wrote that after this Alhambra visit, "I spent a large part of my time puzzling with animal shapes" [1, p. 55]. By carefully studying the Alhambra sketches and noting the geometric relationships of the tiles to one another, he was able to make a dozen new symmetry drawings of interlocked motifs.[2] One of these showed interlocked Chinese boys. In spring 1937 he produced his first print that used a portion of a plane-filling to produce a metamorphosis of figures. In *Metamorphosis I*,[3] the buildings of the coastal town of Atrani morph into cubes which in turn evolve into the Chinese boys [53, pp. 19, 286]. The print was a fantasy, linking his new interest in plane-fillings with his love of the Amalfi coast, but Escher never liked it because it didn't tell a story—how do you link Chinese boys to an Italian town?

In July 1937 the Escher family moved to a suburb of Brussels, where a third son was born. That October Escher showed his meager portfolio of symmetry drawings to his older half-brother Beer, a professor of geology, who immediately recognized that these periodic patterns would be of interest to crystallographers, since crystals were defined by their periodic molecular structure. He offered to send Escher a list of technical papers that might be helpful. There were ten articles in Beer's list, all from *Zeitschrift für Kristallographie*, published between 1911 and 1933, by F. Haag, G. Pólya, P. Niggli, F. Laves, and H. Heesch [53, pp. 24, 337]. Escher found only the articles by Haag and Pólya useful.

Haag's article [28] provided a clear definition for Escher of "regular" plane-fillings and also provided some illustrations. In one of his copybooks, Escher carefully wrote Haag's definition of "regular division of the plane" (here translated):

> Regular divisions of the plane consist of congruent convex polygons joined together; the arrangement by which the polygons adjoin each other is the same throughout.

In the same copybook, Escher also sketched several of Haag's polygon tilings. After studying them, he quickly discovered that the word "convex" in Haag's definition was superfluous, for by manipulating the tile's shape, he was able to sketch several examples of nonconvex polygonal tilings. It was probably at this point that he inserted parentheses around the word "convex" in Haag's definition. Of course he also readily discovered that the word "polygon" was far too restrictive for his purposes; it could easily be replaced by "tile" or "shape". Haag's definition (with Escher's amendments) was adopted by Escher and would guide all of his symmetry investigations. He later carefully recorded the definition on the back of his symmetry drawing 25 (1939) of lizards (the drawing is depicted in Escher's lithograph *Reptiles*).

Pólya's article [43] would have a great influence on Escher. Escher carefully copied (by hand, in ink) the full text that outlined the four isometries of the plane and announced Pólya's classification of periodic planar tilings by their symmetry groups. Pólya was evidently unaware that this classification had been carried out by Fedorov more than thirty years earlier. Escher was already intuitively aware of the congruence-preserving transformations Pólya spoke of but probably didn't understand any of the discussion about symmetry groups. What struck him was Pólya's full-page chart that displayed an illustrative tiling for each of the seventeen plane symmetry groups (Figure 1). Escher carefully sketched each of these seventeen tilings in a copybook and studied them,

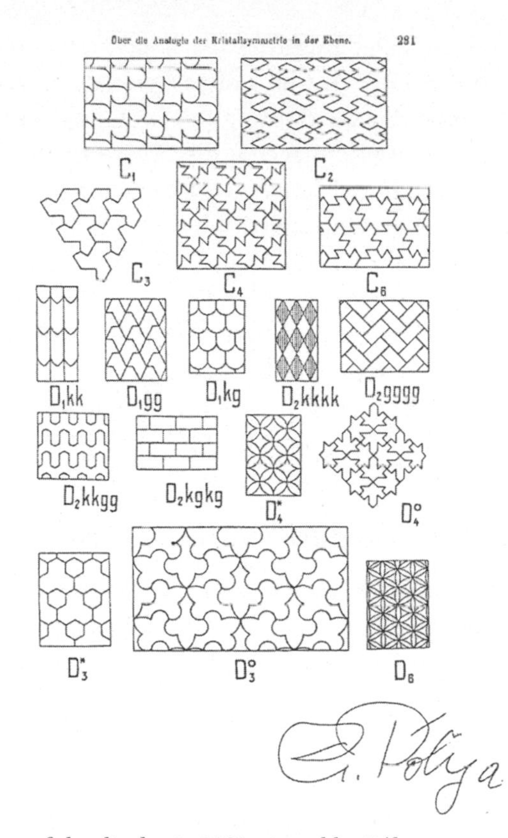

Figure 1. A copy of the display in [43], signed by Pólya.

map-coloring some of them [47]. Among these, there were tilings that displayed symmetries he had not recorded in the Alhambra; for example, tilings whose only symmetries other than translations were glide-reflections or fourfold (90°) and twofold (180°) rotations. Within one month of studying these, Escher had completed his first symmetry drawings displaying fourfold rotation symmetry: squirming lizards interlocked four at a time, pinwheeling where four feet met [53, p. 127]. He featured a portion of one of these drawings at the center of his woodcut *Development I*, completed in the same month.

Escher was so grateful for the help that Pólya's paper provided that he wrote to the mathematician to thank him. He sent Pólya the print *Development I* and asked the mathematician whether or not he had written a book on symmetry for "laymen" as his article indicated he had hoped to do. Although a writer once characterized Pólya's reply as polite but formal, indicating he hadn't written the hoped-for book [53, p. 22], Pólya wrote to me in 1977 that he and Escher had corresponded more than once and that he regretted losing the correspondence in his haste to come to America in 1940. A recent discovery of a forgotten suitcase full of Pólya's notes and other collected letters and papers, now in the Pólya archives at Stanford University, shows that Pólya even sent Escher his own attempt at an Escher-like tiling. Among these papers is Pólya's drawing of a tiling by snakes, inscribed "sent to MCE", at the address where Escher resided from 1937 to 1940. Also, there is an outline of Pólya's never-completed book *The Symmetry of Ornament* and many sketches of tilings, both for the planned book and for the 1924 article that so influenced Escher [53, pp. 335–6].

Escher as a Mathematical Researcher

From 1937 to 1941 Escher plunged into a methodical investigation that can only be termed mathematical research. Haag's article had given him a definition of "regular division of the plane", and Pólya's article showed him that there were many tile shapes that could produce these. He wanted to find more and characterize them. The questions he pursued, using his own techniques, were:

(1) What are the possible shapes for a tile that can produce a regular division of the plane, that is, a tile that can fill the plane with its congruent images such that every tile is surrounded in the same manner?
(2) Moreover, in what ways are the edges of such a tile related to each other by isometries?

The only isometries that Escher allowed in order to map a tile to an adjacent tile were translations, rotations, and glide-reflections—a reflection would require a tile's edge to be a straight segment, not a natural condition for his creature tiles. In 1941–1942 he recorded his many findings in a definitive *Notebook* that was to be his private encyclopedia about regular divisions of the plane and how to produce and color them [46], [48], [53]. The *Notebook* had two parts: its cover inscribed (here translated) "Regular divisions of the plane into asymmetric congruent polygons; I Quadrilateral systems MCE 1-1941 Ukkel; II Triangle systems X-1942, Baarn".

Escher's study of "quadrilateral systems" was extensive. He represented these tilings symbolically with a grid of congruent parallelograms in which each parallelogram represented a single tile. He shaded the grids checkerboard style, so that each parallelogram shared edges only with parallelograms of the opposite color. He was interested in asymmetric tiles (after all, his creature tiles were to be primarily asymmetric), and in order to indicate the asymmetry, placed a hook inside each parallelogram. The hook provided orientation, while small circles and squares on the tile's boundary indicated twofold and fourfold centers about which the tile could rotate into an adjacent tile. Escher was aware that certain symmetries required special parallelogram grids and so considered five different categories: arbitrary parallelogram, rhombus, rectangle, square, and isosceles right triangle (a grid of squares in which the diagonals have been drawn). He labeled these A–E, respectively. As he sought to answer his two questions, he filled the pages of several school copybooks with his sketches of marked grids representing tilings, scratching out those that didn't work out or that duplicated an earlier discovery. Each time he discovered a marked grid that represented a regular division of the plane, he recorded it and made an example of a tiling with a "shaped tile", its vertices marked by letters.

To quickly record how each edge of a tile was related to another edge of the same tile or an adjacent tile, Escher devised his own notation: = meant "related by a translation" and || "related by a glide-reflection". An S on its side meant "related by a 180° rotation" and L meant "related by a 90° rotation". Figure 2 shows one copybook page with five different "rhombus systems" on the left and shaped tilings for two of these systems on the right. Note Escher's "voorbeeld maken!" at the bottom of the page—"make an example!" His results were recorded entirely visually, with no need for words. Ultimately he found ten different classes of these tilings and numbered the classes I–X. His *Notebook* charts giving both visual and descriptive versions of the classes are in [53, pp. 58–61].

To discover still other regular divisions, those for which three colors would be required for map-coloring, Escher employed a technique that he called "transition". Figure 3 recreates one of his examples. He would begin with a two-color regular division from one of his ten categories (Figure 3 begins with type II^A). Each of these categories had four tiles meeting at every vertex and required only two colors. He would then choose a tile and a segment of the boundary that connected one of its

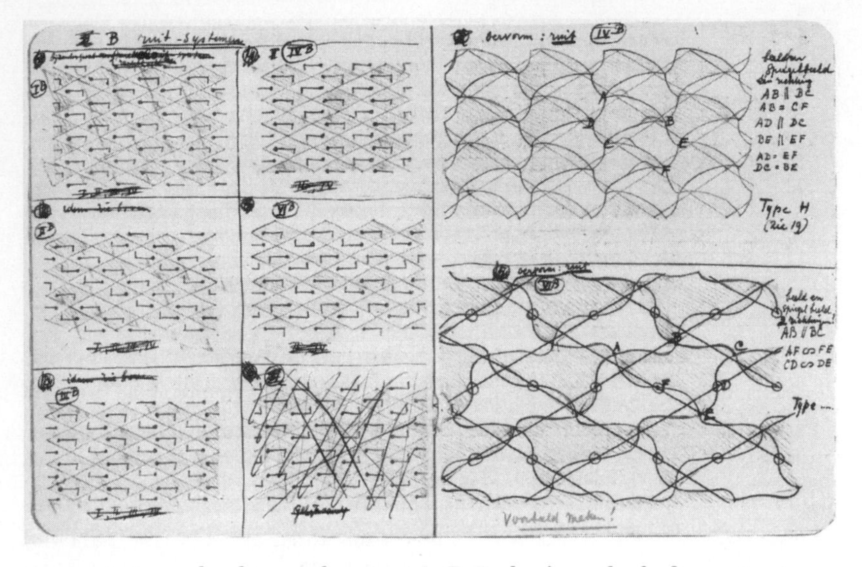

Figure 2. A copybook page showing M. C. Escher's method of investigation of regular divisions of the plane. His symbolic notation is explained in our text.

vertices (say B) to another carefully chosen boundary point (say A) that was not a vertex of the tiling (Figure 3a). Using A as a pivot point, he would then pivot the boundary segment connecting A and B (stretching it if necessary) so that vertex B slid along the boundary of the tile, stopping at a new position (say C). Repeating this on the corresponding segments of the boundaries of all tiles produced a new tiling with vertices at which three tiles met, requiring three colors for map-coloring (Figure 3b). The process could be continued with the new segment AC, sliding C along the boundary until it reached a vertex D of the original tiling. This produced a new tiling that again required only two colors (Figure 3c). At the intermediate (three-color) stage, the network of tile edges was certainly not homeomorphic to the original, but surprisingly, at the end (two-color) stage, the new network of tile edges might also not be homeomorphic to that of the original tiling. Escher thought of the intermediate (three-color) tiling as having components of both the beginning and ending two-color tilings, and so labeled it with both types. In Figure 3, his type II^A system is transformed to II^A–III^A, and that is transformed to system III^A. In this instance, the tiles in the final tiling

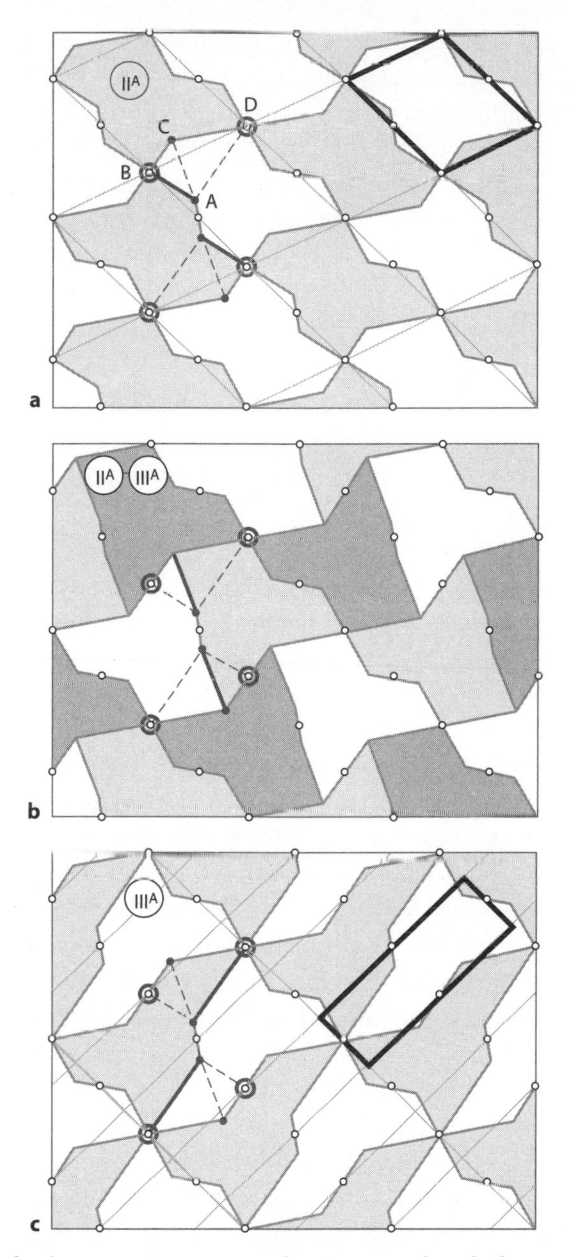

Figure 3. Escher's transition process takes (a) a 2-colored tiling with 180° rotation centers at tile vertices and midpoints of edges, to (b) a 3-colored tiling, to (c) a 2-colored tiling with 180° rotation centers at midpoints of tile edges.

have three, not four, edges and meet six at a vertex; Escher noted that this was an exceptional case [53, p. 62].

Escher did not record these discoveries with words, but in his *Notebook* he displayed sixteen pages of carefully drawn illustrations of transitions that cover all of his ten categories [53, pp. 62–69]. In many cases, he discovered more than one distinct transition of the same tiling. Using today's terminology, he discovered how to produce tilings of different isohedral types beginning with a single isohedral tiling. And he also recorded in a chart (a digraph!) which of his ten categories led to others. This chart makes clear that his process of transition can change the topological and combinatorial properties of a tiling but not change its symmetry group [48], [53, p. 60].

The last section of Escher's "quadrilateral systems" study summarizes in ten pages his investigations of what he called "2-motif" tilings [53, pp. 70–76]. He would begin with a regular division, map-colored with two or three colors, then split each tile (in the same manner) into two distinct shapes, so that the resulting tiling could be colored with two colors. This investigation was spurred by his fascination with what he called "duality"—many of his prints play with the idea of interchanging the role of figure and ground, or juxtaposition of opposites. *Sky and Water I* and *Circle Limit IV* (*Angels and Devils*) are famous examples.

For example, in *Sky and Water I* (Figure 4) a horizontal row of interlocked flat tiles at the center alternates white fish and black birds, dividing the print into upper and lower halves (sky and water). The fish in this center row can serve as figure and the birds as ground, or vice versa. But as the eye moves upward from this row of tiles, the creatures separate and take on distinct roles. The black birds become three-dimensional as they rise, while the white fish melt to become sky. The fish become the background against which figures of birds fly. As the eye moves downward from the center row of tiles, the opposite transformation takes place. Now the fish gain three-dimensional form and the black birds dissolve to become water in which the fish swim. In mathematics, the essence of dual objects is that each completely defines the other, such as a set and its complement, a statement and its negation. In addition to the figure/ground duality, there are other kinds of duality represented in this single print: black and white, sky and water. And opposites: bird and fish often denote opposites (think "neither fish nor fowl"), and in the print, each bird is placed exactly opposite a fish, with the invisible surface of the water acting as a compositional mirror.

Figure 4. M.C. Escher's Escher's *Sky and Water I*, 1938. Woodcut, 435 cm × 439 cm. © 2011 The M.C. Escher Company-Holland. All rights reserved. www.mcescher.com

Part II of Escher's *Notebook* is brief, devoted to what he called "triangle systems"—regular divisions having 120° rotation centers (system A) or 60°, 120°, and 180° rotation centers (system B). After explaining the necessary placement of rotation centers, he records only twenty different tilings, several with two motifs, and all carefully map-colored to respect symmetry. Most require three colors. Unlike in his quadrilateral systems, some of his tiles have rotation symmetry, and from these he derives other tiles with one or two motifs [53, pp. 79–81].

In 1941, as he was nearing the end of these investigations, Escher and his family moved to Baarn, Holland, where he would spend all but the last two years of his life. In the years following, he produced more than 100 regular divisions of the plane, each final version numbered and carefully drawn on graph paper, its creature tiles outlined in ink, map-colored using watercolors, respecting the symmetries of the tiling.

As his portfolio of symmetry drawings grew, he referred to it as his "storehouse". Fragments of these drawings would be featured in many prints, notecards, exhibit announcements, painted and tiled public works, and even carved on the surface of a ball. In all, there are 134 numbered symmetry drawings and many unnumbered sketches.

Escher carried out several other minor mathematical investigations in order to achieve certain effects in his art. Some of these results were recorded in a notebook entitled *Regular Division of the Plane: Abstract Motifs, Geometric Problems*, and others were gathered in small folios. He studied several Moorish-like tilings and investigated linked rings (seen in his last print, *Snakes*). He enumerated several tilings by congruent triangles while designing bank notes. He also recorded two theorems he evidently discovered but did not prove. One was about concurrent lines in a triangle, and the other about concurrent diagonals in a special tiling hexagon [53, pp. 82–93]. At my request, the first theorem was verified by A. Liu and M. Klamkin [35] and the second by J. F. Rigby [44].

An investigation in 1942 that was an amusement, shared with his children and grandchildren [18, pp. 9–11], [50], was combinatorial—to determine how many different patterns could be generated by following this algorithm:

Decorate a square with an asymmetric motif and use four copies of the decorated square (independently chosen from any of four rotated aspects) to fill out a 2×2 larger square, then translate the larger square in the direction of its edges to fill the plane.

With a methodical search of the 4^4 possible 2×2 filled squares, eliminating obvious duplications, and by sketching examples with a simple motif for the rest, he ultimately found twenty-three distinct patterns. That is, no two of these twenty-three patterns were identical, allowing rotations. He also asked the combinatorial question in two special cases in which reflected aspects of the decorated square were also allowed. In these cases, the choices of the four copies of the decorated square were restricted as follows:

Case (1)—two choices must be the same rotated aspect and independently, the other two choices the same reflected aspect.
Case (2)—two choices must be different rotated aspects and independently, the other two choices different reflected aspects.

Escher's results were sketched in copybooks and later printed with inked carved wooden stamps using a motif that produced patterns resembling knitted or crocheted pieces. He also made a "ribbon" design, outlining crossing bands in a square, and carved four wooden stamps—one of the original design, one of its reflection, and two others that reversed the crossings in the first two stamps. He did not attempt to find the number of patterns produced by the 4^{16} possible 2×2 squares filled with aspects of these, but he did produce several patterns with them and colored them with a minimum number of colors so that continuous ribbon strands had the same color and no two bands of the same color ever crossed [53, pp. 44–52], [17, p. 41].

Escher's Interactions with Mathematicians

Until 1954 few mathematicians outside of Holland knew of Escher's work. That year the International Congress of Mathematicians (ICM) was held in Amsterdam, and N. G. de Bruijn arranged for an exhibit of Escher's prints, symmetry drawings, and carved balls at the Stedelijk Museum [11]. He wrote in the catalog, "Probably mathematicians will not only be interested in the geometrical motifs; the same playfulness which constantly appears in mathematics in general and which, to a great many mathematicians is the peculiar charm of their subject, will be a more important element" [2].

When Roger Penrose visited the exhibit, he was amazed and intrigued. Escher's print *Relativity* especially caught his eye. It shows three prominent staircases in a triangular arrangement (and some smaller staircases), as seen from many different viewpoints, with several persons simultaneously climbing or descending them in an impossible manner, defying the law of gravity. Penrose was inspired to find a structure whose parts were individually consistent but, when joined, became "impossible". After returning to England he came up with the idea of the now-famous Penrose tribar in which three mutually perpendicular bars appear to join to form a triangle (Figure 5). Following that, his father devised an "endless staircase", another object that can be drawn on paper but is impossible to construct as it appears [41, pp. 149–50]. Penrose then closed the loop of discovery by sending the sketches of these impossible objects to Escher, who in turn used them in crafting the perpetual motion in his print *Waterfall* and the never-ending march of the monks in *Ascending and Descending.*

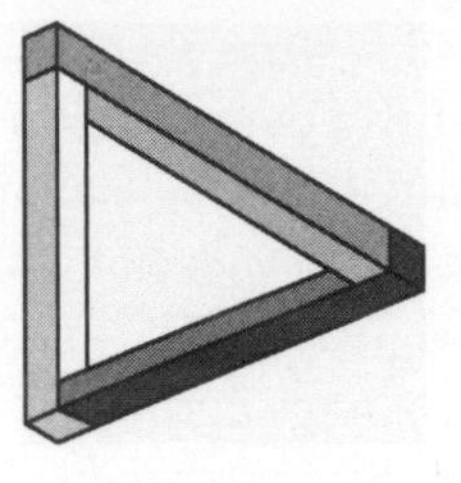

Figure 5. Penrose's tribar.

Penrose also visited Escher's home in 1962 and brought a gift of identical wooden puzzle pieces derived from a 60° rhombus. Escher soon sent Penrose the puzzle's solution, enclosing a sketch of the unique way in which the pieces fitted together. Here, congruent tiles were surrounded in two distinct ways. In 1971 Escher produced his only tiling with one tile that was not a regular division (today it would be called 2-isohedral). It was the last of his numbered symmetry drawings, with a little ghost that filled the plane according to the rules of Penrose's puzzle [41, pp. 144–5; 53, p. 229].

H.S.M. Coxeter also saw Escher's work for the first time during that ICM in 1954, and upon returning to Canada, he wrote Escher a letter to express his appreciation of the artist's work. Three years later, he wrote again to ask if he might use two of Escher's symmetry drawings to illustrate an article based on his presidential address to the Royal Society of Canada. The article discussed symmetry in the Euclidean plane and also in the Poincaré disk model of the hyperbolic plane and on a sphere surface [3]. Escher readily agreed, and when he later received a reprint of the article, he wrote to Coxeter, "some of the text-illustrations and especially figure 7, page 11, gave me quite a shock" [5, p. 19]. The figure's hyperbolic tiling, with triangular tiles diminishing in size and repeating (theoretically) infinitely within the confines of a circle, was exactly what Escher had been looking for in order to capture infinity in a finite space.

Escher worked over the figure with compass and straightedge and circled important points (Figure 6). From this, he managed to discern enough of the geometry to produce his print *Circle Limit I*. But he wanted to know more, and sent a large diagram to Coxeter showing what he had figured out, namely, the location of centers of six of the circles (Figure 7). In his letter, he politely asked Coxeter for "a simple explanation how to construct the [remaining] circles whose centres approach [the bounding circle] from the outside till they reach the limit." He also

Figure 6. Coxeter's Figure 7, with Escher's markings (here computer-enhanced for visibility).

asked, "Are there other systems besides this one to reach a circle limit?" [5, p. 19; 54, p. 263]. Coxeter replied with a minimal answer to Escher's first request:

> The point that I have marked on your drawing (with a red o on the back of the page) lies on three of your circles with centres 1, 4, 5. These centres therefore lie on a straight line (which I have drawn faintly in red) and the fourth circle through the red point must have its centre on this same red line. [54, p. 264]

From this, Escher was supposed to construct the complete scheme. By contrast, Coxeter answered the second question at length, beginning,

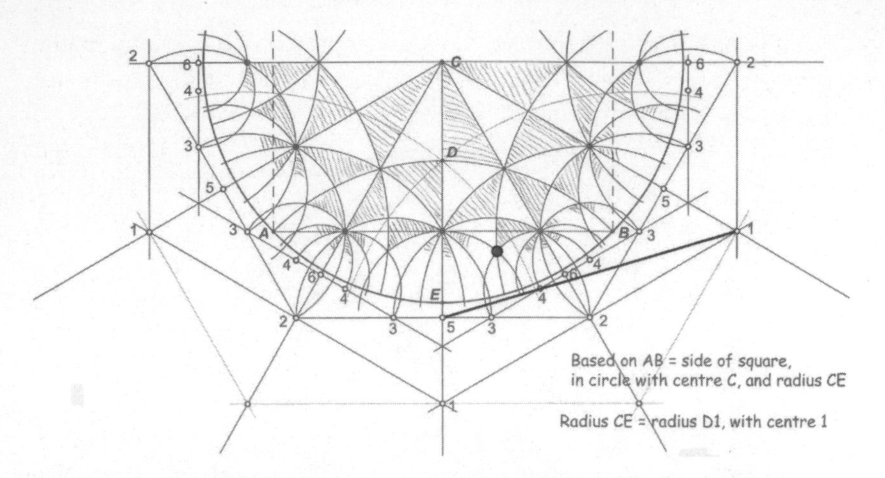

Figure 7. Escher's diagram sent to Coxeter, exhibiting what the artist had figured out. The original drawing is faint, drawn in pencil on tracing paper. This reconstruction by the author shows Coxeter's red markings as a filled black circle and thick black line.

"Yes, infinitely many! This particular pattern is denoted by [4, 6]" and then explained for which p and q patterns $[p, q]$ exist, referring to the text *Generators and Relations*, and enclosing a "spare copy of $\left\{ \begin{smallmatrix} 3 \\ 7 \end{smallmatrix} \right\}$" [54, p. 264].

Escher was disappointed with this reply, yet it only increased his determination to figure things out. He wrote to his son George:

> [Coxeter] encloses an example of using the values three and seven, of all things! However this odd seven is of no use to me at all; I long for two and four (or four and eight) . . . My great enthusiasm for this sort of picture and my tenacity in pursuing the study will perhaps lead to a satisfactory solution in the end. . . . it seems to be very difficult for Coxeter to write intelligibly to a layman. Finally, no matter how difficult it is, I feel all the more satisfaction from solving a problem like this in my own bumbling fashion. [1, p. 92; 54, pp. 264–5]

Escher did successfully carry out his "Coxetering", as he called his work with hyperbolic tilings, and in 1959–1960 he produced three other *Circle Limit* prints. Upon earlier receiving *Circle Limit I*, Coxeter had praised Escher for his understanding of the conformal pattern, and in 1960, when he received the complex *Circle Limit III*, Coxeter wrote Escher a three-page letter sprinkled with symbols explaining the print's mathematical content, with references to several technical texts, and the implications for coloring seen in the "compound {3, 8} [6{8, 8}] {8, 3} of six {8, 8}'s inscribed in a {3, 8}" [54, p. 265]. And Escher despaired to George, "Three pages of explanation of what I actually did . . . It is a pity that I understand nothing, absolutely nothing of it . . . " [1, pp. 100–1; 54, p. 265].

In 1960 Coxeter arranged for Escher to give two lectures at the University of Toronto about his work, and the Coxeters hosted the artist at their home. The Coxeter-Escher correspondence continued for several years, with two letters of note. In March 1964 Coxeter wrote, "After looking again and again at your *Circle Limit III* on my study wall, I finally realized that my remark about its 'impossibility' was based on my own misunderstanding, as you will see in the enclosed," which was his review of Escher's book [20] for *Mathematical Reviews.* He added, "The more I look at your work, the more I admire it" [9]. That review [4] was the first time Coxeter revealed that the white arcs forming the backbones of fish in Escher's *Circle Limit III* were not, as he and others had assumed, badly rendered hyperbolic lines but rather were branches of equidistant curves. In 1979 and again in 1995 he published articles [5, 6] devoted to those white arcs, explaining, "they 'ought' to cut the circumference at the same angle, namely 80° (which they do, with remarkable accuracy). Thus Escher's work, based on his intuition, without any computation, is perfect . . . " [5, pp. 19–20].

In other articles Coxeter gave mathematical analyses of Escher's work and indicated that the artist had anticipated some of his own discoveries [54]. In May 1964 Escher sent Coxeter his print *Square Limit* and explained with a diagram its underlying geometric grid of self-similar triangles (see reconstruction in Figure 8). Escher's explanatory sketch was on graph paper, in red and blue colored pencil. It showed the first three rings surrounding the center square to indicate how the division process can continue forever. He had devised this fractal structure himself, and while a Euclidean construction with straight segments, it possessed the desired property of his *Circle Limits*—figures diminished as they approached the bounding square [17, pp. 104–5; 53, p. 315; 59, pp. 182–3]. A 90° rotation about the center of the diagram is a color symmetry, sending red tiles to blue, blue to red, and white to white. In Escher's print, the triangles are replaced by fish.

Figure 8. Escher's geometric grid for the print *Square Limit*.

Escher had only brief interactions with other mathematicians; none would influence his work as did Pólya, Penrose, and Coxeter. Edith Müller, who had been A. Speiser's Ph.D. student, wrote to me that Escher had learned of her dissertation (a symmetry analysis of the Alhambra tilings) and visited her in 1948 in Zurich to discuss her (and his) work. She told him about how Speiser had learned to make lace in order to better understand symmetry.

Heinrich Heesch, another student of Speiser, carried out extensive research on tilings in the mid-1930s but did not publish until the 1960s. He, too, defined "regular" tilings as plane-fillings with congruent tiles in which every tile was surrounded in the same manner. Also, like Escher, he was interested in characterizing the conditions on edges of asymmetric tiles that could tile in this manner and for which the tiling had no reflection symmetries. He proved there were exactly twenty-eight types of these tiles and displayed a visual chart of them in his 1963 book with Otto Kienzle [29]. He assigned each edge of a tile a letter—T, G, or C_n—according to how it related to another edge by translation, glide-reflection, or $360°/n$ rotation. He sent the book to Escher, who at that time was very ill; for Escher, this information came more than twenty years after his own discoveries of all but one of those twenty-eight types [53, pp. 324–6].

In the last two years of Escher's life, mathematics teacher Hans de Rijk (a.k.a. Bruno Ernst) collaborated with Escher to write a book that would interpret the body of the artist's work, with special attention to the mathematical underpinnings of many prints. Every Sunday without fail they would spend time together as the manuscript took shape. This book [17] and a definitive catalog of Escher's graphic work [37] were both published in 1976, four years after the artist's death, and were the first to show many of Escher's painstaking preliminary drawings for his prints, some of them geometric marvels. A shorter version of de Rijk's analysis of Escher's work is in [1], pp. 135–54.

Escher's Work Used to Teach Mathematical Ideas

Escher enjoyed the role of teacher, giving lectures to diverse audiences—scientific gatherings, school students, museum audiences, even Rotary clubs. His lecture poster (Figure 9) shows in five different illustrative tilings how he explained the actions of translations (*verschuiving*), rotations (*assen*), and glide-reflections (*glijspiegeling*) that would carry a tile into an adjacent tile. Numbers identify the various aspects of a tile, circles and squares identify twofold and fourfold rotation centers, respectively, and adjacent dashed lines act as rails along which a tile glides and then reflects (in a line equidistant from the rails). Escher used large, brightly colored cardboard cutouts in the shapes of these tiles, mounted on straightened wire hangers, to demonstrate the motions of the isometries.

When Escher's book [20] was published in Holland in 1960, it included a short essay in the introduction by crystallographer P. Terpstra, to teach about symmetry and the seventeen plane symmetry groups. When the British translation was published, the essay appeared as a separate pamphlet; it never appeared with the American edition. Evidently the publisher, like Escher, thought it too technical.

Caroline MacGillavry, a crystallographer at the University of Amsterdam, was the first scientist to see the possibility of using Escher's art as a teaching tool in a text. When she first visited his studio in the late 1950s, she marveled: "The notebook in which he wrote his 'layman's theory' has been a revelation to me. It contains practically all the 2-, 3-,

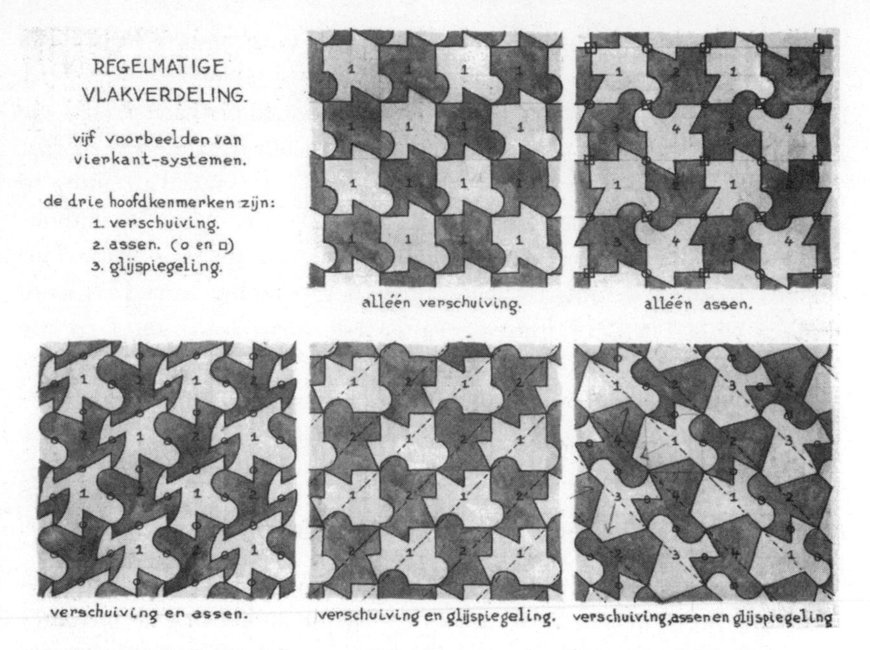

Figure 9. M.C. Escher's lecture slide about regular divisions of the plane, showing five "quadrilateral systems." © 2011 The M.C. Escher Company-Holland. All rights reserved. www.mcescher.com

and 6-colour rotational two-dimensional groups, with and without glide-reflection symmetry" [39, p. *x*]. That visit gave birth to her idea of collaborating with Escher to use his symmetry drawings in a text for beginning geology students, to teach the classification of colored periodic tilings according to their symmetries. The International Union of Crystallography agreed to sponsor the publication. In the book's introduction, she notes,

> Escher's periodic drawings . . . make excellent material for teaching the principles of symmetry. These patterns are complicated enough to illustrate clearly the basic concepts of translation and other symmetry, which are so often obscured in the clumsy arrays of little circles, pretending to be atoms, drawn on blackboards by teachers of crystallography classes. On the other hand, most of the designs do not present too great difficulties for the beginner in the field. [39, p. *ix*]

In reviewing Escher's store of periodic drawings (by then, more than 100), she noted that one of the simplest symmetry groups, type *p2* with no color symmetries, was not represented. At her request, Escher produced a new symmetry drawing to fill the gap [39, plate 2; 53, p. 210]. He also produced another requested type [39, plate 34; 53, p. 211] and refreshed or redrew some others for the publication.

Coxeter may have been the first mathematician (outside of Holland) to use Escher's work to illustrate a mathematics text. His *Introduction to Geometry* was unusual when it was published in 1961, with many nonstandard topics, including symmetry and planar tessellations, which he illustrated with Escher's symmetry drawings used earlier in [3]. Martin Gardner devoted a Mathematical Games column in *Scientific American* to a review of the book and republished the drawings, bringing Escher's symmetry work to the attention of the wider scientific world [21]. It was not long before scores of math texts (at all levels) and articles on teaching displayed Escher's periodic drawings and prints. While the elementary concepts of planar isometries, similarities, and symmetry are obvious ones for which Escher's symmetry drawings and prints provide wonderful illustrations, the drawings can also be used in teaching higher-level concepts of abstract algebra and group theory. In her article [57], Marjorie Senechal discusses how, by studying the color symmetry groups of Escher's periodic drawings, students can better understand the definition of a group, commutativity and non-commutativity, group action, orbits, generators, subgroups, cosets, conjugates, normal subgroups, stabilizers, permutations and permutation representations, and group extensions.

Teachers (and texts) of mathematics and science also use Escher's prints for artful depictions of mathematical objects (knots, Möbius bands, spirals, loxodromes, fractals, polyhedra, divisions of space) and to provide intriguing visual metaphors for abstract mathematical concepts (infinity, duality, reflection, relativity, self-reference, recursion, topological change) [49]. In his Pulitzer-Prize-winning book *Gödel, Escher, Bach: An Eternal Golden Braid*, Douglas Hofstadter uses Escher's work in essential ways to convey ideas of recursion and self-reference, and several authors have used Escher's prints to illustrate complex ideas of perception and illusion.

Often those who view art impose on it their reading of the artist's intention, and mathematicians' use of Escher's work to illustrate the idea of infinity and other mathematical concepts might be questioned. But it should be noted that Escher was intrigued by these concepts and set out to embody their essence in many of his prints. His fascination with infinity and how to capture it was a theme he returned to again and again. He spoke eloquently of this quest in his essay "Approaches to Infinity":

> Man is incapable of imagining that time could ever stop. For us, even if the earth should cease turning on its axis and revolving around the sun, even if there were no longer days and nights, summers and winters, time would continue to flow on eternally. . . .
>
> Anyone who plunges into infinity, in both time and space, further and further without stopping, needs fixed points, mileposts, for otherwise his movement is indistinguishable from standing still. There must be stars past which he shoots, beacons from which he can measure the distance he has traversed. He must divide his universe into distances of a given length, into compartments recurring in an endless sequence. Each time he passes a borderline between one compartment and the next, his clock ticks. . . . [37, pp. 37–40]

For Escher, mathematical concepts, especially infinity and duality, were a constant source of artistic inspiration.

Mathematical Research Related to or Inspired by Escher's Work

Several aspects of Escher's work anticipated by decades theoretical investigations by members of the scientific community. And some of his work has directly inspired mathematical investigations. We note here (necessarily briefly) many of these investigations.

Classification of "regular" tilings using edge relationships of tiles was Escher's method and also that of H. Heesch, but it was limited to asymmetric tiles and tilings with symmetry groups having no reflections. In the 1970s Branko Grünbaum and Geoffrey Shephard undertook a systematic classification of several kinds of tilings having transitivity properties with respect to the symmetry group of the tiling—isohedral (tile-transitive), isogonal (vertex-transitive), isotoxal (edge-transitive). Their method relied on using adjacency symbols and incidence symbols that recorded how (in the case of isohedral tilings) each tile was surrounded; the transitivity condition implied that every tile was surrounded in the same way. Their book [25] remains the fundamental reference on all aspects of tilings.

Two-color and 2-motif tilings were Escher's way of expressing duality. It is interesting to note that the first classification of two-color symmetry groups was carried out in 1936 (at almost the same time Escher was making his independent investigations) by H. J. Woods, who was interested in these black-white mosaics for textile designs [10; 62]. When a monohedral (one tile) tiling was colored in two colors, and a symmetry of the tiling interchanged the tiles and interchanged their colors, he called it "counterchange symmetry". (For example, in a checkerboard-colored tiling of the plane by squares, a reflection of the tiling in an edge of one column of squares would be a counterchange symmetry.) The scientific community and Escher were unaware of Woods's work. Later this kind of symmetry, so prevalent in Escher's work, was called "antisymmetry" by Russian crystallographers; that terminology is not used today. Escher noted that some crystallographers had trouble accepting the idea of antisymmetry; he said that he couldn't work without it [1, p. 94].

Escher's method of splitting tiles to produce 2-motif tilings has been shown to be a powerful one. Today, the term 2-isohedral is used to describe tilings in which the symmetry group of the tiling produces two orbits of tiles—there are two distinct congruence classes of tiles with respect to the symmetry group of the tiling. It has been proved that every 2-isohedral tiling can be derived beginning with an isohedral tiling and applying the processes of splitting and gluing [13] and that this same process extends to produce k-isohedral tilings [32]. Andreas Dress [14] and I [55] have studied other aspects of these tilings.

Color symmetry was not a serious concern of crystallographers until the 1950s; even then, it was not easily embraced, and it took many years before color symmetry groups were studied systematically. When crystallographers and mathematicians did begin to investigate color symmetry groups, they (like Caroline MacGillavry) turned to Escher's work for illustrations and discoveries. Even today, there are competing notations for color symmetry groups [7; 25; 60; 61].

Metamorphosis, or *topological change*, was one of Escher's key devices in his prints. His interlocked creatures often began as parallelograms, squares, triangles, or hexagons, then seamlessly morphed into recognizable shapes, preserving an underlying lattice, as in his visual demonstration in Plate I in [19]. At other times the metamorphosis of creatures changed that lattice, as occurs in his *Metamorphosis III*. William

Huff's design studio produced some intriguing examples of "parquet deformations" that preserve lattice structure [30], and, more recently, Craig Kaplan has investigated the varieties of deformation employed by Escher [34].

Covering surfaces with symmetric patterns was Escher's passion—the Euclidean plane, the hyperbolic plane, sphere surfaces, and cylinders—and always these coverings represented nontrivial symmetry groups of the patterned surface. Douglas Dunham has explored many families of Escher-like tilings of the hyperbolic plane and how to render them by computer [15; 52, pp. 286–96]. Others have studied how to cover different surfaces with periodic designs and sometimes asked, "What symmetry groups do these coverings represent?" See [7; 45; 56].

Escher's algorithm to produce patterns with decorated squares has inspired mathematicians and computer scientists to use combinatorial techniques (Burnside counting) and computer techniques to check his work and to answer more questions. Escher's results of twenty-three patterns for his simplest case and ten for his case (1) are exactly right. The correct answer for his case (2) is thirty-nine; for this case, Escher missed three patterns and counted one pattern twice [50]. Other questions have been asked and answered: How many patterns are there with two stamps (the original and its reflection) if Escher's restrictions on choice are removed [12]? How many with two stamps and translation only in one direction [42]? How many with the four "ribbon pattern" stamps and the additional action of under-over interchange added to the group of symmetries [22]? Can Escher's algorithm be computer-automated [38; 40]? Can allowable coloring of the ribbon patterns be automated [23]?

Creating tile shapes was almost an obsession with Escher. He would begin with a simple tile (often a polygon) that he knew would produce a regular division, then painstakingly coax the boundary into a recognizable shape. Who but Escher could conjure the polygon in Figure 10 into a helmeted horseman? [53, pp. 110–1] He explained,

> The border line between two adjacent shapes having a double function, the act of tracing such a line is a complicated business. On either side of it, simultaneously, a recognizability takes shape. But the human eye and mind cannot be busy with two things at the same moment and so there must be a quick and continual jumping from one side to the other. [39, p. *vii*]

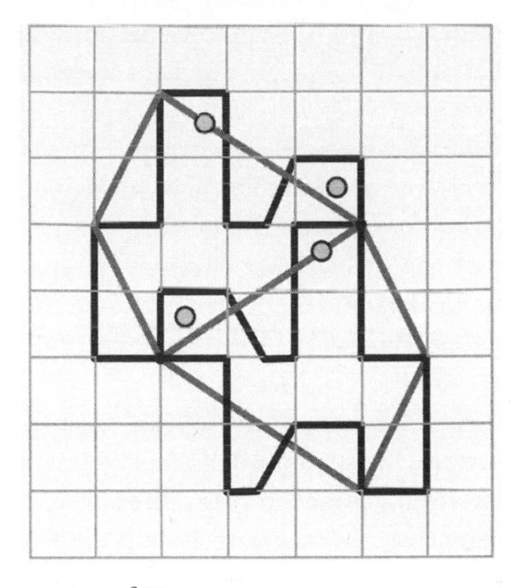

Figure 10. The beginning of *Horseman.*

Kevin Lee was the first to implement Escher's process with a computer program [36]. Craig Kaplan and David Salesin devised a computer program to address a complementary question—beginning with any shape, can it be gently deformed (still being recognizable) into a tile that will produce an isohedral tiling [33]?

Local vs. global definition of "regularity" was not Escher's concern; he followed the local rule that every tile be surrounded in the same way. But every one of Escher's "regular divisions" is an isohedral tiling; it satisfies the global regularity condition that the symmetry group is transitive on the tiles. An isohedral tiling necessarily has local regularity, but are the two definitions equivalent? In the Euclidean plane, yes, at least for asymmetric tiles and edge-to-edge tilings by polygons, but not so in the hyperbolic plane or in higher dimensions [51]. P. Engel also addresses this question in [16].

Symmetry of a tile inducing symmetry of its tiling was encountered and noted by Escher. When he used a tile with reflection symmetry (such as a dragonfly), it always induced reflections as symmetries of the tiling. He would note the tile was symmetric, and add an asterisk * to his classification symbol. But in a couple of instances, he created a tiling in which the tile was almost symmetric (and with slight modification can be made symmetric), yet the reflection line for the tile is not a reflection line for any of its tilings. In [24], Branko Grünbaum calls such tiles "hypersymmetric" and asks if they can be characterized. This is an open question.

Orderliness not induced by symmetry groups occurs at least twice in Escher's work: in his fractal construction of squares of diminishing size (Figure 8) and in his combinatorially perfect but not color-symmetry perfect coloring of one of his most complex designs with butterflies [1, p. 76]. Branko Grünbaum and others have asked for serious studies of other kinds of "orderliness" in tilings and patterns, not only that defined by symmetry groups [26; 27].

"Completing" Escher's lithograph "Print Gallery" recently posed a mathematical challenge to H. Lenstra and B. de Smit—how they came to understand the underlying geometric grid, "unroll" it, complete missing bits of the unrolled print, and roll it up again is described in [58].

In 1960 Escher wrote, "Although I am absolutely innocent of training or knowledge in the exact sciences, I often seem to have more in common with mathematicians than with my fellow artists" [20, Introduction]. Although he struggled with mathematics as a school student, when he became a graphic artist he was driven to pursue mathematical research, learn new geometric ideas, depict mathematical concepts, and pose mathematical questions. He could not have imagined the scope of influence his work would have for the scientific community.

Acknowledgments

The author thanks the M. C. Escher Company for permission to reproduce works by M. C. Escher. Bill Casselman took the photo of Coxeter's article used in Figure 6. Figures 3, 5, 6, 7, 8, and 10 were created by the author using *The Geometer's Sketchpad*.

Notes

1. The title of a 1995 exhibit of Escher's work at the National Gallery of Canada in Ottawa was titled "M. C. Escher: Landscapes to Mindscapes."

2. Escher's colored plane-fillings have been called tessellations, periodic drawings, tilings, and symmetry drawings. I prefer to use the last term.

3. All of Escher's prints that are named in this essay can be found in the catalogs [1] and [37], and many of them can also be found in [20], [59], and on the official website www.mcescher .com.

References

[1] F. H. Bool, J. R. Kist, J. L. Locher, and F. Wierda, *M. C. Escher: His Life and Complete Graphic Work*, Harry N. Abrams, New York, 1982; Abradale Press, 1992.

[2] N. G. de Bruijn, Preface, in *Catalog for the Exhibition M. C. Escher*, cat. 118, Stedelijk Museum, Amsterdam, 1954.

[3] H.S.M. Coxeter, Crystal symmetry and its generalizations, in A Symposium on Symmetry, *Trans. Royal Soc. Canada* **51**, ser. 3, sec. 3 (June 1957), 1–13.

[4] ———, Review of The Graphic Work of M. C. Escher, *Math. Rev.* MR0161210 (28:4418), 1964.

[5] ———, The non-Euclidean symmetry of Escher's picture *Circle Limit III, Leonardo* **12** (1979), 19–25, 32.

[6] ———, The trigonometry of Escher's woodcut *Circle Limit III*, in [52] pp. 297–305. Revision of *Math. Intelligencer* **18** no. 4 (1996), 42–6 and *HyperSpace* **6** no. 2 (1997), 53–7.

[7] ———, Coloured symmetry, in [8], 15–33.

[8] H.S.M. Coxeter, M. Emmer, R. Penrose, and M. L. Teuber, eds. *M. C. Escher: Art and Science*, North-Holland, Amsterdam, 1986.

[9] H.S.M. Coxeter and M. C. Escher, Correspondence, M. C. Escher Archives, Haags Gemeentemuseum, The Hague, The Netherlands.

[10] D. W. Crowe, The mosaic patterns of H. J. Woods, *Comput. Math. Appl.* **12B** (1986), 407–11, and in *Symmetry: Unifying Human Understanding*, I. Hargittai, ed., Pergamon, New York, 407–11.

[11] J. Daems, Escher for the mathematician, Interview with N. G. de Bruijn and Hendrik Lenstra, *Nieuw Archief voor Wiskunde* **9**, no. 2 (2008), 134–7.

[12] D. Davis, On a tiling scheme from M. C. Escher, *Electron. J. Combin.* **4**, no. 2 (1997), #R23.

[13] O. Delgado, D. Huson, and E. Zamorzaeva, The classification of 2-isohedral tilings of the plane, *Geom. Dedicata* **42** (1992), 43–117.

[14] A W.M. Dress, The 37 combinatorial types of regular "heaven and hell" patterns in the Euclidean plane, in [8], 35–46.

[15] D. J. Dunham, Creating repeating hyperbolic patterns—old and new, *Notices of the AMS* **50**, no. 4 (April 2003), 452–5.

[16] P. Engel, On monohedral space tilings, in [8], 47–51.

[17] B. Ernst (J. A. F. de Rijk), *The Magic Mirror of M. C. Escher*, Random House, New York, 1976; Taschen America, 1995.

[18] G. Escher, M. C. Escher at work, in [8], 1–11.

[19] M. C. Escher, The regular division of the plane, in [1], pp. 155–73 and in [31], 90–127.

[20] ———, *Grafiek en Tekeningen M. C. Escher*, J. J. Tijl, Zwolle, 1960. *The Graphic Work of M. C. Escher*, Duell, Sloan and Pearce, New York, 1961; Meredith, 1967; Hawthorne, 1971; Wings Books, 1996.

[21] M. Gardner, Concerning the diversions in a new book on geometry, *Sci. Amer.* **204** (1961) 164–75. In *New Mathematical Diversions*, MAA, 2005, pp. 196–209.

[22] E. Gethner, D. Schattschneider, S. Passiouras, and J. Joseph Fowler, Combinatorial enumeration of 2 × 2 ribbon patterns, *European J. Combin.* **28** (2007), 1276–311.

[23] E. Gethner, Computational aspects of Escher tilings, Ph.D. dissertation, University of British Columbia, 2002.

[24] B. Grünbaum, Mathematical challenges in Escher's geometry, in [8], 53–67.

[25] B. Grünbaum and G. C. Shephard, *Tilings and Patterns*, W. H. Freeman, New York, 1987.

[26] B. Grünbaum, Levels of orderliness: Global and local symmetry, in *Symmetry 2000*, I. Hargittai and T. C. Laurent, eds., Portland Press, London, 2002, 51–61.

[27] ———, Periodic ornamentation of the fabric plane: Lessons from Peruvian fabrics, in *Symmetry Comes of Age: The Role of Pattern in Culture*, D. K. Washburn and D. W. Crowe, eds., U. of Washington Press, Seattle, 2004, 18–64.

[28] F. Haag, Die regelmässigen Planteilungen und Punktsysteme, *Z. Krist.* **58** (1923), 478–88.

[29] H. Heesch and O. Kienzle, *Flächenschluss. System der Formen lückenlos aneinanderschliessender Flachteile*, Springer, Berlin, 1963.

[30] D. Hofstadter, Parquet deformations: Patterns of tiles that shift gradually in one dimension, *Scientific American* (July 1983): 14–20. Also in *Metamagical Themas: Questing for the Essence of Mind and Pattern*, Basic Books, New York, 1985, 191–212.

[31] W. J. van Hoorn and F. Wierda, eds., *Escher on Escher: Exploring the Infinite*, Harry N. Abrams, New York, 1989.

[32] D. H. Huson, The generation and classification of *k*-isohedral tilings of the Euclidean plane, the sphere, and the hyperbolic plane, *Geom. Dedicata* **47** (1993), 269–96.

[33] C. S. Kaplan and D. H. Salesin, Escherization, in *Proc. 27th Inter. Conf. Computer Graphics and Interactive Techniques (SIGGRAPH)*, ACM Press/Addison Wesley, New York, 2000, 499–510.

[34] C. S. Kaplan, Metamorphosis in Escher's art, *Bridges Leeuwarden Conf. Proc. 2008*, Tarquin, 39–46.

[35] M. S. Klamkin and A. Liu, Simultaneous generalizations of the theorems of Ceva and Menelaus, *Math. Mag.* **65** (1992), 48–52.

[36] K. D. Lee, Adapting Escher's rules for "regular division of the plane" to create *TesselMania!®*, in [52], 393–407.

[37] M. C. Escher, Approaches to infinity, in *The World of M. C. Escher*, J. L. Locher, ed., Harry N. Abrams, New York, 1972, 37–40.

[38] R. Mabry, S. Wagon, and D. Schattschneider, Automating Escher's combinatorial patterns, *Mathematica in Ed. and Res.* **5**, no. 4 (1996–97), 38–52.

[39] C. H. MacGillavry, *Symmetry Aspects of M. C. Escher's Periodic Drawings*, Oosthoek, Utrecht, 1965. Reprinted as *Fantasy and Symmetry*, Harry N. Abrams, New York, 1976.

[40] S. Passiouras, *Escher Tiles*, http://www.eschertiles.com/.

[41] R. Penrose, Escher and the visual representation of mathematical ideas, in [8], 143–57.

[42] T. Pisanski, B. Servatius, and D. Schattschneider, Applying Burnside's lemma to a one-dimensional Escher problem, *Math. Mag.* **79**, no. 3 (2006), 167–80.

[43] G. Pólya, Über die Analogie der Kristallsymmetrie in der Ebene, *Z. Krist.* **60** (1924), 278–82.

[44] J. F. Rigby, Napoleon, Escher, and tessellations, *Math. Mag.* **64** (1991), 242–6.

[45] D. Schattschneider and W. Walker, *M. C. Escher Kaleidocycles*, Ballantine, New York, 1977; Pomegranate Communications, Petaluma, 1987; Taschen, Berlin, 1987.

[46] D. Schattschneider, M. C. Escher's classification system for his colored periodic drawings, in [8], pp. 82–96, 391–2.

[47] ———, The Pólya–Escher connection, *Math. Mag.* **60** (1987), 292–8.

[48] ———, Escher: A mathematician in spite of himself, *Structural Topology* **15** (1988), 9–22. Reprinted in *The Lighter Side of Mathematics*, MAA, 1994, 91–100.

[49] ———, Escher's metaphors: The prints and drawings of M. C. Escher give expression to abstract concepts of mathematics and science, *Sci. Amer.* **271**, no. 5 (1994), 66–71.

[50] ———, Escher's combinatorial patterns, *Electron. J. Combin.* **4**, no. 2 (1997), #R17.

[51] D. Schattschneider and N. Dolbilin, One corona is enough for the Euclidean plane, in *Quasicrystals and Discrete Geometry*, J. Patera, ed., Fields Inst. Monographs, v. 10, AMS, 1998, 207–46.

[52] D. SCHATTSCHNEIDER and M. EMMER, eds., *M. C. Escher's Legacy: A Centennial Celebration*, Springer Verlag, 2003.

[53] D. SCHATTSCHNEIDER, *M. C. Escher: Visions of Symmetry*, W. H. Freeman, 1990, new edition Harry N. Abrams, 2004.

[54] ———, Coxeter and the artists: Two-way inspiration, in *The Coxeter Legacy: Reflections and Projections*, C. Davis and E. W. Ellers, eds., Fields Inst. Comm., ser. no. 46, AMS, 2006, 255–80.

[55] ———, Lessons in duality and symmetry from M. C. Escher, in *Bridges Leeuwarden Conf. Proc. 2008*, Tarquin, 2008, 1–8.

[56] M. SENECHAL, Escher designs on surfaces, in [8], 97–110.

[57] ———, The algebraic Escher, *Struct. Topology* **15** (1988), 31–42.

[58] B. DE SMIT and H. W. LENSTRA JR., The mathematical structure of Escher's *"Print Gallery"*, *Notices of the AMS* **50**, no. 4 (2003), 446–51. Also http://escherdroste.math.leidenuniv.nl.

[59] E. THÉ (design), *The Magic of M. C. Escher*, Harry N. Abrams, New York, 2000.

[60] D. K. WASHBURN and D. W. CROWE, *Symmetries of Culture: Theory and Practice of Plane Pattern Analysis*, U. of Washington Press, Seattle, 1988.

[61] T. W. WIETING, *The Mathematical Theory of Chromatic Plane Ornaments*, Marcel Dekker, New York, 1982.

[62] H. J. WOODS, Counterchange symmetry in plane patterns, *J. Textile Inst.* **27** (1936), T305–T320.

Celebrating Mathematics in Stone and Bronze

HELAMAN FERGUSON AND CLAIRE FERGUSON

Motives

I celebrate mathematics with sculpture and sculpture with mathematics. Eons-old stone strikes me as a perfect medium through which to celebrate timeless mathematics.

I used to simplify life by putting science in one room and art in another. This avoided exposing my entangled soul, lest someone think me not sufficiently dedicated to one discipline or the other. Keep science and art separated, my generation was informed. "You can't do both." Parents advised, "If you can do science and have a lick of sense, you'd better. Artists starve."

Now we live in a golden age of both art and science. More career choices. I am grateful that I have had the opportunity to combine and do both [22].

Mathematicians have their own aesthetic, a sense of beauty and elegance difficult to communicate [2]. My design process starts with some inspiring mathematics. My sculpture convolves mathematical abstractions and fundamental forms shared by everybody.

Humans are toroidal. This feature links our anatomy and abstract topology. For example, the abstract mathematical idea could be a quotient space. An easy example is a torus. The shared form could be a handshake, a less easy example of a triply punctured torus. My convolution includes geometry, topology, and humanity in forms I can express in paper, computer, clay, bronze, and stone.

I design sculpture to be touchable. Art museum curators warn us to "Look, don't touch." My sculpture succeeds when touched, held, fingered, crawled through, and then thought about.

Although I do not make models, I do make the invisible visible, touchable, and occasionally knowable. Each of my sculptures involves a circle of beautiful mathematical theorems. I love mathematics for reasons difficult for me to articulate in words, but which I can articulate in sculpture. I want some wonders of our beautiful art and science to have a life in a larger world [4].

Theorems in Bronze

Lost-Wax Bronze

Like most sculptors, early on I cast my own bronzes. This process involves complex steps of positive primary, negative mold, positive wax, negative ceramic, and positive bronze. These precede chasing, patina, or polish of the final surface. Some pieces I cast solid, then carved, and then polished [4].

Bronze is very like cast iron and usually comes in ingots. Many industrial copper alloys go by the name of bronze, and bronzes throughout the world can differ in many respects. Manganese bronze is typically found in faucets. Fine art bronze in North America is an alloy of copper with silicon specialized for pour consistency and polishability. A typical "molecule" of silicon bronze is

$$9438\mathrm{Cu} + 430\mathrm{Si} + 126\mathrm{Mn} + 4\mathrm{Fe} + \mathrm{Zn} + \mathrm{Pb}.$$

Division of the coefficients of this "molecule" by 10,000 gives a partition-of-unity recipe for fine art silicon bronze. Bronze has had industrial and military value through the ages; bronze artifacts have always been vulnerable, because we humans dream of peace but engage in war. Nonetheless, polished bronze collects light like no other medium, and it makes dazzling awards, e.g., [11, 14, 18, 19, 20].

Umbilic Torus NC

Some twenty years ago, I created *Umbilic Torus NC* in an elegant faience antique verde bronze [4].

I had already carved in stone a number of these twisted tori, for example, the umbilic toroid in a polishable Deseret limestone I pulled out of the Rocky Mountains. The NC refers to numerical control, at a time when milling machines were served by paper tape.

The *Umbilic Torus NC* form intrigues me most among all twisted toroids because of its explicit connection with the representation theory

Figure 1. Umbilic toroid in Deseret limestone. Reprinted by permission of Claire and Helaman Ferguson.

of $GL(2, R)$, the 2×2 invertible real matrices. This group acts on homogeneous binary quadratic forms (two variables, three coefficients) to give the familiar stratification of ellipses, hyperbolas, and parabolas. This group also acts on homogeneous binary cubic forms (two variables, four coefficients) to give the stratification by elliptic umbilics, hyperbolic umbilics, parabolic umbilics, and pure cubics [5].

I wanted to carve the 1/3 twisted torus with radial cross-section deltoid (hypocycloid with three cusps) and sagittal cross-section cardioid (epicycloid with one cusp) with some numerical accuracy. I also wanted to articulate the surface with a surface-filling Peano-Hilbert curve. The three-axis milling machine presented a rapidly rotating carbide ball cutter to the material and received its point-to-point movements from paper tape. I solved the tool-path problem by discovering that the choice of a surface-filling curve for the tool path was more efficient and more aesthetic. The offsets and tool moves all had to be preprogrammed. At the time all this data in the G-code of the milling machine would generate enough paper tape to fill the manufacturing lab. Fortunately, the lab people found a hard disk drive to interface with the milling machine in lieu of the paper tapes.

In Figure 2 we see the tool path curve in three-space as a ghost trajectory of the path for the physical cutter to follow. This machine carving

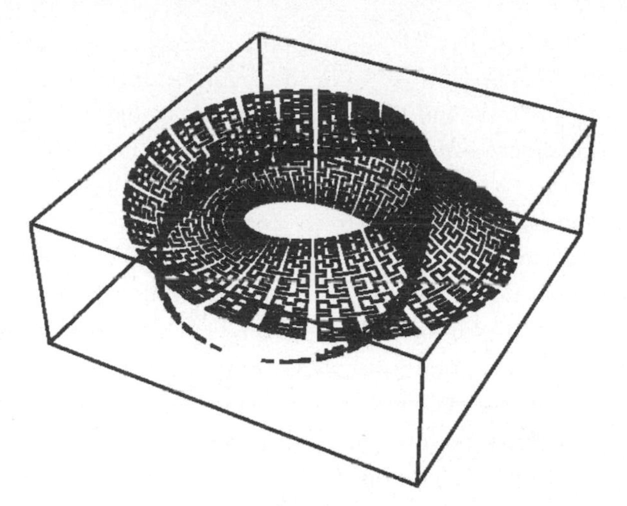

Figure 2. *Umbilic Torus NC*. Reprinted by permission of Claire and Helaman Ferguson.

only gives the first primary positive curvilinear waffle form cut into a dense styrofoam. To create the primary form for the *Umbilic Torus NC*, I used applied mathematics, computer science, and engineering. After the many positive-negative steps of the lost-wax bronze casting process I finished the *Umbilic Tori NC* with antique verde patinas.

The umbilic torus image and others like it have appeared on the covers of calculus books. I encounter students who tell me they have spent a lot of time looking at those covers and the description of the image. One young woman told me she thought her calculus instructor awful and that the umbilic torus image and its description were what got her through the class.

Theorems in Stone

On Carving Stone

Stone is one of my favorite media. Maybe I choose stone because I was raised by a stone mason who saw beauty in common field stones. My aesthetic choices include geological age, provenance, and subtraction. We learn addition and then we learn subtraction. Subtraction is harder, isn't it?

Traditionally, one originates sculptures by either addition or subtraction. Addition is popular: model clay, add clay to an armature, or weld pieces of metal together. These are operations with modules. Most art schools do not teach subtraction in the form of carving stone. To begin with a block of stone and take away what is not desired, to leave the desired, is considered really old-fashioned. It is difficult to do and even more difficult to teach. But I think subtraction is more interesting than addition, especially if I subtract from stone myself.

Mathematicians are notorious for wanting to do things themselves, prove their own theorems, or prove other people's theorems without looking at the known proofs. Sculptors tend to the opposite. Most stone carving today is like a glamorous rock-music recording production; artists with enough money job it out—outsource. The question for me is "job out what?" How do I job out C^{∞} functions? Negative Gaussian curvature? More important, having done so, what have I learned?

But there may be no practical reason to outsource simply because we live in amazing times. I can go almost anywhere and put together a modest stone carving studio from the inventory of a local hardware store, complete with diamond saws. That would have been impossible when I began pulling my own studio together some forty years ago. Most of the credit for this goes to the recent development of diamond-cutting technology [8]. Because of synthetic diamond production, sawn and polished stone can be found everywhere now, while even a short time ago it was prohibitively expensive.

Usually my sculpture goes into colleges or universities. I can expect my work to be there for generations of faculty, colleagues, students, employees, and their families to enjoy. My work may well perturb their preconceptions about why creative math and creative art possess the unique and inspiring vitality that they do possess.

If a stone has been around for millions of years, it will probably last another few thousand years, especially after I carve it into something of no obvious military or industrial value. I start with a relatively worthless piece of stone, and when I get done with it, from a functional point of view, it is even more worthless. This suits me because my stone sculpture will probably last longer as a result.

If someone digs up my theorems in stone in a few thousand years, I expect that the excavator can decode what I have encoded and continue celebrating mathematics. I'd like my sculpture to be big and strong enough for a few beautiful mathematical theorems to survive outside of books, classrooms, and the here and now, e.g., [9, 10, 12, 13, 16, 17] .

Figure 3. Reprinted by permission of Claire and Helaman Ferguson.

Next I give a mathematical discussion of subtraction and relate that to a block of stone. Then I conclude with a discussion of two of my large stone sculptures.

Subtraction

One of our oldest mathematical algorithms is the Euclidean algorithm as found in Euclid's Book VII for pairs of integers and Book XI for pairs of real numbers. The algorithm is described in Euclid as continually subtracting in turn. In 1977 Rodney Forcade and I discovered, with proofs, uncountably many generalizations of Euclid's algorithm to n-tuples of reals, complexes, or quaternions [6]. Now known as PSLQ, these are subtraction algorithms. I begin with a list of real numbers $x \in \mathbb{R}^n$ and construct a list of integers $m \in \mathbb{Z}^n$, relating them in a linear combination, $x \cdot m = 0$, if such a relation m exists. If no such relation is constructed, then at least PSLQ constructs a lower bound on the size of any possible relation. PSLQ will discover an underlying relation of dimension n in time that is a polynomial function in n and the logarithm of the Euclidean norm of the coefficients; I consider our algorithm a subtraction process with $GL(n, \mathbb{Z})$ matrix inverse operations. This subtraction algorithm has led to many new discoveries [1].

I give an example of one of these discoveries: a new formula for π was discovered with PSLQ [3]. This formula has the remarkable feature that permits binary digits of π beginning at some arbitrary position to be computed directly, without any need to compute any of the preceding digits.

Figure 3 looks like subtraction: the large positive black area is a disk with three smaller disks subtracted so that the figure has area π. This partially deleted disk is reasonably accurate as printed here, like the one I cut from black acrylic to 1,000th of an inch with a Cartesian laser robot.

The area π of this disk is given by an inner product of a real vector $x = (x_1, x_4, x_5, x_6) \in R^4$ and an integer lattice point $m = (4, -2, -1, -1) \in Z^4$. The three negatives correspond to the three deleted disks. In the theorem,

$$\pi = x \cdot m,$$

where

$$x_j = \sum_{k \geq 0} \frac{1}{16^k} \cdot \frac{1}{8k + j},$$

we only need $j = 1, 4, 5, 6$ to express π. These x_j are real numbers, given by the geometrically convergent series above; we can compute them to arbitrary precision.

The discovery of this new π formula [3] was made by taking the input for PSLQ to be

$$y = (-\pi, x_1, x_2, x_3, x_4, x_5, x_6, x_7, x_8) \in R^9.$$

Let a basis for $y^\perp$ be given by 9×9 rank 8 matrix H, so that $yH = 0$. Then PSLQ iterates on the pair y, H presented as sufficiently long decimal strings. Each iteration constructs an integer matrix $A \in GL(9, \mathbb{Z})$, where $A^{-1}H$ becomes small. In the process yA becomes small, too, but more important, a coordinate of yA may become 0 so that a column of A is a relation for y.

For this example, taking 32-place decimal presentations of y and H, after 50 iterations PSLQ constructed the $GL(9, \mathbb{Z})$ matrix $A =$

$$\begin{pmatrix}
1 & 0 & 0 & 0 & 0 & 0 & 0 & 0 & 0 \\
4 & -11 & -36 & -68 & 0 & 61 & -68 & -175 & -3 \\
0 & -3 & 4 & 2 & 1 & -4 & -6 & -2 & 0 \\
0 & 11 & 32 & 69 & 0 & -58 & 73 & 177 & 3 \\
-2 & 17 & 75 & 134 & -4 & -110 & 127 & 348 & 6 \\
-1 & 21 & 42 & 86 & 1 & -80 & 93 & 234 & 4 \\
-1 & -3 & -40 & -64 & 3 & 56 & -52 & -170 & -3 \\
0 & 12 & 22 & 40 & 3 & -46 & 52 & 118 & 2 \\
0 & -8 & -6 & -21 & -5 & 15 & -27 & -61 & -1
\end{pmatrix}$$

which has determinant ± 1. The first column of this 9×9 matrix A is a relation for y, a fairly small lattice point. The rows of the inverse matrix A^{-1} actually give good rational approximations to the vector $y \in \mathbb{R}^9$. Note that some of the coordinates of this first column of A are 0. Hence the formula for π needs only four of the x_j's.

Where is the block of stone in all this? The block is in nine dimensions and is very tiny. In this example the block has volume on the order of 10^{-288}.

The vector $y \in \mathbb{R}^9$ is presented for this experiment by thirty-two decimals in each of the nine coordinates, not as real numbers. In any computer we know of, all the arithmetic is in terms of truncated rational numbers, bit strings most likely. The PSLQ experiment does all these subtractions and in the end only reveals that in this tiny block there is a real vector with that relation. That is the discovery, and it is as empirical as carving stone.

PSLQ gives no proof. The proof, as for all the discoveries made by PSLQ, must be found elsewhere. Of course, the form once discovered may suggest a proof. Once we see this formula for π we can easily imagine that it could have been discovered by Newton or Euler, but evidently it was not. There are many more PSLQ discoveries than we have proofs for at this time, and I suppose that is as it should be.

Fibonacci Fountain, Essential Singularity II

A rainbow occasionally appears in my *Fibonacci Fountain*. I completed this sculpture, quite coincidentally, eight centuries after Leonardo of Pisa introduced addition and subtraction with Arabic numerals on paper instead of operating with Roman numerals or an abacus (it should be noted that decimal positional arithmetic was discovered in India at least seven centuries before Leonardo and was later developed by the Arabs). We remember him for giving us that rabbit problem, leading to the Fibonacci numbers:

$$0,1,1,2,3,5,8,13,21,34,55,\ldots,$$

from which descended linearly recurrent sequences [21]. It would seem that linear recurrent sequences are even more ubiquitous than rabbits.

As I walk about the 2/3-mile circumference of the 15/2-acre Lake Fibonacci at the Maryland Science and Technology Center in Bowie, Maryland, I see an amazing assortment of plants and trees. The area is a shrunken plantation; only 500 acres remain. If I nose about a bit, I observe Fibonacci phyllotaxis in scale, seed, cone, sprout, leaf, and twig of these ratios (where the subscript denotes the relevant ratio): $\text{elm}_{1/2}$, $\text{balsam}_{2/3}$, $\text{oak}_{2/5}$, $\text{cherry}_{2/5}$, $\text{hemlock}_{3/8}$, $\text{poplar}_{3/8}$, $\text{pear}_{3/8}$, $\text{pine}_{5/8}$, $\text{willow}_{5/13}$, $\text{daisy}_{8/13}$, $\text{sunflower}_{34/55}$. The plantation owners collected and planted every sort of flower, bush, and tree available over at least two centuries. The collection is now neglected, and the collectors and their slaves lie buried on a forested hill near the Fibonacci Fountain.

My *Fibonacci Fountain* contains over 45 tons of billion-year-old Texas granite. It stands 18 feet above the water, supported underwater by concrete and steel to a depth of 14 feet, which is supported in turn by 28 pilings in 40 feet of mud. When the test cores were drilled no bedrock was found.

Fourteen water cannons spurt a mathematical profile over thirty-six feet into the air, recycling Lake Fibonacci with freshly oxygenated water. The view is especially refulgent with the interplay of stone, water, sunlight, and fog. The profile of the fountain in the usual x, y, z coordinates approximates

$$z = z(x, y) = \left| \tau^{\frac{1}{x + y\sqrt{-1}}} \right|, \quad \tau = \frac{1 + \sqrt{5}}{2} \approx \left(f_n \sqrt{5} \right)^{\frac{1}{n}}$$

for the usual Fibonacci two-term recursion sequence, $f_n = f_{n-1} + f_{n-2}$, $n > 1$, $f_0 = 0$, $f_1 = 1$. I located water cannons at Fibonacci number intervals along the x-axis in the $y = 0$ vertical plane. Restricted to that plane, the profile gives an essential singularity curve which tends to infinity and zero as x tends to zero from right and left, respectively: a smooth, infinitely differentiable, C^∞, but not analytic at $x = 0$, function [4].

This distinction between smooth and analytic, allowing C^∞ functions of compact support, is worth celebration by itself.

To do architectural-size sculpture, I find friends with huge cranes. Assembling the fountain required a 75-ton crane. The 75 tons does not refer to the weight being lifted nor the weight of the crane. The 75 tons refers to the lift capacity of the crane as determined by the angle and length. The angle is that which the stick (extendable boom) makes relative to the crane platform; the length is how much stick is out. The crane had to stand on solid ground 100 feet from the fountain. This arrangement created a balance problem: A 1500-pound granite slab hanging from 140 feet

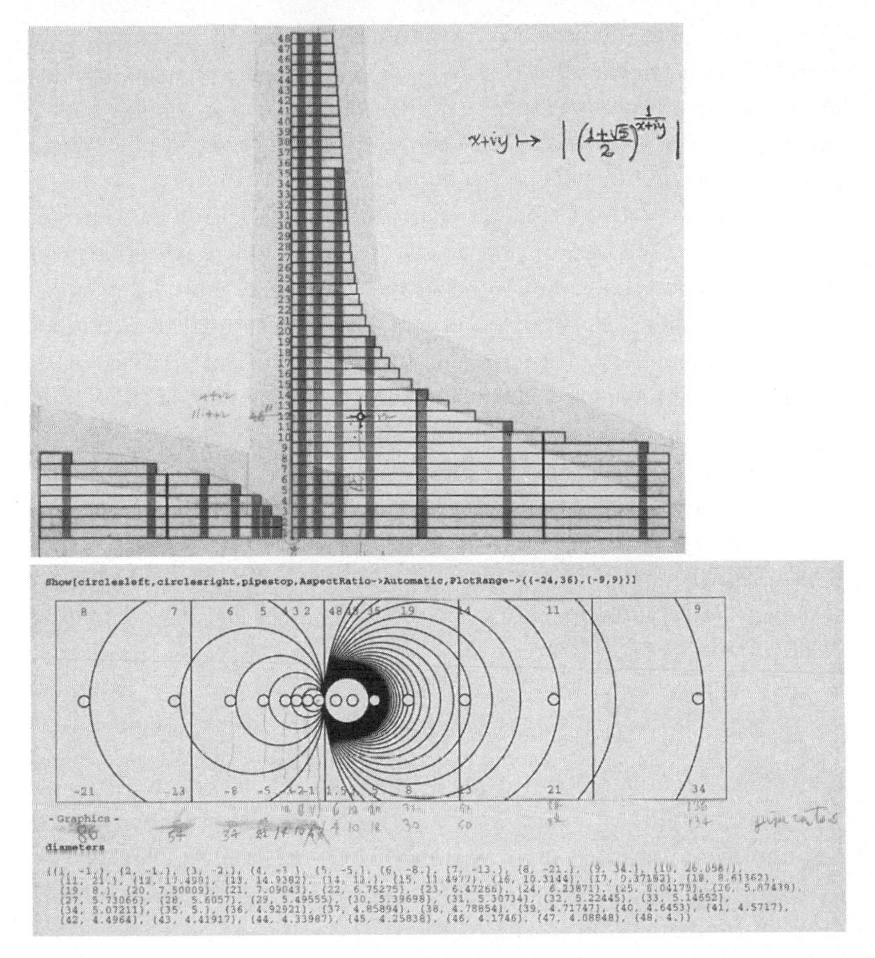

Figure 4. Reprinted by permission of Claire and Helaman Ferguson.

of stick, at a low angle, could have tipped the crane into the squishy mud of the lake bed. We avoided this disaster by loading a 2-ton block of quartz diorite (which I happened to have in my studio) onto the back of the crane. There were dicey moments; fortunately, no one was injured.

Pigeonholes and Gauss

I composed the fountain of layers of 1500-pound slabs, each nominally 4 inches thick.

In Figure 4 are side and plan drawings of the fountain, the x–z plane section and the projection on the x–y plane, respectively. The nominal 4 inches meant that the slabs came out of the quarry mill no thinner than 4 inches and no thicker than 4-1/2 inches, so any two slabs could differ in thickness by as much as a half inch. Stacking up half-inch differentials could have been disastrous. What to do? Use the powerful pigeonhole principle from number theory, of course.

I had eighty slabs of two colors of billion-year-old Texas granite, forty of each color from which to choose. I knew that at least two of these 1500-pound "pigeons" were within 1/64 of an inch difference. I calipered all these "pigeons" and found the usual normal distribution, from which I could select compatible subsets of "pigeons" for each layer.

Figure 5 gives a comparison and a conflict. On the bottom, the sheet records the tally marks for the distribution of the eighty slabs within the half inch. There are two, one for the forty red slabs and the other for the forty beige slabs. You can see each has the expected normal bulge in the middle. On the top we have the lovely abstraction, a Gaussian distribution, the familiar bell-shaped, unimodal curve and probability measure on the real numbers,

$$t \mapsto e^{-t^2}, \quad \frac{1}{\sqrt{\pi}} \int_{-\infty}^{\infty} e^{-t^2}\, dt = 1.$$

In the bottom image we see messy discrete reality. In the top image we see clean continuous theory with the transcendental π thrown in for good measure. I created this sculpture as I create all my sculptures, by balancing the tension between these two: ugly or beautiful, truth or ideal, or sometimes the other way round.

Note that I have not drawn the normal curve as a narrow inky line on a blank background. In fact, I have not drawn the curve at all. Instead I painted a red region and a blue region. This is more like a sculpture for me. It reminds me of the stone and air interface I create when I make a sculpture, where I spend so much time and energy to get it just right.

Lightning Strikes!

I spent three years working on this Fibonacci Fountain. Then, three years after it was installed, a lightning bolt obliterated several of the granite-enclosed water cannons. I was close enough to hear the monster thunderclap and to see the flash. Such an event might have struck me in

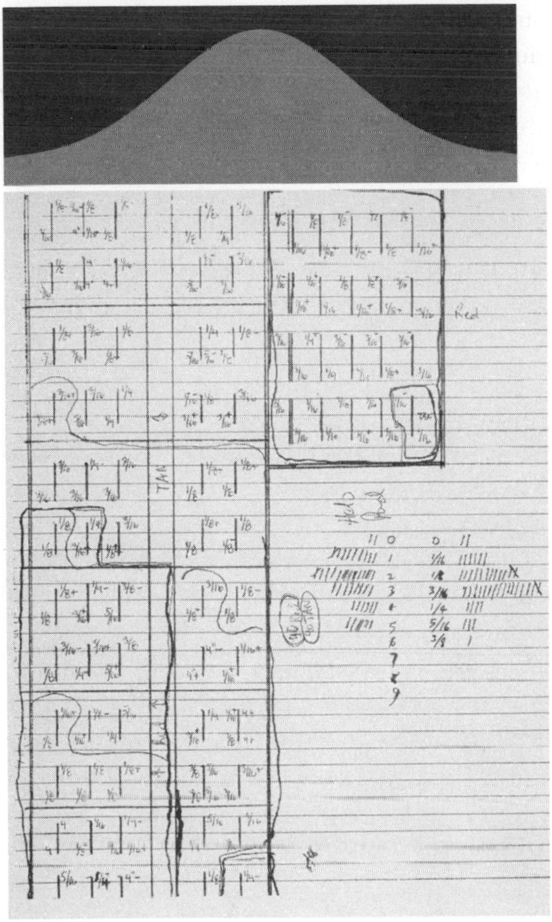

Figure 5. Reprinted by permission of Claire and Helaman Ferguson.

particular as ill-fated or portentous, for, at the age of three, I had seen my natural mother killed by lightning. Fortunately, I have grown out of feeling superstitious about threes. I in fact consider three a very helpful first odd prime number.

This lightning bolt reminded me that even if I work in billion-year-old granite as a hedge against the future, I only fob off nature for a little while. She moves on indifferent to our little interruptions.

After the strike, a scuba-diver/mathematician friend brought up three hundred pounds of granite shards, very sharp, like broken glass, which had been blown off the fountain. To prevent future strikes, I put a charge diffuser on the fountain of the sort engineers put on antennae to protect them from environmental high voltages and currents. The reconstructed fountain once again rejuvenates the lake with fresh oxygen and gathers geese under its mists.

Invisible Handshake I

For this negative Gaussian curvature carving, I started with a twenty-four-ton block of two-billion-year-old black quartz diorite from South Africa.

Minimal vs. Negative Gaussian Curvature

Geometer Alfred Gray inspired this negative curvature direction in my sculpture when he introduced me to Celso Costa's minimal surfaces with torus topology [7].

About each point (x, y, z) on a smooth surface S in $\mathbb{R}^3$, there is a little circle image on the surface. Each point of the circle has a plane through the normal to the (x, y, z) point. The plane intersects the surface S in a curve. Each curve has its own signed curvature. This mapping from the little circle to curvatures is bounded, so there will be a maximum and minimum curvature at the point (x, y, z), the two principal curvatures $\kappa_{max}(x, y, z)$ and $\kappa_{min}(x, y, z)$, respectively. The product of these two curvatures

$$\kappa = \kappa(x, y, z) = \kappa_{max}(x, y, z) \cdot \kappa_{min}(x, y, z)$$

is the Gaussian curvature at the point (x, y, z). Gauss proved that this product curvature is an intrinsic invariant of the surface S. Isometric deformations of the surface S will have the same product curvature.

A surface S is a *minimal surface* if the two principal curvatures are equal in absolute value and opposite in sign at every point (x, y, z),

$$\kappa_{max} = -\kappa_{min}, \quad \kappa = -\kappa_{max}^2.$$

Weierstrass parametric representations [7] of minimal surfaces are given by the real or imaginary part of the mapping

$$z \mapsto \int_{z_0}^{z} f(w)\left(1 - g(w)^2, \sqrt{-1}(1 + g(w)^2), 2g(w)\right) dw,$$

where f and g are suitable meromorphic functions of w, and z, z_0 are in a suitable region of the complex plane.

These Weierstrass parametric representations provide uncountably many immersions of minimal surfaces in three dimensions. What can I do with uncountably many virtual sculptures? I can at least sample them with stereo-pair computer graphics and pick ones I particularly like. It is possible to compute images from integrals with modern tools such as *Mathematica*.

What does Gaussian curvature mean to me sculpturally? The robust physical idea is not the equality of the two principal curvatures. The robust sculptural idea is that the Gaussian curvature must satisfy an *inequality* and be negative,

$$\kappa(x, y, z) < 0.$$

This observation has far-reaching consequences for my sculpture: every point on a negative Gaussian curvature surface has a saddle neighborhood. Translated into stone, the idea means that locally every point is the keystone of a fabric of arches with the warp bending one way and the weft bending the opposite way. Such stone carvings should have great structural strength, and they do. Translated to anatomy, the idea lets us recognize many epithelial saddle forms on our skin surface. For instance, in the common handshake two people press together their matching negative Gaussian-curvature parts between the thumb and palm.

For me, the consequences of negative Gaussian curvature became solid in sculptures of Costa's embeddings and Weierstrass's immersions of triply punctured tori in $\mathbb{R}^3$. I related this to elliptic curves $y^2 = x^3 + ax + b$ and a Kepler law (Jupiter shaking hands with the Sun).

I carved my early negative Gaussian curvature sculptures in Carrara marble using virtual-image projection systems. These were based on Stewart platform and cable metrology ideas. The old concept is subtraction again: carve away what is not supposed to be there. New technology allows me to virtually project the image given by parametric equations into a block of stone. Most importantly, in the midst of all this technology, I have been able to learn new negative Gaussian curvature forms and carve them directly [4]. Mathematics can be my invisible model.

Carving Quartz Diorite

The *Invisible Handshake* block of quartz diorite was unusually large as quarry blocks go. Upright, this block stood nine feet high, six feet wide,

Figure 6. Reprinted by permission of Claire and Helaman Ferguson.

and five feet thick. My block came by boat into the port of New Orleans from South Africa, was barged up the Mississippi River to Minnesota, and then trucked to my studio. At twenty-four tons, this block was a single semi-trailer truckload barely below the limit for a legal load on the highway. I used a seventy-ton crane to lift my block off the truck and ease it down into my studio, where I could finally sink my diamonds into it.

I cored the first hole and changed forever the topology, transforming the block from simple connectivity into a torus. I built and rebuilt scaffolding to carve the torus further into a triply punctured torus.

I do not suggest that such a carving project is easy, even with new diamond cutting technology. Carving uses many tools, my diamond chainsaw principal among them.

Final Setting and Footprint

A ten-foot-diameter granite disk provides the setting for my *Invisible Handshake I*, tiled by hyperbolic pentagons. The footprint of the sculpture is the center right-angled pentagon of 2-1/2-foot radius.

Figure 7. *Fibonacci Fountain*. Reprinted by permission of Claire and Helaman Ferguson.

I created a computer-graphics color image of the hyperbolic checkerboard of right-angled pentagons in the Poincaré hyperbolic disk, with the *figure five* moving, tense, but heeded through the hyperbolic plane. This was my response to the imagist poem "The Figure Five" by William Carlos Williams and the subsequent painting by his friend Charles Demuth.

On the right side I've transferred these right-angled hyperbolic pentagons into a fabric sculpture quilt, a negative Gaussian-curved checkerboard of identical panel pentagons. This quilt adapts itself well to the human form and, like the human form, it can be flattened locally but not globally. The finished stone carving *Invisible Handshake* weighs 7-1/2 tons and has a precisely carved pentagonal footprint. This corresponds to the central pentagon of a Poincaré disk ten feet in diameter hyperbolically tiled in two colors of granite. On the left is a small bronze analog of this setting with a hyperbolic pentagon footprint.

The sculpture fits into the space of this central tiling of a conformal Poincaré disk. The closed necklace of corner-connected pentagons is in number 5 times every other Fibonacci number (those ubiquitous rabbits) [15].

The Nine-Ton Bell Reveals Itself

I had carved the twenty-four-ton block down to a little over nine tons. To prepare to carve the right angled pentagon footprint, as shown with

Figure 8. Reprinted by permission of Claire and Helaman Ferguson.

attached water-jet precision-cut template (Figure 8 on right), I laid the block on one side. To do this simple move, I hired a sixty-ton crane, a crane operator, and a rigger man. We had everything rigged up for the tricky rotation. The nine-ton sculpture hung horizontally about four feet off the concrete floor. While I was changing the timber cribbing, I heard a crackle. The sculpture had slipped about an inch, which created enough friction to melt the thick six-inch-wide nylon rigging strap. The nine tons struck the floor, with one corner crushing a six-inch hole in the concrete. It rang like a bell! No evidence of this mishap showed on the piece, which was not what I would have expected given a big block. I would have assumed that such a force on a point would split the block.

I learned later that the velocity of sound in this quartz diorite is greater than the velocity of sound in steel, even though the density of steel is much greater than any stone. The negative Gaussian curvature form has saddle (double-arch) forms everywhere. An upright arch is quite strong in relation to gravity. Because a negative Gaussian curvature form presents a web of arches, every point is a keystone with a double arch regardless of the orientation relative to gravity. This gives the sculpture great strength even though the curved form is practically hollow compared to its original solid form. Evidently the energy from the force of impact is radiated almost instantly and uniformly throughout the sculpture.

I should not have been too surprised about this, because I had been carving negative Gaussian curvature forms in stone and had noted their bell-like aspects. I had also noted the peculiar strength of the negative

Gaussian curvature form in a large snow (not ice) carving I did that softened in the sun but did not collapse.

Negative Gaussian curvature surfaces, especially minimal surfaces, must have infinite extent in $\mathbb{R}^3$. To create a sculpture I have to terminate the surface. Rather than follow the quarry faces of the raw block, I now choose to compute an appropriate wave front boundary by solving the Gauss-Christoffel equations for geodesics on a surface in $\mathbb{R}^3$ [7]. The primary pressure wave will radiate along geodesics.

Future

My current sculpture studio is in an industrial park in Baltimore, Maryland. My studio volume is 45,500 cubic feet. My "tool box" is a shipping container, which when filled with hand tools weighs 14,000 pounds. As I sit here, I feel in my mind my thirteen-ton block of beautiful billion-year-old Texas red granite, and my fingers sweat. This raw granite block compels me to think of the right timeless theorems. Its time is now.

References

[1] David H. Bailey, Integral relation detection, *Communications in Science and Engineering*, Top 10 Algorithms of the Century: 24–8, January/February 2000.

[2] James A. Cannon, Mathematics in marble and bronze: The sculpture of Helaman Rolfe Pratt Ferguson, *The Mathematical Intelligencer* **13** (1991), no. 2, 30–9.

[3] Simon Plouffe, David H. Bailey, and Peter B. Borwein, On the rapid computation of various polylogarithmic constants, *Mathematics of Computation* **66** (1997), no. 218, 903–13.

[4] Claire Ferguson, *Helaman Ferguson: Mathematics in Stone and Bronze*, Meridian Creative Group, Erie, Pennsylvania, 1994.

[5] Helaman R. P. Ferguson, Two theorems, two sculptures, two posters, *American Mathematical Monthly* **97** (1990), no. 7, 589–610.

[6] Helaman R. P. Ferguson, Analysis of PSLQ, an integer relation finding algorithm, with David H. Bailey and Steve Arno, *Mathematics of Computation* **68** (1999), no. 225, 351–69.

[7] Alfred Gray, *Modern Differential Geometry of Curves and Surfaces with* Mathematica, xxiv+1053 pages. CRC Press, Boca Raton, second edition; Chapter 22.2 is Gauss's Theorema Egregium; Chapter 32, pages 735–60 is on minimal surfaces via the Weierstrass representation (not in the first edition), 1998.

[8] Howard Tracy Hall, Ultrahigh pressure research: Tetrahedral anvil press, *Science* **128** (1958), 445–9.

[9] Sculptor Helaman Ferguson, *Eightfold Way*, volume Carrara White Marble and Albemarle Virginia Serpentine, Mathematical Sciences Research Institute, 17 Centennial Way, Berkeley, California, 1993.

[10] Sculptor Helaman Ferguson, *Four Canoes: Two Linking Klein Bottles*, volume 12 tons with plaza, billion-year-old Texas red granite, half-billion-year-old Academy Black California

quartz diorite. University of St. Thomas, Sabo Square, University of St. Thomas, corner of Cretin and Summit Avenues, St. Paul, Minnesota, 1995.

[11] SCULPTOR HELAMAN FERGUSON, *Clay Mathematics Institute Award*, Clay Mathematics Institute, Cambridge, Massachusetts, 1998–present.

[12] SCULPTOR HELAMAN FERGUSON, *Fibonacci Fountain: Essential Singularity II*, volume 42 tons of red and beige Texas billion-year-old granite, concrete, and steel. Dean Morehouse, Lake Fibonacci, Maryland Science and Technology Center, Bowie, Maryland, 2000.

[13] SCULPTOR HELAMAN FERGUSON, *Invisible Handshake I*, volume 9 ft × 5 ft × 6 ft, negative Gaussian curvature carving; base is 10-ft diameter hyperbolic disk tiled by right-angled pentagons forming a checkerboard in two colors of granite. Merck Pharmaceutical, Upper Gwynedd, Pennsylvania, 2002.

[14] SCULPTOR HELAMAN FERGUSON, David and Bessie Borwein Award, CMS/SMC Career Award. Canadian Mathematical Society, 2004–present.

[15] SCULPTOR HELAMAN FERGUSON, *Five-Fold Negative Gaussian Curvature*, volume two-part bronze, triply punctured torus with right-angled pentagon footprint, cuneiform-inscribed Poincaré disk, 2005.

[16] SCULPTOR HELAMAN FERGUSON, *SYZYGY: Venus and Mars redux*, volume billion-year-old Texas red and beige granite, articulated Poincaré discs with Mayan Mars and Venus pyramids. Hamilton College, in front of the Science Center, Hamilton College Campus, Clinton, New York, 2006.

[17] SCULPTOR HELAMAN FERGUSON, *Invisible Handshake II*, volume 3-ton quartz diorite, half-billion-years-old, negative Gaussian curvature, Macalester College, Olin-Rice Science Center, Macalester College, St. Paul, Minnesota, 2008.

[18] SCULPTOR HELAMAN FERGUSON, AMS Public Service Award, American Mathematical Society, 2009–present.

[19] SCULPTOR HELAMAN FERGUSON, Gauss Society Donor Award, Mathematical Sciences Research Institute Gauss Society, 2009–present.

[20] SCULPTOR HELAMAN FERGUSON, *Stephen Anson Coons Award*, career award. ACM/SIGGRAPH, biennial, 1999–present.

[21] IVARS PETERSON, "Fibonacci Fountain," *Science News*, 2002.

[22] KATHERINE UNGAR, Helaman Ferguson: Carving his own unique niche, in symbols and stone, *Science* **314** (2006), no. 373, 412–3.

Mathematics Education:
Theory, Practice, and Memories over 50 Years

John Mason

Having been invited to look back over my life in mathematics education, I take the liberty of recalling some of the most stimulating moments as they come back to me, in an attempt to analyze what mathematics education has been about for me.* In particular I want to suggest that while the field has maintained and even widened the gap between theory and practice, it is incumbent upon us to remain steadfast that the purpose of our work is to understand and contribute to student learning of mathematics. One way I have consistently attempted to do this is to try to *preach what I practice* (and find effective) rather than the other way around.

One of my great amusements is that in the U.K. I am, I believe, seen mainly as a theorist, someone who works with ideas and tries to make them available to others. By contrast, in other countries I believe I am seen as immensely and intensely practical. In my defense in the U.K. I can point to numerous publications which have offered teachers practical actions to initiate when working with learners. But as I have often said to staff on arrival at the Open University, you have 12 to 18 months to establish a reputation in the university; after that it is very hard to alter. So too in the academic world more generally. For example, it seems that when people attend one lecture-presentation, they assume that that is all the person can talk about, when actually most of us are willing to engage with a wide range of issues at different phases of mathematics education.

*An earlier version of this article was included as the opening chapter in Lerman, S., and Davis, B. (Eds.) (2009) *Mathematical action and structures of noticing: Studies on John Mason's contributions to mathematics education,* Dordrecht, Netherlands, Sense Publishers. It is reprinted with the kind permission of Sense Publishers.

Some Historical Accounts

I say "memories over 50 years" in the title, because I started tutoring at the age of 15 at the request of my mathematics teacher Geoff Steele. It was only later that I realized just how profoundly he had influenced me through his stimulation and challenge. Many years later still I discovered that he had had no training in teaching nor much in mathematics: he worked hard to keep just ahead of me, writing out in long-hand theorems in projective geometry, leading me through Hall and Knight on continued fractions, and engaging me in Susanne K. Langer's symbolic logic. I came to value very highly the contact I had with topics that did not appear in the formal curriculum until late university, and on this I base my recommendations that students be challenged sideways, in breadth and depth, rather than accelerated through the curriculum. It was true that I twice skipped ahead, but this provided space to consolidate and explore broadly, and I am ever so grateful for it.

My first attempts at tutoring were of course completely naïve. I explained things when students I was tutoring got stuck. I tutored high school students in my first year at university, and then started supporting students in my college in the years below me. I discovered later that they used to go first to my friend across the hall who knew how to do the problems; then they would come to me and watch me struggle, asking them questions about what theorems they knew and so on! In my third and fourth years I tutored for the university. It was here that I discovered the effectiveness of *being mathematical with and in front of learners*, although I didn't formulate this slogan until much later. I would face a class of students stuck on a problem about which I knew almost nothing. So I would ask them to read the problem out loud, then to tell me from their notes what the technical terms meant, and what theorems they had in their notes concerning them. In every case I eventually "saw" how to resolve the problem (and to my retrospective regret) I would then show them how to do the problem. At least they saw a slightly more experienced learner struggling publicly, so they could pick up some practice for themselves.

In graduate school I was shown George Pólya's film *Let Us Teach Guessing* (1965) on a Friday afternoon before teaching a semester class at 7:45 each morning starting on the Monday. As I later realized, the film released in me many of the practices used with me by Geoff Steele, and resonated vibrantly with practices I had developed spontaneously so as not to look a complete fool in tutorials. I got agreement with the class that they would work "my way" on Mondays; on Tuesdays to Thursdays we would then work "their way" in order to finish the chapter of the week, and on Fridays I would do revision problems with and for them. Within a week or two we were working "my way" until at least part way through Wednesdays, finishing the chapter Thursdays, and revising on Fridays.

So what did it mean to "work my way"? My recollection is that I would ask them questions to get them thinking. I would construct examples and generally cajole them. I would summarize in technical language what I thought they had begun to "see" or appreciate. I felt that I was engaging them in the thinking. I am sure that an observer would have seen me getting them to "guess what was in my mind," but the class was interactive and I hope challenging.

On my arrival at the Open University I realized that distance education was the antithesis of what I thought learning was about, but I settled down to write material. It was then that I realized how problematic teaching mathematics is, and that I had been enculturated as a structuralist through the influence of Bourbaki on many of my lecturers. One advantage was that, unlike the practice in a face-to-face institution where each lecture is an event to be survived, I frequently found myself at my desk wondering what example to use, what definition to use, what theorems to put in what order, and how best to interlace examples with abstractions and generalities.

As I was one of the few people in the faculty who had worked with Pólya's ideas, I was asked to organize the summer schools. I instituted sessions such as investigations (inspired by what I discovered going on in primary schools in the U.K. and based on Pólya's film), *mental calisthenics* (reproducing sessions I had experienced myself in primary school), *surgeries* (where students could come and ask for help on any aspect of the course), *tutor revelations* (in which a tutor would work through some typical questions while exposing the inner thoughts, choices, and incantations accompanying the solving) as well as lectures and tutorials. I even tried sessions called *tutor bashing*, in which a tutor poses a question to another tutor who then works at it from cold in front of the students before posing a question to the next tutor in sequence. This was an attempt to show students that tutors were mortal and fallible and that mathematics does not flow perfectly out of the pen in its completed form.

It took me several years to realize that what was obvious to me about how the mathematical practices described by Pólya, such as specializing and generalizing, imagining and expressing, conjecturing and convincing, were not obvious to many of our tutors. Thus some tutors adopted a "don't let them leave for coffee break without giving them the formula, because they'll just get it from others anyway" approach, and recommended this to tutors new to the summer schools. On probing this it transpired that students in tutorials where the tutor did not like investigations usually came out not liking them, and students with tutors who did like them usually came out enjoying them. The stance, beliefs, and attitudes of the tutor could be highly influential. This later resonated with the pioneering work of Alba Thompson (1984, 1992) working with teachers.

In 1973–74 I spent nearly a year with some 125 others in a house in Gloucestershire under the direction of J. G. Bennett, mathematician, scientist, linguist, and seeker. It was a pivotal year for me, crystallizing many awarenesses and awakening me to many others. Ever since then, I have been reconstructing the ideas and practices I encountered. In particular, I experienced deeply an experiential approach to enquiry, which for me extended through mathematics to every aspect of life and thought. Many years later I decided to devote time to reconstructing those practices, expressing them in the domain of mathematics education in particular, but applicable in any caring profession. I called it the Discipline of Noticing (Bennett, 1976; Mason, 1996, 2002). My idea was to provide teachers and other carers with a philosophically well-founded method and theoretical framework for researching their own practice, that is, for working on themselves.

I began my research life in combinatorial geometry, and achieved a certain notoriety within the rather small community of like-minded scholars as someone with a practical, example-rich approach to tackling really difficult mathematical problems from a structural perspective. It was after an event run by a colleague, Johnny Baker, in which teachers from secondary mathematics and science departments reported on their experience of teaching and using mathematics in schools, that I realized that although the mathematical problems I worked on were difficult, very few people cared, whereas the problems in mathematics education are essentially unsolvable, but a very large number of people care. So I turned my attention explicitly to mathematics education, and set myself three years to establish a reputation in the field. Despite this change of direction, I have never lost my interest in, no, my addiction to, working on mathematics. I always have some mathematical problem in the back of my mind that I work on in otherwise idle moments. This serves to keep me in touch with my own experience and sensitizes me to struggles that others may have with different topics, concepts, or problems.

It was while designing a course for mathematics teachers that I persuaded my colleagues that it would be a good idea for teachers to engage in mathematical thinking for themselves at their own level each week. But then a decision had to be made as to how to choose what problems to offer them, and how to structure that experience. We found that we had plenty of advice to offer, in my case gleaned from Pólya, Bennett, and awareness of my own experience. Together Leone Burton and I decided to write a book about problem solving, in order to assist us in developing a structural framework for making suitable selections for the course. Leone introduced me to Kaye Stacey, and *Thinking Mathematically* (Mason et al., 1982) was born. Surprisingly, perhaps, the only opportunity I ever had to teach a course like it was before it was conceived, when I taught summer courses in Toronto, though others continue to use the book more than 25 years on. A new extended edition has appeared just this year (Mason et al., 2010).

My nonacademic activities in the '70s brought me into contact with a wide range of group activities exploring sensitivity training and body-mind connections. I encountered a wide range of practices and authors such as Abraham Maslow (1971). I immediately recognized from him that I was really interested in what is possible, what *could* be, rather than *what is the case currently*. Since at the Open University there were no students on campus to use as subjects, nor easy contact with schools, it matched my situation to be more interested in what is possible through examining my own experience while also working with teachers in in-service professional development sessions. One effect was that, with only a few exceptions, I only ever worked with teachers and students on a one-off basis rather than over an extended period of time.

Soon after I arrived at the Open University I discovered the Association of Teachers of Mathematics, and began to go to their meetings. I encountered there a group working on mental imagery, and this fired my imagination, literally. I adopted and adapted practices that Dick Tahta and others used with posters and animations, to use with videotape of classroom interactions, and this was the basis for what our group constructed as an approach to using video. At ICME 5 in Adelaide, I discovered this was called or at least akin to "constructivism," which then evolved into *simple* or *psychological* constructivism, *radical* constructivism (much my preference), and later *social* constructivism. Combined with my experiential orientation, this set me up to resonate strongly with Ernst von Glasersfeld when I met him in Montreal in 1984.

In Adelaide I also met Guy Brousseau and extended contact with Nicholas Balacheff. Impressed by several of the constructs they used, in particular *transposition didactique* (Chevallard, 1985), *situation didactique,* and the *didactic tension* (Brousseau, 1984, 1997) and *epistemological obstacles* (Bachelard, 1938), it took me a long time to appreciate even the most surface features of the deeply analytic frame that informed their impressive research. However, my own experience in working with groups of teachers experientially had shown me that offering results of research enquiries to others and expecting changes in practice is in itself highly problematic.

When I was a graduate student there were frequent calls for research on the most effective way to teach mathematics to undergraduates: 3 lectures and 2 tutorials per week, or 5 lectures per week, or 3 tutorials and 2 lectures, or what? It was evident to me that the issue depended on too many factors connected with the setting, the individuals, the expectations, and the practices within lecturing and tutorials to be able to declare one better than another universally. Whereas most mathematicians that I knew were seeking a mathematical type of theorem with definitive conclusions, I was convinced that any value system would be situation dependent. I found statistical findings deeply unsatisfying, because either they would agree with my own prejudices, in which case they told me nothing, or they would contradict those biases, in which case I would reject them as being unsuitable or irrelevant to my situation. I felt perfectly at home with the impossibility of mathematical-like theorems in mathematics education, because of the presence of human will, intention, and ideals.

When I started working with teachers in the U.K. I soon realized that making use of "findings" was problematic for others as well. Proposing a research finding (qualitative or quantitative) that is close to current practice is likely to be responded to by assimilation without noticing any subtle differences. Proposing a research finding that challenges thinking, or which is not immediately compatible with practices, is unlikely to stimulate people to try the idea, much less adopt it simply because it is a research finding. There has to be something which catches attention, either because it seems implausible and so motivates checking, or because it appears to match perceived current needs.

In my own case, even if I do try something, I am most likely to modify it to fit with my perspective and approach. Indeed, there is no "it" as such, only my reconstruction. This corresponds with my view of classroom incidents, indeed incidents and events generally: there is no "event" as such, merely the stories told about it, whether at first, second, or later hand. On the other hand, if the finding fits with my experience, it is likely to seem "obvious," so I am likely to pay no attention to any slight differences that might in fact be significant. Instead, I feel reassured and carry on. Thus for me a successful professional development session is one in which participants can actually imagine themselves acting differently in some situation in the future which they recognize. This statement is much more significant than it may appear at first sight. I emphasize *imagine themselves*, for one thing I learned from Bennett was the immense power of mental imagery for preparing actions to take place in the future.

What does seem to be helpful is prompting people to experience something which sheds light on their past experience and offers to inform their future choices. I see professional development as personal enquiry, stimulated and supported by work with colleagues, but essentially a psychological issue with a sociocultural ecology. I have however always resisted pushing this as far, for example, as my one-time colleague Barbara Jaworski (2003, 2006, 2007) has done. I am content to indicate possibilities to others rather than trying to maximize efficiency and efficacy. For me, change is such a delicate matter that it must be left to the individual within their various communities, to the extent that "I cannot change others; I can work on changing myself", and even that is far from easy!

My interests have always been in supporting others in fostering and sustaining mathematical thinking in their students. I have at various times concentrated on mental imagery, modeling, problem solving, and language, but these have been only byways in getting to grips with the nature and role of attention. I have, for example, found it convenient to shift my discourse from *processes of problem solving* to *exploiting natural powers*, finding that the same ideas (imagining and expressing, specializing and generalizing, conjecturing and convincing, among others) continue to be potent as long as they are expressed in a context which resonates with people's experience and a discourse that they recognize. Early in my reading of mathematics education books and papers, I recognized that each generation has to re-express insights in their own vernacular, even though these insights have been expressed before. Indeed, each person has to re-experience and reconstruct for themselves. This contrasts with mathematics, in which it is possible to be directed along a "highway" toward problems at the boundary without traversing all of the country in between. Mathematics education is not like that, and perhaps never will be, at least until we establish common ways of working. I take up this theme in the next section.

In the early 1980s I had the chance to attend a number of seminars led by Caleb Gattegno when he tried to revivify his science of education (Gattegno, 1987, 1990) in the mathematics education community in England. I found his approach attractive, with a good deal in common with what I had learned from Bennett, but leading to a rather different cosmology. I began to get a taste of what it is like when an experienced "graybeard" assembles their to-them-coherent-and-comprehensive framework or theory. Whereas when the fragments are being worked on and described there is often considerable interest among colleagues, once the whole is assembled, people don't really want to know. I ran into this phenomenon again when reading Richard Skemp's later book (1979), where again my experience was one of interest in some of his distinctions, without appreciating all of them, or the way they all fit together. Reading Jean Piaget, Zoltan Dienes, Hans Freudenthal, David Ausubel, Frédérique Papy, and Humberto Maturana all had similar effects on me, partly perhaps, because I came across their work after or near the end of their careers. I found many specific distinctions of great value, but resisted taking on board their overarching theories.

I assume that the issue is one of subordination. Philosophers are trained to suppress their own thinking in order to "think like" the philosophers they are studying. In mathematics education, the intention is to improve the experience of learners, and this pragmatic dimension may contribute to a reluctance to let go of one's own stance in order to enter, absorb, and fully appreciate the stance of someone else. Several French colleagues have given me the impression that in France they are more used to subordinating to an established theory, whereas Northern

European Anglo-Saxon cultures appear to be more pragmatic and less theory-oriented in assembling their own personal framework or theory that "works for them."

The Rise of Mathematics Education

Others are more scholarly at researching the ebb and flow, the waxing and waning of salient constructs in mathematics education. My memory is that in the 1970s and early 1980s research interest focused on students. I was, naturally, caught up in the Pólya-inspired *problem solving* discourse of *processes*, as manifested in *Thinking Mathematically*. As data accumulated, attention turned to student errors and misconceptions. I recall the breath of fresh air when Douglas McLeod and Verna Adams (1989) edited a book on affect and problem solving, and Alba Thompson championed devoting attention to teachers' beliefs as influencing both how people teach and what students learn. As I look back now, it seems to me that one of the reasons for each generation revisiting and reconstructing classic insights and awareness is that, as well as participating in a process of personal reconstruction, each generation finds itself dissatisfied with the explanatory and/or remedial power of the current discourse and foci. The discourse seems somehow to be drained of its power to inform choices, partly through over- and misuse perhaps. Each generation seeks fresh fields for explanation as to why students, on the whole, do not learn mathematics effectively or efficiently, and latterly, why teachers do not teach what they know and why they know so little of what it is necessary to know in order to teach effectively.

On the one hand we have Henri Poincaré's position of being mystified as to why perfectly rational people can fail to succeed at the perfectly rational discipline of mathematics, and on the other hand we have generations of students convinced that either fractions or algebra was a watershed for their involvement in mathematics. Clearly rationality is not the central feature of most people's psyches. One of the many things that has impressed me about Open University students over the years is that when I used to ask students in our mathematics courses why they were going to all that effort, I almost always got the reply "always liked mathematics at school; never could do it, mind, but always wanted to know more." Something touches people, even if it remains dormant.

Attention in mathematics education research has shifted variously between the structure of and inherent obstacles in specific topics, psychological aspects of learning mathematics, psychological aspects of teaching mathematics, sociological aspects of teaching and learning mathematics, acts of teaching, teachers' beliefs and how they influence learners, the historical-sociocultural forces at work in and through institutions, and the content and format of teacher education courses, not to say the obstacles encountered by novice teachers due to weak mathematical background, and the destructive forces of school practices and government policies on the ideals and aspirations of novice teachers emerging from teacher education courses, to name but a few. Identity, agency, and collaboration are currently popular, and multiple selves may be just over the horizon. Most of what is accepted as *research* involves making observations of others (what I call *extra-spective*) whether associated with deliberate interventions or not. Observations and transcripts are turned into data by being selected for analysis. Analysis then applies a framework for making distinctions, or generates or modifies such a framework. The data becomes the object being analyzed, not the original phenomenon.

It is tempting to say that we (the community of mathematics educators, scholars, and researchers) have accumulated a great deal of data. We have individually, though perhaps not collectively, drawn a multitude of distinctions and formulated a plethora of constructs to analyze and account for what has been observed. But what have we really got to show for all this effort? Publications proliferate faster than I for one can read them, much less take them in and integrate them. Rarely do we get evidence that the framework has enabled teachers to modify their practice and so influence student learning. So what is the mathematics education enterprise?

The Enterprise of Mathematics Education

On the surface, it is reasonable to expect that those engaged in mathematics education research and scholarship have as their aim the improvement of conditions for learning (and hence for teaching) mathematics more effectively, at every age and stage. Consequently evidence of effectiveness must lie ultimately in improvement in learner experience and performance, in both the short term and long term.

Of course there is an immediate obstacle, for there is little or no agreement as to what constitutes evidence of learner experience, much less evidence of improvement. Since it is easiest to gauge by scores on tests, national and international studies administer tests and pronounce on the results. Questionnaires and even interviews with selected subjects can be carried out. But there is a fundamental difficulty. Test results only indicate what subjects did on one occasion under one set of circumstances with or without specific training in preparation. Interviews at best reveal only what the interviewer probes and selects due to their sensitivities, and questionnaire responses are highly dubious indicators of what lies beneath surface reactions to specific questions. As soon as you identify an indicator of mathematical thinking or other mathematical competence or success and incorporate it into a test item, it should take a competent teacher at most two years to work out how to train students to answer those types of questions. It all comes back to Guy Brousseau's notion of the *didactic contract* as manifested by the *didactic tension* and its parallel, the *assessment tension*:

> The more clearly and specifically the teacher (assessor) indicates the behaviour sought, the easier it is for the learner to display that behaviour without generating it from themselves (understanding).

Put another way in a discourse derived from Caleb Gattegno, training behavior is important and useful, but it tends to be inflexible and even dangerous if it is not paralleled with educating awareness. It is awareness (what enables you to act, what you find "comes to mind" in the way of actions) that guides and directs (en)action, using the energy arising from affect (by harnessing emotions). It is ever so tempting to train someone's behavior by giving them rules and mnemonics to memorize, and quantities of exercises on which to rehearse. But real learning only occurs when these form the basis for reflection and integration so that awareness is educated (as in the Confucian culture approach to teaching and learning). Alternatively, one can work on educating awareness and training behavior together, through harnessing emotion, and this is the approach that I have endeavored to practice, and through practicing, to articulate for myself and others so as to make the process more efficient over time. Because I am interested in what is possible, and because the only way of directing other people's attention is through being aware of the focus of my own attention, I use *intra-spection* (between selves or between people) as distinct from *intro-spection*, which garnered a negative connotation through the indulgences of people trying to develop phenomenological research methods in the early part of the twentieth century.

One of the underlying tensions in mathematics education that I am aware of is that between a "scientific stance" and a "phenomenological stance." Editors want their journals to contribute to the scientific development of knowledge. Journals have recently become so obsessed with theoretical frameworks that papers get longer and longer, without any growth in substance. I suspect that colleagues, especially editors, want to see mathematics education build a coherent and well-founded structure of knowledge. They like to see people building on one another's work, adding to and refining rather than starting afresh. I wonder however whether this is even possible, much less desirable, given the nature and focus of mathematics education, working as it does with human beings placed in institutional settings of various sorts, and exercising their wills and intentions through their dominant dispositions. I want to put a different case, the case of working with lived experience.

Structured Awareness

I have often thought and sometimes said, that when I am engaged in my enquiries, I enjoy it most when I am at the overlap between mathematics, psychology and sociology, philosophy, and religion. There is something about working on a mathematical problem which is for me profoundly spiritual; something about working on teaching and learning that integrates all three traditional aspects of my psyche (awareness, behavior, and emotion, or more formally, cognition, enaction, and affect) as well as will and intention, which themselves derive from ancient psycho-religious philosophies such as expressed in the Upanishads (Rhadakrishnan, 1953) and the Bhagavad Gita (Mascaró, 1962; see also Raymond, 1972). I associate this sense of integration with an enhanced awareness, a sense of harmony and unity, a taste of freedom, which is in stark contrast to the habit and mechanicality of much of my existence. Even a little taste of freedom arising in a moment of participating in a choice, of responding freshly rather than reacting habitually is worth striving for.

One way to summarize such experiences is that, in the end, what I learn most about, is myself. This observation is not as solipsistic, isolating, and idiosyncratic as it might seem, for in order to learn about myself I need to engage with others (who may, as is the case for hermits, be virtual), and I need to be supported and sustained in those enquiries. A suitable community can be invaluable, though an unsuitable community can be a millstone! I reached this conclusion through realizing that when a researcher is reporting their data, and then analyzing it, the distinctions they make, the relationships they notice, the properties they abstract all tell me as much about their own sensitivities to notice and dispositions to act as they do about the situation data being analyzed. Indeed I proposed an analogy to the Heisenberg principle in physics: the ratio of the precision of detail of analysis to the precision of detail about the researcher is roughly constant (Mason, 2002, p. 181).

The seeds of this observation were in working with teachers on classroom video, informed by techniques for working on mathematical animations (e.g., those of Jean Nicolet and Gattegno's reworking of them). The technique is to get participants to reconstruct as much of the film as they can after seeing it just once. When they have made a good attempt, they have specific questions about portions they only partly recall, or where different people have different stories. So a second viewing makes sense, but only because there are specific questions. Applied to classroom video, we adopted a similar stance in order to counteract the common reaction of "I wouldn't let that teacher in my classroom," or "my low attainers are lower attaining than those low attainers" (Jaworski, 1989; Pimm, 1993). It seemed that teachers saw classroom video as a challenge to their identity and practices. By getting them to recount specific incidents briefly but vividly from the video with a minimum of judgment, evaluation, and explaining, we found that they soon recognized incidents as being similar to incidents they had met in their own experience. So the videos became an entry into participants' own past experience, and hence gave access to their lived experience. This makes so much more sense than critiquing the behavior of some unknown teacher whose class may already have left school, so there is no way that their behavior could be altered!

Incidents which strike a viewer usually resonate or trigger associations with incidents recalled from the past. Describing these to others briefly but-vividly so as to resonate or trigger their own recollections provides a database of rich experiences which can be accessed through the use of pertinent labels. Often sameness and difference between reconstructed incidents has to be negotiated among colleagues, and this is what prompts probing beneath the surface. As Italo Calvino (1983, p. 55) said, "It is only after you come to know the surface of things that you venture to see what is underneath; but the surface is inexhaustible."

I have come to recognize that Bennett (among many others over the centuries) was right when he highlighted the fundamental act of making distinctions. It is, after all, how organisms at all levels of complexity operate. Change is the experience of making distinctions over time; difference is the experience of distinction making in time. Evaluation is the experience of distinguishing relative intensities (as a ratio or scaling). Bennett went much further, amplifying Gurdjieff's observation that "man is third force blind" (see Orage, 1954). In other words, distinguishing things, this from that, is important, but locks you into tension or evaluation, and is just the beginning of what is possible. In order to appreciate how the world works (whether material, mental, symbolic, or spiritual) it is necessary to become aware that actions require three impulses: something to initiate, something to respond, and something to mediate between these, to bring them into or hold them in relationship. The product of actions can then go on to serve to initiate, respond to, or mediate a further action. Bennett continued this neo-Pythagorean analysis into the quality of numbers from 1 to 12 in his monumental four-volume work *The Dramatic Universe*, which he called *systematics*, long before "systems theory" became a slogan. Perhaps because of my structural upbringing, I found myself resonating with his approach, to the extent that I could sometimes hear in the structure of his talks the ways in which he was systematically employing systemic qualities of a particular number.

Precision and Replication

There is another issue concerning transforming observations into data and the degree of precision presented in research reports. Over the years there has been an evident growth in the length and complexity of papers in mathematics education. It used to be that some detailed transcripts along with some analysis stimulated colleagues to investigate the phenomenon in their own setting. A classic example would be the paper by Stanley Erlwanger (1973, reprinted in Carpenter et al., 2004) about Benny's encounter with fractions. Nowadays this paper would probably be rejected by journals as failing to present an adequate theoretical framework and discussion of method and ethics. Many recent papers are so heavily theory laden in the opening sections that by the time I get to the substance I have forgotten exactly which parts of which theory are actually being employed, and indeed sometimes it is not even very easy to detect this. It seems to me that often only tiny fragments of theoretical frameworks are called upon. Indeed, I have no problem with this at all, because of my eclectically cherry-picking approach to understanding and practice: all I can ever do is be stimulated or sensitized to notice, that is to discern details not previously attended to, and through that discernment, raise questions deserving of enquiry. But if authors are selecting fragments, why not be straightforward about it? I go so far as to suggest that experience itself is fragmentary, despite consciousness and the collection of selves that make up consciousness and personality trying to develop stories to make it look continuous and coherent (Mason, 1986, 1988). This is the one detail on which I disagree with William James' notion of a stream of consciousness (James, 1890). My own observations agree with Tor Nørretranders (1998) that these stories are a fabricated illusion.

I realize that editors have a commitment to building the scientific foundations of mathematics education, but I don't see that present practices are actually furthering the field, in the main. What we do have is a plethora of distinctions, sometimes several labels for at best subtly distinct distinctions, and sometimes the same label used for different distinctions. What the field really needs is some agreement on ways of working, rather than on theoretical frames and stances. We need to build up a vocabulary for how we compare observations, turn them into data, and negotiate meaning among ourselves. This would then make it easier to offer similar distinctions to others, including teachers, teacher educators, and policy makers, and to negotiate similarities, differences, and intensities. Caleb Gattegno offered his *science of education,* but this is too radical for most to agree to; in the Discipline of Noticing I tried to offer a less radical and more practical foundation for ways of working; I am sure that others feel they have done the same. The problem in my view lies not in the fact that everyone discerns slightly differently, but that we don't have established ways of negotiating similarities and differences in what is noticed and in what triggers that noticing, and in what actions might then be called into play.

Despite the developments in style (I hesitate to use the word *improvement*) it is still rarely if ever possible to imagine, much less actually carry out, a replication of a study reported on in a mathematics education research paper in a journal. There is simply never enough detail. I happen to suspect that it would never be possible to replicate a study exactly, precisely because of the complex range of factors comprising the traditional triad of student, teacher, and mathematics all embedded in an institutional environment or milieu.

If it is either impossible or not necessary to be able to replicate the conditions of a study, what is it that we are gaining by reporting on our studies? My radical response to such a question is that what matters most is *educating awareness* by alerting me to something worth noticing because it then opens the way to my choosing to respond rather than react with a more creative action than would otherwise be the case. I don't usually need all sorts of detailed data, because the more precise and fine-grained the detail, the less likely I am to pay attention to the overall phenomenon being instantiated, and so the less likely I am to recognize it again in the future and so choose to act differently.

Reprise

I am genuinely perplexed about the role and nature of structure in a domain such as mathematics education. On the one hand, with my structural background, I find it really helpful to be able occasionally to invoke one or other structure in order to inform my thinking. But I have colleagues who resist such an approach, just as I have resisted accommodating the whole of other people's structured frameworks. It is too simplistic to say that each could be expounded and then tested by experiment to see which is best. I am reminded of a sequence of lectures I was required to attend in my first year at university on the leap of faith. My recollection is that they were about the philosophical conundrum of how you cannot investigate or enquire into what it is like to believe something without actually believing it. Put in an overly extreme form perhaps, "if you can critique it, you haven't experienced it fully." Of course this is anathema to many in Western society, but increasingly popular to fundamentalists the world over.

My own experience is that I do not usually use my own frameworks systematically or mechanically, because they have been integrated into how I perceive the world and how my thinking progresses. Every so often it is useful to ask myself if I have taken all aspects of an action, an activity, a potentiality, a moment, a transformation into account, and this is when it can be fruitful to remind myself of the pertinent number and its structural qualities. More specifically, the Structure of a Topic framework (Griffin and Gates, 1989; Mason and Johnston-Wilder, 2004a, 2004b), based on Bennett's "present moment" system associated with qualities of six is particularly useful when preparing to teach a topic. The six modes of interaction (expounding, explaining, exploring, examining, exercising, and expressing) arising from the qualities of three alert me to possible forms of interaction and bring different interventions to mind (Mason, 1979).

I suspect that each of us does something similar. We act in the world; when some tension or disturbance arises, we resort to accustomed modes of thinking using whatever discernments of distinctions come to mind, and then carry on. We have habitual but slowly evolving forms of activity with which we feel comfortable; we are sensitized to certain aspects of potentiality in a situation; we stress certain aspects of the present moment; and so on. Calvino (1983, p. 107) said something similar: "The universe is a mirror in which we can contemplate only what we have learned to know in ourselves," which in turn resonates with a North American shaman Hyemeyohsts Storm (1985), who phrased it as "the Universe is the Mirror of the People." The universe, whether material, imagined, or symbolic, provides a mirror for seeing ourselves and so bringing possible actions to mind. Indeed, all we can see is in fact ourselves, in the sense that what we discern and relate is a reflection of ourselves. What professional development means is ongoing work to extend sensitivities, striving for a greater balance in the interplay of component features, so as to participate more fully in the evolution of awareness.

References

Bachelard, G. (1938; reprinted 1980) *La Formation de l'esprit scientifique*, Paris, J. Vrin.

Bennett, J. (1956–1966) *The dramatic universe* (four volumes), London, Hodder & Stoughton.

Bennett, J. (1976) *Noticing* (The Sherborne Theme Talks, Series 2), Sherborne, UK, Coombe Springs Press.

Brousseau, G. (1984) "The crucial role of the didactical contract in the analysis and construction of situations in teaching and learning mathematics," in Steiner, H. (Ed.), *Theory of mathematics education* (Paper 54), Bielefeld, Germany, Institut für Didaktik der Mathematik der Universität Bielefeld, pp. 110–9.

Brousseau, G. (1997) *Theory of didactical situations in mathematics: didactiques des mathématiques, 1970–1990*, in Balacheff, N., Cooper, M., Sutherland, R., and Warfield, V. (Trans.), Dordrecht, Netherlands, Kluwer.

Calvino, I. (1983) *Mr Palomar*, London, Harcourt, Brace Jovanovich.

Carpenter, T., Dossey, J., and Koehler, J. (Eds.) (2004) *Classics in mathematics education research*, Reston, Va., National Council of Teachers of Mathematics.

Chevallard, Y. (1985) *La transposition didactique*, Grenoble, France, La Pensée Sauvage.

Erlwanger, S. (1973) "Benny's conception of rules and answers in IPI mathematics," *Journal of Mathematical Behaviour* **1**(2), 7–26.

Gattegno, C. (1987) *The science of education, part I: theoretical considerations*, New York, Educational Solutions.

Gattegno, C. (1990) *The science of education*, New York, Educational Solutions.

Griffin, P., and Gates, P. (1989) *Project Mathematics UPDATE: PM753 A,B,C,D, Preparing to teach angle, equations, ratio and probability*, Milton Keynes, U.K., Open University.

Hall, H. S., and S. R. Knight. (1940) *Higher algebra. A sequel to elementary algebra for schools*, London, Macmillan.

James, W. (1890, reprinted 1950) *Principles of psychology* **1**, New York, Dover.

Jaworski, B. (2006) "Theory and practice in mathematics teaching development: critical inquiry as a mode of learning in teaching," *Journal of Mathematics Teacher Education* **9**(2), 187–211.

Jaworski, B. (1989) *Using classroom videotape to develop your teaching*, Milton Keynes, U.K., Centre for Mathematics Education, Open University.

Jaworski, B. (2003) "Research practice into/influencing mathematics teaching and learning development: towards a theoretical framework based on co-learning partnerships," *Educational Studies in Mathematics* **54**(2–3), 249–82.

Jaworski, B. (2007) "Developmental research in mathematics teaching and learning: developing learning communities based on inquiry and design," in *Proceedings of 2006 Annual Meeting of the Canadian Mathematics Education Study Group*, Calgary, Alberta, University of Calgary, 3–16.

Langer, J. (1957) *An introduction to symbolic logic*. New York, Dover.

Mascaró, J. (Trans.) (1962) *The Bhagavad Gita*, Harmondsworth, U.K., Penguin.

Maslow, A. (1971) *The farther reaches of human nature*, New York, Viking Press.

Mason, J. (1979) "Which medium, which message," *Visual Education* (February), 29–33.

Mason, J. (1986) "Probes and fragments," *For the Learning of Mathematics* **6**(2), 42–6.

Mason, J. (1988) "Fragments: the implications for teachers, learners and media users/researchers of personal construal and fragmentary recollection of aural and visual messages," *Instructional Science* **17**, 195–218.

Mason, J. (1996) *Personal enquiry: moving from concern towards research*, Milton Keynes, U.K., Open University.

Mason, J. (2002) *Researching your own practice: the discipline of noticing*, London, RoutledgeFalmer.

Mason, J., Burton, L., and Stacey, K. (1982) *Thinking mathematically*, London, Addison Wesley.

Mason, J., Burton, L., and Stacey, K. (2010) *Thinking mathematically (2nd ed.)*, London, Pearson.

Mason, J., and Johnston-Wilder, S. (2004a) *Designing and using mathematical tasks*, Milton Keynes, U.K., Open University (2006 reprint, St. Albans, U.K., QED).

Mason, J., and Johnston-Wilder, S. (2004b) *Fundamental constructs in mathematics education*, London, RoutledgeFalmer.

McLeod, D., and Adams, V. (Eds.) (1989) "Affect and mathematical problem solving: a new perspective," New York, Springer-Verlag.

Nørretranders, T. (1998) *The user illusion: cutting consciousness down to size*, Sydenham, J. (trans.), London, Allen Lane.

Orage, A.(1954) *Essays & aphorisms*, New York, Janus.

Pimm, D. (1993) "From should to could: reflections on possibilities of mathematics teacher education," *For the Learning of Mathematics* **13**(2), 27–32.

Pólya, G. (1962) *Mathematical discovery: on understanding, learning, and teaching problem solving, combined edition*, New York, Wiley.

Pólya, G. (1965) *Let us teach guessing* (film), Washington, D.C., Mathematical Association of America.

Raymond, L. (1972) *To live within*, London, George Allen & Unwin.

Rhadakrishnan S. (1953) *The Principal Upanishads*, London, George Allen & Unwin.

Skemp, R. (1979) *Intelligence, learning and action*, Chichester, U.K., Wiley.

Storm, H. (1985) *Seven arrows*, New York, Ballantine.

Thompson, A. (1984) "The relationship of teachers' conceptions of mathematics and mathematics teaching to instructional practice," *Educational Studies in Mathematics* **15,** 105–27.

Thompson, A. (1992) "Teachers' beliefs and conceptions: a synthesis of the research," in Grouws, D. (Ed.), *Handbook of research in mathematics teaching and learning*, New York, MacMillan, 127–46.

Thinking and Comprehending in the Mathematics Classroom

Douglas Fisher, Nancy Frey, and
Heather Anderson

Literacy—reading, writing, speaking, and listening—is a critical foundational skill that provides individuals access to information in all other disciplines and domains. As teachers and researchers, we know that literacy impacts every aspect of a person's life, from success in school and work to living a productive life. As Shanahan (2007) noted in his International Reading Association keynote, low levels of literacy put people at risk in all kinds of ways, from being taken advantage of by scam artists to not understanding health information.

Literacy involves more than learning how to read. Breaking the code and developing fluency are important aspects in the development of a literate life. But that's not our focus in this chapter. We're interested in how literacy, broadly defined to include thinking in words and images, impacts content learning and achievement. This book has examples of successful content literacy initiatives in every subject area. Our focus in this chapter is mathematics and the ways in which literacy and numeracy can be integrated such that students learn more content.

Mathematics Knowledge Is Critical

Failing a year of mathematics is highly correlated with failures in future years of school and difficulty in finding gainful employment (Nichols, 2003; Thompson and Lewis, 2005). Math, specifically algebra, is a gatekeeper course. Haycock (2003) says, "Just as we educators have learned that courses like Algebra II are the gatekeepers to higher education, we must now come to understand that they are gatekeepers to well-paying jobs, as well."

Failure in mathematics is also a common cause of college dropouts (Heck and Van Gastel, 2006). Colleges spend significant resources remediating students in mathematics, most commonly college-level algebra. The largest higher educational system in the world, the California State University (CSU), has established goals to reduce the number of students who require remedial instruction upon entering college. As part of the CSU effort, an Early Assessment Program (EAP) was developed. This is an optional assessment given to 11th graders that provides students and their families with feedback about readiness for college algebra, the first in a sequence of required mathematics courses for undergraduates. Of 141,648 students who took the math test in 2007, only 77,870 (55%) demonstrated proficiency. Remember that these are the students who chose to take the exam thinking that they were ready for college. Data from the ACT is even worse. Of the 1.2 million students tested, just 40% were ready for their first course in college algebra (ACT, 2004).

Although it can be argued that students are doing better today than they have in the past, there is a clear and immediate need to continue to raise mathematics achievement. Thankfully, there is evidence for how to do this. Before we explore three ideas for using literacy to improve understanding in mathematics, let's recall how many students experience math class. The following comes from Doug's experience in a ninth-grade algebra class.

A Common Experience in Math Class

The day starts like every other so far this semester. As we enter the room, our teacher calls off odd numbers. By now, we know that when we're called on like this, we have to solve the assigned homework problem on the board. Our teacher watches the group of students assigned to complete the problems and offers periodic criticisms and compliments. When everyone is finished and the answers are correctly posted on the board, we check our homework and then pass it forward to the teacher (of course, we've all checked our homework on the bus to make sure we've all got the same answers because homework counts for 25% of the grade).

With the homework review complete, now 15 minutes into the period, our teacher rolls out the overhead projector. It's the kind with rollers on both sides so that the transparency paper slides across as he finishes writing. He solves algebra problems for us for about 15 minutes. It's the last period of the day and his hands have blue stains from the number of times he's spit on them to erase.

Our task during this time is to take notes exactly as he presents them. Our notebooks must have specific page numbers that match his and are worth 25% of our grade. He provides us with a table of contents for our notes during the first week of school, and we are to keep the page numbers current. For example, as noted in Figure 1, page 13 focuses on "inverse of functions." If we want to take notes on examples, we are told to add pages such as 13a, 13b, etc., so that we do not make mistakes with the numbering system. We do not summarize our notes or organize them in any systematic way; we copy them exactly as they are presented in class.

When he finishes the lecture and note-taking component of the class, we have the remainder of the period available to start our odd-numbered problem set for the day. If we do not finish the problems in class, we are to take them home and complete the rest. If we talk during class, our teacher will call out an even-numbered problem for which there is no answer in the back of the book. As punishment for talking, we have to go to the board and attempt to demonstrate our prowess in front of our peers. Obviously, we quickly learned to be quiet in this class! We also learned to talk with one another outside of the class to check answers and get help.

Every fourth week, we have a test. The tests comprise the remaining 50% of our grade and include long lists of problems to solve, selecting from the correct multiple-choice answer. We never had to explain our thinking, either orally or in writing. We just needed to select the correct answer from the choices provided. As our teacher said many times during the year, "This is what you'll have to do on the standards test, so you might as well get used to it now.'"

You're probably asking yourself, "How will anyone learn anything in this class?" Yet our experiences, and probably yours, bear out the fact that this type of instruction is very common in mathematics classrooms. To analyze the scenario a bit further, it is clear that the teacher values

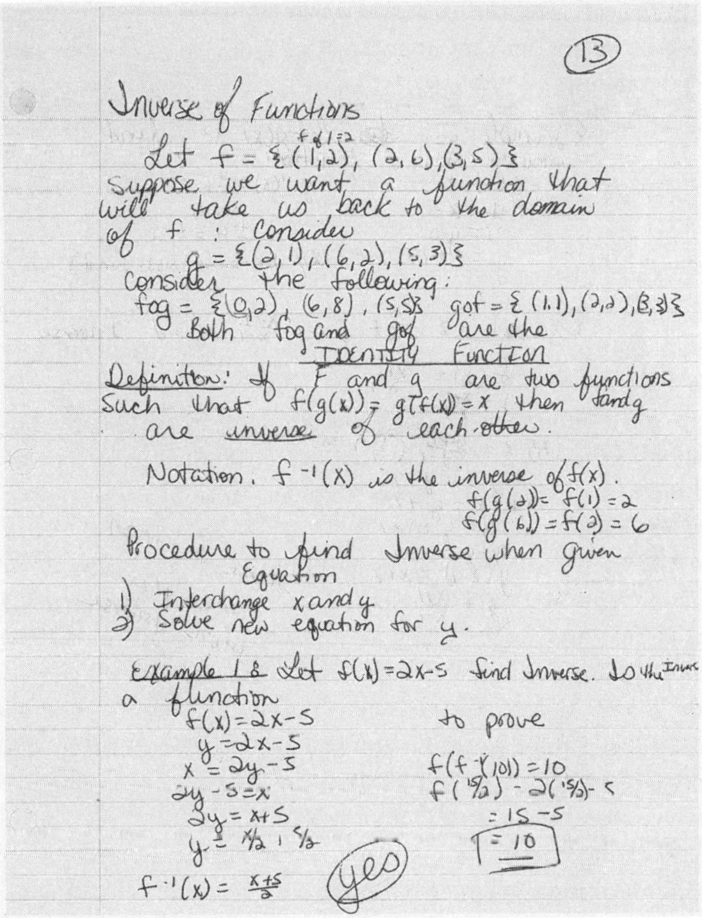

(13)

Inverse of Functions

Let f = {(1,2), (2,6), (3,5)}

Suppose we want a function that will take us back to the domain of f: Consider

g = {(2,1), (6,2), (5,3)}

Consider the following:

fog = {(2,2), (6,8), (5,5)} gof = {(1,1), (2,2), (3,3)}

Both fog and gof are the IDENTITY FUNCTION

Definition: If f and g are two functions such that f(g(x)) = g(f(x)) = x, then f and g are inverse of each other.

Notation. f⁻¹(x) is the inverse of f(x).

f(g(2)) = f(1) = 2

f(g(6)) = f(2) = 6

Procedure to find Inverse when given Equation

1) Interchange x and y

2) Solve new equation for y.

Example 1: Let f(x) = 2x − 5 find Inverse. Do the Inverse a function

f(x) = 2x − 5

y = 2x − 5

x = 2y − 5

2y − 5 = x

2y = x + 5

y = x/2 + 5/2

f⁻¹(x) = (x+5)/2

yes

to prove

f(f⁻¹(10)) = 10

f(15/2) − 2(15/2) − 5

= 15 − 5

= 10

Figure 1. Sample note page.

practice—he provides his students with lots of opportunities to engage in independent learning. He also values students having correct information and right answers. He wants their notes to be exact, and he wants students to practice testing formats. Unfortunately, this teacher has no way of understanding his students' thinking and the types of errors they make. Although he explains information, he doesn't let his students in on his thinking—the thinking of an expert.

With a few adjustments to the structure of the classroom, students would likely develop a deeper understanding of mathematics and begin to see the relevance of this content in their lives. The remainder of this chapter focuses on three areas that we know to be effective ways to engage students in thinking about mathematics (e.g., Fisher and Frey, 2007): modeling, vocabulary development, and productive group work.

Engaging Students in Thinking about Mathematics

Modeling

There exist decades of evidence that teacher modeling positively impacts student performance and achievement (Afflerbach and Johnston, 1984; Duffy, 2003; Olson and Land, 2007). Modeling provides students with examples of the thinking required, as well as the language demands, of the task at hand. In essence, the student gets to peer inside the mind of an expert to see how that person thinks about, processes, and solves a problem.

Unfortunately, there is significant confusion between modeling and explaining. Think of a lecture you've attended. It was probably full of explanations. And explanations aren't all bad. We all need things explained to us sometimes. But we also need modeling, which personalizes the experience for the learner as the teacher uses "I" statements to share his or her thinking. Modeling also provides information about the cognitive process that went on in the mind of the expert; it's the *why* that we're after here. But as Duffy (2003) pointed out, "The only way to model thinking is to talk about how to do it. That is, we provide a verbal description of the thinking one does or, more accurately, an *approximation* of the thinking involved" (p. 11).

Accordingly, mathematics teachers must model their thinking by talking and thinking aloud for their students. Some of the common areas of thought that math teachers model include:

- Background knowledge (e.g., "When I see a triangle, I remember that the angles have to add up to 180°.").
- Relevant versus irrelevant information (e.g., "I've read this problem twice, and I know that there is information included that I don't need.").
- Selecting a function (e.g., "The problem says 'increased by,' so I know that I'll have to add.").
- Setting up the problem (e.g., "The first thing that I will do is . . . because . . . ").

TABLE 1.
Math signal words

Function	*Sample terms*	*Examples*
Words that signal addition	and, made larger, more than, in addition, sum, in excess, added to, plus, add, greater, increased by, raised by	Forty-five and twenty-two are what? Translation: $45 + 22 =$
Words that signal subtraction	decreased by, subtract, difference, from, made smaller by, diminished by, reduce, less than, minus, take away	If you take away 3 from 29, what do you have left? Translation: $29 - 3 =$
Words that signal multiplication	product, multiplied by, times as much, of, times, doubled, tripled, etc., percent of interest on	What is the product of fourteen and sixteen? Translation: $14 \times 16 =$
Words that signal division	per, quotient, go(es) into, how many, divided by, contained in	How many times can 5 go into 100? Translation: $100/5 =$

- Estimating answers (e.g., "I predict that the product will be about 150 because I see that there are 10 times the number.").
- Determining reasonableness of an answer (e.g., "I'm not done yet as I have to check to see if my answer makes sense.").

Let's listen in on Heather's modeling of her thinking relative to the algebra problem: The sum of one-fifth p and 38 is as much as twice p. In her words:

> "Okay, I've read the problem twice, and I have a sense of what they're asking me. I see the term *sum*, so I know that I'm going to be adding. I know this because *sum* is one of the signal words that are used in math problems [for a list of signal words, see Table 1]. I also know that when terms are combined, like *one-fifth p*, they are related because they make a phrase 'one-fifth of p' so I'll write

that *1/5p*. The next part says *and 38*, so I know that I'll be adding 38 to the equation. Now my equation reads $1/5p + 38$. But I know that's not really an equation. I know from my experience that there has to be an equal sign someplace to make it an equation. Oh, they say *as much as*, which is just a fancier way of saying *equal to*. So, I'll add the equal sign to my equation: $1/5p + 38 =$. And the last part is *twice p*. And there it is again, one of those combined phrases like *one-fifth p*, but this time *twice p*. So I'll put that on the other side of the equation: $1/5p + 38 = 2p$. That's all they're asking me to do. For this item, I just need to set up the equation. But I know that I can solve for *p*, and I like solutions. I know that you can solve for *p* as well. Can you do so on your dry erase boards?"

Heather clearly understands the task and expectations. But more importantly, she understands her own thinking on the subject. To be effective modelers, teachers have to move beyond their expert blind spots. Gladwell (2005) notes that even brilliant experts have biases and blind spots that prevent them from seeing the problem as it really is. Expert blind spots prevent teachers from recognizing content that would be helpful to students. Too often teachers are unaware that much of their subject matter knowledge, while second nature to them, is very difficult for their students to learn. Nathan and Petrosino (2003) noted in their study of new secondary math teachers that this blind spot was prevalent because they lacked the experience to recall how a new concept is acquired by a novice.

Our experience suggests that there is a good reason for these expert blind spots. The goal of instruction is for students to reach automaticity such that they no longer have to pay conscious attention to every aspect of the problem at hand. As students develop their understanding of mathematics, components become automatic. For example, by mid-elementary school, multiplication facts should have moved from conscious thought to automatic execution. This process frees up working memory such that the brain can work on other parts of the problem. Many math teachers have reached automaticity with mathematics in general and, as a result, have lost the awareness of their cognitive problem-solving processes. The key is to slow down thinking such that the process once again becomes clear. When you know what you think, it's easier to model for students. And simply said, students desperately need to witness experts in action.

Vocabulary Development

Returning to the modeling provided by Heather in the example above, it's hard not to notice the vocabulary that she used. In every academic endeavor, words matter. We use words to communicate with one another, and our selection of specific words is intended to convey specific information. The problem is that students often don't know what the words mean, especially in a mathematical context.

In response to this problem, teachers often identify words for students to learn. Of course, learning words requires much more than a list. It is also more than providing definitions for students to memorize, such as, "A polygon is a simple closed figure comprised of line segments." Although definitional meaning is an aspect of learning a word, when it comes too early in the process it can confound rather than clarify. Consider the vocabulary embedded in that definition—you need to understand what a *closed figure* is, be able to identify a *line segment*, and know the meaning of *comprised*. Students have to engage with the words multiple times to get a sense of their meaning and usage. A number of instructional routines are useful in mathematics word learning (e.g., Fisher and Frey, 2008b), including the following:

- *Word walls*, on which teachers post 5–10 words on a wall space that is easily visible from anywhere in the room. The purpose of the word wall is to remind teachers to look for ways to bring words they want students to own back into the conversation so that students get many and varied experiences with those words (Ganske, 2006, 2008).
- *Word cards*, in which students analyze a word for its meaning, what it doesn't mean, and create a visual reminder. A sample card for the word *rhombus* can be seen in Figure 2.
- *Word sorts*, in which students arrange a list of words by their features. Word sorts can be open (students are not provided with categories) or closed (students are given categories in which to sort). An example of a word sort in which words could be used more than once can be found in Table 2.

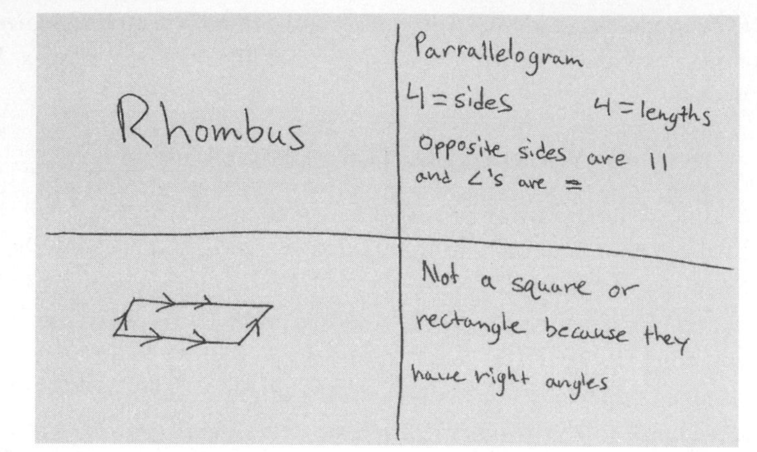

Figure 2. Sample word card.

TABLE 2.
Sample word sort

Circle	*Square*	*Triangle*
Radius	Area	Angle
Diameter	Perimeter	Area
Circumference	Quadrilateral	Altitude
Area	Angles	Sides
	Sides	
	Length	
	Height	

- *Word games*, in which students play with words and their meanings. For example, this might involve a bingo game of sorts (Pat Cunningham calls it *Wordo*), wherein students write words from the class in various squares and then the teacher randomly draws definitions until someone gets bingo. We also like games such as Jeopardy, Who Wants to Be a Millionaire?, or $25,000 Pyramid because they allow students to review words while having a bit of fun. A great website that provides information about vocabulary games is http://jc-schools.net/tutorials/vocab/ppt-vocab.html.

These instructional routines are useful, especially for technical words, that is, those words that are specific to a discipline or content area. In mathematics, it's important that students learn the accepted meanings of words and phrases such as *square root, polygon, linear equation,* and *Fibonacci sequence*. Of course, those aren't the only words students need to know. In addition to technical words, students in mathematics classrooms must learn the intended mathematical meanings of common words. These are known as the specialized vocabulary terms as they tend to change their meaning in different contexts. For example, the word *prime* has a common meaning related to the best in quality, as in *prime beef*. However, in mathematics the term takes on a specialized meaning: a number that is only divisible by itself and 1. Our experience suggests that students need these differences made explicit to them, especially English language learners who may know one meaning of a term, but a meaning that does not help them understand the mathematics. For example, one of our students understood the term *expression*, having heard it in her English language development (ELD) class. That teacher talked about facial expressions and reading social clues through expressions. The newcomers in the ELD class developed an appreciation of the term *expression* and were able to read nonverbal clues in their new environment. However, when this student enrolled in algebra, the knowledge she had about the term failed her. She had no idea what the teacher meant when he said, "Let's write an expression for the information we have" or "Evaluate the expression $5 \times z + 12$ when $z = 3$."

The best way we've found to teach and reinforce specialized vocabulary is through a mathematics journal such as the one in Table 3. Of course this requires that teachers notice the specialized vocabulary in their speaking and in the texts they use. It also requires that teachers take the time to provide instruction on the difference between the common meaning of the word and the math-specific definition or usage. But as the student who did not understand the mathematical use of *expression* taught us, focusing on specialized vocabulary terms is time well spent.

The first two areas that we've presented have focused a lot on the teacher: The teacher provides expert modeling, the teacher determines which words are worthy of instruction, and then the teacher provides time for students to engage effectively with the words. The final area of attention focuses on the role of students. To significantly raise achievement

Table 3.
Sample mathematics journal

Word	*Common meaning*	*Math meaning*	*Where I found it*
Prime	Best or high quality	A number that can only be divided by 1 or itself	Math book, page 34
Expression	Something someone says or the feelings on a person's face	A group of symbols that make a mathematical statement	Wikipedia
Set	To put something someplace	A collection of specific numbers	After-school tutor

in mathematics, students have to have opportunities to interact with one another in regard to content.

Productive Group Work

Unlike the classroom scenario that we used to start this chapter, classrooms that work well are filled with student talk, student interaction, and meaningful work. Simply listening to math instruction and then doing math problems will not result in learners who understand and use mathematical concepts in their daily lives, much less in their college classrooms. As noted in the opening scenario, students want, and need, to talk about the content. Our experience suggests that the productive group-work phase of instruction allows students to consolidate their thinking about the content. In that respect, it's a critical aspect of learning. Importantly, we have evidence that students use the information modeled by the teachers when they are working with their peers. And even more importantly, independent work is of higher quality when students have first had an opportunity to collaborate with others. In fact, this aspect is so important that it is one of the foundational principals of *Working on the Work* (Schlechty, 2002), a reform effort with the aim of improving student achievement by focusing on the tasks students are asked to complete.

The following five features should be considered in any productive group-work task (Fisher and Frey, 2008a; Johnson, Johnson, and Smith, 1991).

1. *Positive interdependence.* Members must see how their efforts contribute to the overall success of the group. The task cannot be one that individuals could have completed independently. Rather, the task has to have at least an aspect of interdependence such that students need each other to complete their work successfully.
2. *Face-to-face interaction.* As part of the task, group members must have time to interact live. While they can also interact in virtual and electronic worlds, our experience suggests that the opportunity to interact on the physical level encourages accountability, feedback, and support.
3. *Individual and group accountability.* As we have noted, productive group work is not simply a matter of having a group of students complete a task in parallel with peers that they could have done alone. Having said that, we also know that the risk of productive group work lies in participation. In nearly every group, there are likely members who would allow their peers to complete the required tasks. To address this issue, each member of the group must be accountable for some aspect of the task. Of course, this is a perfect opportunity to differentiate based on students' needs and strengths. In addition to the individual accountability, the group must be accountable for the overall product. This also ensures that overly involved students will not monopolize the conversations during productive group work.
4. *Interpersonal and small-group skills.* One of the opportunities presented during productive group work is social skill development. Wise teachers are clear about their expectations related to interpersonal skills and communicate these to students. For example, during a group brainstorming session about ways to represent the concept of slope, Heather reminds her students that "put downs for ideas are not allowed, especially during a brainstorming session."
5. *Group processing.* As part of the learning associated with productive group work, students need to learn how to think about,

and discuss, their experiences. The goal of the discussion is for students to consider ways in which they can improve their productivity and working relationships.

Following her modeling in which she thinks aloud about a problem and its solution, Heather provides groups of three students (triads) with a problem. Each triad has the same problem as one other triad, and the two triads with the same problem will discuss their results at the end of the class session. Before doing so, students in each triad must solve the problem with words, numbers, and pictures. They are working on reasonableness of their answers. One of the groups received the problem: If Esme cuts an apple into 8 equal pieces and gives Kaila 1 piece, how much of the apple is left? Is it reasonable to suggest that there is more than 50% of the apple left?

The group members go into action, first talking about the problem. Andrew asks how Esme got the pieces to be equal. Maria responds, "That's extraneous to the problem. We can ask that as a follow-up, but first we have to solve the problem." Their individual accountability is widened by their separate ability to explain the solution to their teacher and another group. The group accountability includes a presentation of the solution in words, numbers, and a picture. This group decides that each member will take one of the required representations and work alone for a minute before sharing the results with the team.

Maria takes writing, Jamal takes numbers, and Andrew takes the picture, and they begin to work. As they finish, they trade papers for a quick review and an opportunity to ask clarifying questions. Jamal suggests that Maria add a sentence about subtracting fractions with common denominators. Andrew asks Jamal if the answer is reasonable, "You have the problem worked out, but I think you forgot the second part. Is it reasonable to suggest that there is more than 50% of the apple left?"

The students then discuss the answer and the various ways that they solved the problem. Using a timer, they each get 60 seconds to explain their thinking. When the 3 minutes are up, triads with the same problem meet. They each explain their thinking and their solutions to the members of the other triad that solved the same problem while Heather listens in on the conversations. Naturally, there are a number of ways that students solved their assignment problems and they're having a chance to hear about alternatives.

Returning to their triads, the students talk about what they learned from the experience. Heather reminds them to "Talk about what you learned about problem solving and also what you can do to make your triad more effective." Maria talks about adding numbers to her writing like Sophia did. Andrew talks about how all six of them got the answer right, but in different ways. He says, "I like Michael's picture better than mine, but we got the same answer." In response to increasing the productivity of the group, Jamal suggests that next time they each solve the problem two different ways (writing, numbers, or pictures) and then compare all of the different ways, "so that we'll know more ways to figure out what we have to do."

The amount of student talk and student engagement in Heather's class is significant. Students know what is expected of them from the models she provides. They also learn a lot of words and have opportunities to use those words in context with their interactions with peers. And students in Heather's class collaborate on productive tasks that allow them to consolidate their understanding of mathematics. It's no wonder that Heather's students do so well on state assessments—they know the content very well because of the structures in place in the classroom.

Conclusion

Importantly, we are not suggesting that mathematics instructors become reading teachers any more than we are suggesting that reading specialists become math educators. However, we do recognize the value that modeling, vocabulary, and productive group work play in learning mathematics content, regardless of the level of mathematics being taught. Mathematics instructors, K–12, can improve student achievement through the use of key "literacy" instructional routines. We put the word literacy in quotations because the ideas presented in this chapter are not owned by reading teachers; they are ways to get students to think about and understand the content. And that's the goal of every teacher—to ensure that comprehension occurs.

References

ACT, Inc. (2004). *National data release.* Iowa City, Iowa: ACT.

Afflerbach, P., and Johnston, P. (1984). On the use of verbal reports in reading research. *Journal of Reading Behavior, 16*, 307–22.

Duffy, G. G. (2003). *Explaining reading: A resource for teaching concepts, skills, and strategies.* New York: Guilford Press.

Fisher, D., and Frey, N. (2007). A tale of two middle schools: The role of structure and instruction. *Journal of Adolescent and Adult Literacy, 51*, 204–11.

Fisher, D., and Frey, N. (2008a). *Better learning through structured teaching: A framework for the gradual release of responsibility.* Alexandria, Va.: Association for Supervision and Curriculum Development.

Fisher, D., and Frey, N. (2008b). *Word wise and content rich: Five essential steps to teaching academic vocabulary.* Portsmouth, N.H.: Heinemann.

Ganske, K. (2006). *Word sorts and more: Sound, pattern and meaning explorations K–3.* New York: Guilford Press.

Ganske, K. (2008). *Mindful of words: Spelling and vocabulary explorations 4–8.* New York: Guilford Press.

Gladwell, M. (2005). *Blink: The power of thinking without thinking.* New York, Little Brown.

Haycock, K. (2003). Foreword. In A. P. Carnevale and D. M. Desrochers, *Standards for what? The economic roots of K–16 reform.* Princeton, N.J.: Educational Testing Service.

Heck, A., and Van Gastel, L. (2006). Mathematics on the threshold. *International Journal of Mathematical Education in Science and Technology*, 37(8), 925–45.

Johnson, D. W., Johnson, R. T., and Smith, K. A. (1991). *Active learning: Cooperation in the college classroom.* Edina, Minn.: Interaction Book.

Nathan, M. J., and Petrosino, A. (2003). Expert blind spot among preservice teachers. *American Educational Research Journal, 40*, 905–28.

Nichols, J. D. (2003). Prediction indicators for students failing the state of Indiana high school graduation exam. *Preventing School Failure, 47(3)*, 112–20.

Olson, C. B., and Land, R. (2007). A cognitive strategies approach to reading and writing instruction for English language learners in secondary school. *Research in the Teaching of English, 41*(3), 269–303.

Schlechty, P. C. (2002). *Working on the work: An action plan for teachers, principals, and superintendents.* San Francisco, Calif.: Jossey-Bass.

Shanahan, T. (2007, May 15). Presidential address for International Reading Association, Toronto, Canada.

Thompson, L. R., and Lewis, B. F. (2005). Shooting for the stars: A case study of the mathematics achievement and career attainment of an African American male high school student. *High School Journal, 88*(4), 6–18.

Teaching Research: *Encouraging Discoveries*

Francis Edward Su

What does teaching have to do with research or discovery?* What does it take to turn a learner into a discoverer? Or to turn a teacher into a co-adventurer? A handful of experiences—from teaching a middle school math class to doing research with undergraduates—have changed the way that I would answer these questions. Some of the lessons I've learned have surprised me.

The title of this article may seem a little puzzling. After all, the words *teaching* and *research* usually only appear in the same sentence when separated by the word *and*, on a list of a faculty member's obligations. Do these words belong together at all?

One of my favorite quotes about teaching comes from one we normally associate with research:

> The principal aim of mathematical education is to develop certain faculties of the mind, and among these intuition is not the least precious. It is through it that the mathematical world remains in touch with the real world.
>
> —Henri Poincaré [**5**, p. 128]

But Poincaré also said

> It is by logic that we prove, but by intuition that we discover.
>
> —Henri Poincaré [**5**, p. 129]

So we teach to build intuition, and intuition enables our students to make discoveries. Notice that Poincaré does *not* say that the principal aim of teaching is to convey facts or theorems. The principal aim of mathematical teaching is to build qualities of mind that enable students to make discoveries.

* This article is based on the James R. C. Leitzel Lecture, delivered by the author at the MAA MathFest, Knoxville, Tennessee, on Friday, August 11, 2006.

But how does one do that? How does one turn a learner into a discoverer? When I was first starting out as a new professor, I might have given these answers:

- Teach the needed background.
- Cultivate maturity.
- Inspire them!
- Ask good questions.
- Select the smartest students.
- Give open problems.
- Advertise the thrill of research.
- Be an expert in what you advise.
- Encourage independence.

Now, I shall explain why I believe every one of these pieces of advice is either plainly wrong or, at best, inadequate. Along the way, I will mention some concrete teaching ideas that you may find useful as well.

The Principal Aim of Mathematical Teaching

There is one word that is an excellent metaphor for everything I want to say:

YAWP.

What is a *yawp*?

> I sound my barbaric *yawp* over the roof(top)s of the world!
>
> —John Keating, quoting a Walt Whitman poem,
> in the movie *Dead Poets Society*

As Keating explains, a *yawp* is a loud cry or yell. But in the poetry of Walt Whitman, the word *yawp* refers to the inner groaning inside each of us, too deep for words, that is yearning to be expressed and experienced. In the movie *Dead Poets Society*, an English teacher named John Keating (played by Robin Williams) does more than just teach poetry—rather, he breathes poetry, he inhabits poetry, and he inspires his students to do the same. In one scene, his student Todd Anderson (played by Ethan Hawke) has been struggling all weekend to write a poem, but without success. In class, when Todd fails to deliver a poem, Keating encourages Todd to first find his *yawp*, and once he does, Keating helps him transform his *yawp* into poetry in a truly mesmerizing manner [**6**, 55:07–57:57].

Keating was an exuberant teacher who exhorted his students to *live extraordinary lives.* He did not measure this by the usual metrics of success and accomplishment; rather, for Keating, living *extraordinarily* meant becoming who they were created to be: human beings empowered to express their passion through poetry, and to use their souls, spirits, hearts, and minds to seize the day with all its wonders.

As mathematicians, we have experienced one of the greatest human wonders: the thrill of discovery. And the aesthetic pleasure of an elegant proof captures our spirits just as poetry does. So what I seek to cultivate in my students is a *mathematical yawp*:

> **Definition.** A *mathematical yawp* is that expression of surprise or delight at discovering the beauty of a mathematical idea or argument.

Every student is capable of a mathematical *yawp*. The *yawp* may not be a poem when it is first expressed, but I believe *the principal aim of mathematical teaching is to nurture the mathematical yawp, and help transform it into poetry.*

Math Fun Facts

When I first began teaching calculus in graduate school, I lamented the fact that most students would leave college with a mistaken notion of what mathematics is. After all, most of them took math to fulfill a requirement, usually some brand of calculus or something less advanced, and it was the final course they would take. As a result, the view most students develop of mathematics is that it is a cut-and-dried, 400-year-old list of theorems and applications. They do not get the sense that mathematics is an exciting, living, and developing subject.

I wanted to change that impression. So I began to present daily "Fun Facts," by taking 5 minutes at the start of each lecture to show my students some mathematical idea that I thought was fascinating—for instance: the uncountability of the reals, the nine-point circle, tiling a chessboard with opposite corners removed, the ham sandwich theorem, etc. The facts I chose usually had nothing to do with calculus; the main point was to broaden their perspective of mathematics, to show them math is alive and full of interesting questions and new ways of thinking, and to whet their appetite for learning more.

The student reaction was overwhelming. They loved Fun Facts. They clamored for them at the beginning of every class. Students stopped coming late! If I ever forgot to do a Fun Fact, they would remind me of my obligation, as if going through Fun Fact withdrawal. Sometimes they would clamor for more than one (though perhaps to postpone the calculus lecture). I had nurtured their mathematical *yawp*.

And they *yawp*ed for more. They would often come back to me with questions about Fun Facts, or variants of Fun Facts they had studied or solved. Many of my teaching evaluations cited Fun Facts as a reason that they wanted to take more math, even if they didn't like calculus. I had helped them find their *yawp*, even if it wasn't for calculus!

And then it dawned on me that it was more important to nurture their *yawp* than to teach them calculus.

Lesson #1. Teach the needed background?
***No. Nurture the* yawp.**

Of course, we hope that these goals are in harmony, but if there is ever a conflict, we must remember the principal aim. For instance, I make it a policy never to stop nurturing the *yawp* in favor of "getting through the material."

If you'd like to try using Fun Facts in your courses, I have created a website and iPhone application containing two hundred Fun Facts that I have used. (See [7] or Google "Math Fun Facts" to find it.) But the best Fun Facts are the ones that you care about the most! Give your students the Fun Facts that fueled your passion for mathematics!

What I Learned from 8th-Graders

I found a way to use a similar idea in the K–12 arena. One of the most formative experiences of my career occurred in my second year as a faculty member, when I learned that a principal of a local middle school was looking for a professor to teach a math enrichment course to challenge some bright 8th-graders. Intrigued by the idea, I agreed.

Once a week, the 8th-grade math class at this school was divided into two groups by ability. One half stayed with the usual teacher to get some needed help and extra practice, and the other group met with me.

What should I do with them? Instead of teaching them more of the same stuff that they might encounter again in high school, I chose instead to teach them "enrichment material," such as extended Fun Facts, proof techniques like induction and contradiction, fast mental arithmetic, magic tricks using mathematical ideas, elementary knot theory and Reidemeister moves, mind-boggler puzzles, elementary combinatorics, etc.

Many things about this experience surprised me. First of all, teaching 8th-graders is a lot harder than teaching college students! Any issue related to teaching college students is amplified for 8th-graders. If college students can't sit still for more than fifteen minutes, 8th-graders can't sit still for five. College students get bored; 8th-graders noticeably fidget in their seats. College students give nodding winks at each other; 8th-graders pass notes. If college men speak up in class more than college women, 8th-grade boys dominated my class discussions entirely.

As a result, for everything I would normally do to engage a college class, I had to do better with these 8th-graders. If my college teaching style is interactive, my 8th-grade teaching style had to be more so. I had to pay more attention to domineering students, give more reassurance and encouragement, repeat ideas more often, ask more questions to keep them engaged and ignite their curiosity and imagination . . . and it occurred to me that doing these things was great for my college teaching!

Lesson #2. Cultivate maturity in your students?
No. Restore their childlike curiosity and imagination.

I believe there isn't a single concept in a first-year college course that an 8th-grader couldn't understand if developed well. So what will you do to develop it well, to engage your students?

We must find ways to restore the childlike curiosity of college students. I see that 8th-graders are more curious, more willing to probe with a "why?," more willing to ask questions if they don't understand, and more willing to let you know your pen is leaking in your front pocket. Somewhere along the path to adulthood, they lose the ability to ask questions.

Was this enrichment class a success? Did they develop a mathematical *yawp*? For end-of-term evaluations, I asked these 8th-grade students to complete the following sentence:

I now know that mathematics is . . .

I received some responses like this:

- . . . harder than I thought (I thought it was too easy before).

But most of the responses were like these:

- . . . extremely challenging, yet excessively fun. Also, it's a much broader topic than I had originally thought.
- . . . can be fun and not always so boring. I also now know that sometimes, mathematics can have absolutely nothing to do with numbers.
- . . . a broader subject than it was to me before; this class opened up many new, exciting, and challenging topics that have made me more interested in mathematics as a whole.
- . . . a world of mysteries waiting to be discovered.

Exactly! We seek to build intuition to enable them to make discoveries!

Recently, I heard from a student in this class who is now just graduating from college, eight years later. He said, "from what I remember I really enjoyed it . . . in middle school it was the only class that really opened my eyes to different thinking." All these responses indicate that they had developed a *yawp*.

Another lesson I learned from this class was how easy it was to miss hearing a *yawp*. There was a shy girl in this class who hardly ever spoke up and seemed not to be engaged at all. I didn't notice her for several weeks until I assigned some homework on induction. I was curious to see what 8th-graders could do, and I naively expected the best students to be the ones who were answering all the questions.

By and large, the students seemed to grasp induction. But I'd say the homework looked like the typical work you would get from beginning calculus students, not so good on the form and very unclear in the writing. On the other hand, this student's induction proof was astounding. It had perfect form, perfect understanding, perfect writing—in complete sentences with good connective phrases, journal style, with the clarity and maturity that I expect from professional mathematicians.

Throughout the term she continued to impress me, always in her quiet way. What was interesting to me was that her regular 8th-grade math teacher did not seem to think she was exceptional at all! Why?

And then I wondered how many students like her I was missing in my college courses because of the way I was teaching, and what I could be doing to look for their *yawp*.

***Lesson #3. Identify invisible* yawpers.**

As a sad footnote, after the class was over, I told this student's parents that if she should show an interest in math later on, they should encourage it because she was truly exceptional. Their reaction (or lack thereof) was as if to say: why would I want my girl to study math? And I understood then that there are still social pressures that push children out of callings for which they might have a gift.

Nurturing Poets

I surveyed a dozen of my former thesis and research students and asked them to reflect on their experiences doing research as undergraduates, whether with me or in other settings. I refer to them as Poets, since they have found their *yawp*. Nearly all of them are now in graduate school or just beyond.

One question I asked them was what early influences caused them to think research in mathematics was something they wanted to try. One Poet said,

> My parents were a huge influence on me, posing mathematical questions to me when I was young . . . for example, I remember my dad asking me how many steps it would take to get to the wall if on each step, I went halfway.

She had found her *yawp* because her parents nurtured it by asking probing questions, but they let her experience the joy of discovery herself.

More than half the students cited the experience of working puzzles or doing math problem-solving as early influences. Many had been part of organized problem-solving groups, or at summer math programs where they, in the words of one Poet, "caught a fever of math research." In effect, these groups became their "Dead Poets Societies," places and spaces where they could gather to recite poems and *yawp* together.

At Harvey Mudd College, Andy Bernoff and I run a problem-solving seminar that has become something of a phenomenon on campus. Every fall, about 70 students (that's 10 percent of our student body!) gather on Tuesday nights over pizza to have fun with problem-solving. We keep the atmosphere lively, and we allow them to work in groups. We introduce a few problem-solving techniques at the start, then let them work on a slate of five or six problems, break for pizza, then we have students present their solutions [**1, 2**]. They have the space to make discoveries themselves, and by letting them present their solutions, we give them the space to *yawp* as well (just as any good Dead Poets Society would do).

This seminar now attracts a large fraction of students who are not even math majors, but who come for the thrill of *yawp*ing, either when they solve the problems, or when they see a solution and get the "aha!" experience that is at the heart of a true *yawp*.

Would I still say it is important to inspire students? Of course. But I think it is even better to give students space for self-inspiration so they can experience and express their *yawp*.

Lesson #4. Inspire them?
Better: Create spaces for self-inspiration.

Moore-method courses are another example of spaces where students can find their *yawp* and make discoveries themselves. I won't say much about them here, because much has been said about them, but the best course I ever took as an undergraduate was a Moore-method topology course (taught by Mike Starbird), and it had a profound impact on developing my *yawp*.

We all understand how important it is for a professor to ask good questions of her students. After all, a good question can prompt a discovery. But I now believe it is even better if you can teach students how to ask good questions themselves!

Lesson #5. Ask good questions?
Better: Teach how to ask good questions.

Let me give you an example that I learned from my colleague Lesley Ward. She has used "question stems" in her courses, to teach students how to ask good questions (using a suggestion of T. J. Mueller, and adapting questions from [**3**]). For instance, in an analysis course, after giving part of a lecture on some topic, she asks students to form groups and she hands out cards which have questions like these on them, with blanks:

- What is an example of . . . ?
- How does . . . affect . . . ?
- What would happen if . . . ?
- What is the difference between . . . and . . . ?

The students formulate questions about the lecture based on these models, and then they discuss them in class. In addition to the benefit of teaching students the kinds of questions they might think about, it has the added benefit of students often answering each other's questions!

Another way of encouraging students to ask good questions is to reward them for it. On some exams, I put a problem in which I ask students to write down a research question that they would like to answer, based on what they have learned so far in the class. The answer to the question doesn't have to be unknown to the world, it just has to be unknown to them. I am happy if they have just gone through the reflection necessary to ask a question, so I give full credit for any reasonable attempt at formulating a question.

For our students to experience true research, we should certainly give them open problems. But in order to cultivate a *yawp*ing ability, I believe it is much better to give students open-ended problems.

Lesson #6. Give open problems?
Better: Give open-ended problems.

I learned a terrific example of this from my colleague Michael Orrison, who has developed a project that he has used successfully in a discrete mathematics course. In it he asks students to

> Define a measure of the complexity of a graph. Compute it for several examples, and prove some properties about your measure.

This is the main idea; you can find more details in [4]. What is great about this exercise is that it is open-ended, and there are multiple right answers. Students are able to be creative about the solution, and they are excited about investigating a concept that they have defined themselves. At the end of the project, students make presentations.

Looking at some of the slides from past presentations, one sees that students develop a wide variety of notions of complexity. More interestingly, they find creative ways of expressing their ideas, e.g., by describing their complexity measure as answers to questions: "How expensive is it to plow the town?" "How many friends do you have?" etc.

And their *yawps* are often visible as they explore properties of their measures of complexity—for instance, written on one recent memorable slide, a student exclaims "Really Cool!" and "Yowza!" when she realizes that her measure is invariant under taking graph complements. That's a *yawp* if I ever saw one!

My colleague Darryl Yong has even adapted Orrison's project as an exercise for high school teachers, in a conference called "Imagine Math Day" [**8, 9**]. And there are many other project examples that could follow a similar model. For instance, "What is a fair voting procedure?" or "Define a measure of a center of a distribution." Last summer, I gave a couple of my research students the following open-ended problem: "List 10 generalizations of convexity." After they came up with those ten, I asked them to focus on the most interesting ones and try to characterize those notions of convexity in terms of properties that people already understand. My students had a lot of fun generating their list, and eventually took one of their notions and proved necessary and sufficient conditions that would allow someone to tell if a set satisfied their notion of convexity. This resulted in a paper that they wrote completely on their own, and submitted for publication. Because of one open-ended problem, they created a new idea, and they were motivated to study it and explore it—they *yawp*ed!

Research with Undergraduates

But now suppose you have taken your students to the point where they can *yawp*. How do you help them turn it into poetry? If a *yawp* is the thrill of discovery, a poem is a *yawp* that is communicated well. This is where the undergraduate research experience can play a big role.

One major question that every advisor faces is the question of how to choose students. I always thought that the best metric for this was to choose students who were the smartest, or got the best grades. However, I now think that something else is much more important—strong motivation and persistence.

Lesson #7. Select the smartest students? Not necessarily; select motivated students!

One of the best research students that I have ever had was not a student that fit the mold of getting the best grades, and I did not notice her abilities from her coursework. She was a B+ student, and when she approached me about summer research I had already offered my research positions to other students. But she came back several times to ask, so I finally decided to go out of my way to ask the dean for extra funds to support her, and he agreed.

This was one of the most rewarding experiences that I have had as an advisor. The persistence my student showed in inquiring about research opportunities translated into a tenacity in tackling research problems. She was extremely dedicated and motivated, and always came to research meetings having done everything that we discussed at the last meeting, and more. She was the kind of student who was meticulous and took her time to carefully think things out, and so while taking exams was not a place where her gifts were displayed, she really shone in research problems where persistence and meticulousness are rewarded. And she was an excellent writer. By the end of the summer, she had completed a paper that was polished and ready to submit!

So, if I had to give advice about what to look for in a research student, I would say smarts are important, but if you have to choose between a super-smart student and a smart motivated student, I would go for the latter. Some qualities to look for: Are they persistent? Do they excel in something that shows dedication? For instance, we have a couple of undergraduate research students in our department this summer who are B students, but one is a nationally ranked player in the computer game *StarCraft*, and another is a hacky-sack expert. And their advisors report that these students are doing outstanding, publishable research.

Another question I ask about potential research students is: do they write well? I ask for a writing sample, such as a proof they have written for a homework exercise. Even if you don't choose a student who can write, it is nice to know what kind of writer you are working with. Such information can be useful, for instance, if you are choosing students to work in teams.

Lesson #8. Advertise the thrill of research? Better: Set complete expectations.

I used to want my research students to think research was going to be really fun and exciting. But now, I think that they should be aware of the whole picture. I now tell my students in advance that research can be frustrating at times, and I warn them that often it hits about the fourth week of the summer when they are really wrestling with their research problem. If I tell them in advance, then they aren't surprised when it happens, and they understand it is normal to feel that way.

Along these lines, one Poet says,

> The most valuable aspect, in retrospect, was how my research jobs made researching look far less glamorous than I had thought it might be . . . in a way it was demoralizing, but you likely want your bubble burst before you chase the dream too far.

But the same student also notes that understanding the expectations doesn't diminish his desire to *yawp*—he says, "The hope of delightful discoveries and taking part in that same cool elegance I recognized so long ago is still a large part of what keeps me going."

I also set clear expectations for my students about how to conduct their research. For instance, I ask them to keep track of what they are doing in a research notebook. I explain that this notebook is where they write down ideas that they want to follow through with later. I also tell them to start writing often and early in the process. When they get stuck or bored with thinking about their problem, they can write down what they have been doing, or learn LaTeX and LaTeX their work. Another Poet reports,

> One of the best things I learned was organization . . . especially since none of my later advisors mentioned this at all.

We can teach undergraduates research skills that will be useful to them later, even if, as the Poet suggests, graduate schools don't teach those same skills.

In advising research students, I always had the impression that you should be an expert in what you advise. After all, I think it is common for us to worry: how can we nurture a *yawp* in someone else if we don't know how it is we *yawp* ourselves? But now I think differently.

Lesson #9. Be an expert in what you advise?
No, let the student be the expert.

What do I mean by this? I think it is extremely important to give my student ownership of the problem. I tell my students at the start of the summer: "By the time the summer is over, you will be the local expert on this problem. You will know more than anyone else here about this problem, including me." Doing so gives them the excitement of living up to that challenge, and my job the rest of the summer is to probe, and model asking good questions (rather than trying to give them all the answers that I obviously don't yet have). So they are on an adventure to find out more than anyone else in my local neighborhood about their topic, and, by accompanying them on this journey, always asking or suggesting (but never forcing) where to go next, I am a "co-adventurer" with them. I listen to their *yawps* on this journey.

Not surprisingly, my research with students has taken a random walk through many fields; as an example, our papers have been published in a diverse collection of journals: *Random Structures and Algorithms, Journal of Combinatorial Theory Series A, Discrete and Computational Geometry, Journal of Mathematical Analysis and its Applications, American Mathematical Monthly*, and *Journal of Mathematical Biology*, among others.

In all these examples, I was never the expert on these topics, but I picked up what I needed to know in a journey with my students, often with the students teaching me. I am only a co-adventurer. Students also learn from this journey that research often involves learning many new and unfamiliar things.

Building Community

One of the biggest pieces of advice that I hear about advising research is the adage about encouraging students to be independent. While I believe this should be true much later in the research process, I strongly disagree with this in the early stages.

Lesson #10. Encourage independence?
No! Give close guidance, and build community.

For instance, I think it is very important to meet often and regularly, especially in the beginning of the research process. During those meetings, one should always set goals for the research students to work on for the next meeting. Doing so gives them something concrete to do and to report on and gives structure to the following meeting. It is especially important to have students write something up for each meeting. Doing so refines their thinking, and focuses discussion.

When it comes time to write a real paper, I always sit down with my student in front of the computer and write, together with the student, at least a few pages. This not only helps get the paper off the ground, but doing so also models for the student the process of writing. Students are often initially surprised by the care which I take to get definitions right, to make notations consistent, or to ensure that attributions about prior work are correct. They get to see how I think about writing and they can begin to do the same.

Throughout their research, I want my students to see that mathematics really is a social enterprise. We, as a community, decide what's important, our work depends on the work of others, and we work together with others to advance the field. In one project, I included an undergraduate student on a project that I was doing with an economist in Germany. I included my student in all the meetings (some by phone and some when my collaborator visited). By being fully part of this process, she was able to see one way in which collaboration could work, as well as see in this social setting how important it is to learn to communicate ideas well.

Poets report:

- My research experience definitely helped me learn about writing papers, and perhaps more importantly, about working with others to co-write papers.
- Working with other people besides you demonstrated the different possibilities when people with different strengths collaborated.
- Working with another person on a research question is fundamentally different than any of the other group work I had been exposed to.

I want students to see the importance of building community in their mathematical work. I hope that they can see that through collaboration, they can really have a lot of fun *yaw*ping together, whether that be through the sweat of persisting through a problem or the joy of actually solving it. As the Swedish proverb goes,

A joy shared is twice the joy. A sorrow shared is half the sorrow.

To summarize, how would I answer the question: How do you turn a learner into a discoverer?

- Nurture the *yawp*.
- Restore childlike curiosity and imagination.
- Identify invisible *yawp*ers.
- Create spaces for self-inspiration.

- Teach how to ask good questions.
- Select motivated students.
- Give creative, open-ended problems.
- Set complete expectations.
- Let the student be the expert.
- Give close guidance, and build community.

John Keating did not teach his students poetry; he helped them *yawp*, he helped them inhabit the verse. He encouraged playful childlikeness. He created the space for his students to self-inspire in a resurrected Dead Poets Society, even though he himself never attended a meeting. He taught them how to find good subjects for poems. He engaged the students who were motivated, who had ears to hear. He encouraged creativity, and he encouraged community.

Let's help our students find their *yawp* and transform it into poetry.

Dedication

This article is dedicated to Jim Leitzel and Christine Stevens, co-founders of Project NExT, who have nurtured the professional development of more than a thousand mathematicians and given them places and spaces to *yawp*; and to Michael Moody, who will always be my Keating—O captain, my captain!

Acknowledgments

I thank Art Benjamin, Jon Jacobsen, Susan Martonosi, Claus-Jochen Haake, Ulrike Ervig, Mike Starbird, and Bruce Palka for their advice in preparing the Leitzel Lecture. I gratefully acknowledge NSF Grants DMS-0301129 and DMS-0701308 for supporting my research with undergraduates.

References

1. A. J. Bernoff and F. E. Su, PCMI problem-solving references and resources (2003–2009), available at http://www.math.hmc.edu/~ajb/PCMI/problem_solve.html.
2. ———, Putnam, pizza & problem-solving, *Math Horizons* (Sept 2004) 8–9.
3. A. King, Inquiry as a tool in critical thinking, in *Changing College Classrooms*, D. F. Halpern, Ed., Jossey-Bass, San Francisco, 1994, 13–38.

4. M. Orrison, Graph complexity, in *Resources for Teaching Discrete Mathematics: Classroom Projects, History Modules, and Articles*, B. Hopkins, Ed., MAA Notes #74, Mathematical Association of America, Washington D.C., 2009, 159–61.
5. H. Poincaré, *Science et méthode*, E. Flammarion, Paris, 1908; English trans. F. Maitland, *Science and Method*, T. Nelson, New York, 1914.
6. T. Shulman, *Dead Poet's Society* (dir. Peter Weir), Walt Disney Studios Home Entertainment, Burbank, Calif., 1989.
7. F. E. Su, Math Fun Facts (1999–2010), available at http://www.math.hmc.edu/funfacts/.
8. D. Yong, Imagine Math Day (2009), available at http://www.math.hmc.edu/pdo/imd/.
9. D. Yong and M. Orrison, Imagine Math Day: Encouraging secondary school students and teachers to engage in authentic mathematical discovery, *MAA Focus* **30**(6) (2008) 24–7.

Reflections of an Accidental Theorist

ALAN H. SCHOENFELD

Many years ago, David Wheeler asked me to write "Confessions of an Accidental Theorist" (Schoenfeld, 1987), in which I described how I had come to examine the research issues on which I had focused.* The SIG/RME Senior Scholar Award provides me with a wonderful opportunity to reflect once again on those and related issues. I am truly grateful for the opportunity.

The word *accidental* in the articles' titles refers to the fact that when I began doing both mathematical and educational research, I was "theory-neutral." In mathematics, unless one worries about foundations (logic), one just goes about one's work: The rules of the game are so well established that one simply forges ahead, working on what one hopes is the next meaningful and significant problem. After all, a proof is a proof is a proof; people schooled in mathematics know what one is and how to produce one. My work in education started near the dawn of cognitive science, and I happily adapted tools from artificial intelligence to the study of human thinking and problem solving. My arguments at that time were essentially empirical. If I thought X was a factor in problem solving, I helped students learn X and observed whether it made them better problem solvers. If it did, then X was obviously important. This stance did not ignore theory, of course—it depended on an *information processing* perspective rather than a behaviorist perspective, for example—but it made somewhat passive use of it. As I evolved as a researcher, however, I came to realize that being explicit about theory and models helped me clarify what I was trying to understand and to test and refine my ideas. I am now firmly committed to the dialectic between theory and model-based empiricism as a core component of my work.

*This research commentary is derived from my AERA SIG/RME Senior Scholar Award Presentation at the annual meeting of the American Educational Research Association, April 13–17, 2009, San Diego, California.

In what follows, I outline three core principles in my work, as it has evolved. The narrative that follows draws upon my research history for illustrations. I conclude by addressing some issues that the reviewers asked me to discuss.

Core Principles

1. *Theory matters.* If you take theory and models seriously, then (a) you need to elaborate clearly for yourself "what counts" and how things supposedly fit together, and (b) you must hold yourself accountable to data (see Schoenfeld, 2007). From my perspective, theory is—or should be—the lifeblood of the empirical scientist. Conversely, all educational (more broadly, social science) theory should routinely be tested against empirical data.
2. *One makes progress by systematically pushing the boundaries of the problem space in order to see where the theory "breaks."* That is, it is essential to choose cases for analysis that you think you might be able to understand and that have the following property: If you succeed in explaining them, you will have expanded the scope of the theory, and if you fail, you have found a limitation of the theory.
3. *One can make progress by keeping one's eyes open for interesting things.* From my earlier article: "Human problem solving behavior is extraordinarily rich, complex, and fascinating—and we only understand very little of it. It's a vast territory waiting to be explored . . . I'm convinced that . . . if you just keep your eyes open and take a close look at what people do when they try to solve problems, you're almost guaranteed to see something of interest" (Schoenfeld, 1987, p. 38). I note that *interesting* is a theoretically laden term: What turns out to be interesting is often what turns out to not quite jibe with one's theoretical expectations, so (cf. point 1) the more explicit one is about one's theoretical commitments, the more likely it is that something interesting will lead to a productive line of inquiry.

On Keeping Your Eyes Open

I begin autobiographically, moving at breakneck speed because my early problem-solving work (e.g., Schoenfeld, 1985, 1992) is well known and much of this story was encapsulated in the earlier article (Schoenfeld, 1987). In the early 1970s, I read Pólya's (1945) *How to Solve It*. It felt right; Pólya seemed to describe the kinds of problem-solving strategies I was using as a mathematician. But I asked problem-solving experts, and they said, "Pólya doesn't work." This (interesting) contradiction is what got me started in educational research. I took a postdoc at Berkeley, where my mentor, Fred Reif, offered me a wonderful deal: "Read until you get sick of reading, at which point we will assume you are literate. Then you can start to do research."

I read Newell and Simon's (1972) *Human Problem Solving*, in which they looked at human problem solving with an eye toward abstracting regularities in problem-solving performance and implementing those regularities as computer programs. The idea was that computers and humans were information processors. Inspired by Pólya on the one hand and Newell and Simon on the other, I decided to look at people solving problems to see if I could find out what would enable other people (rather than computers) to become more effective at solving problems. This was my first step toward empiricism: I needed to look closely at what people do! Note that this theoretically driven empiricism is a source of inspiration as well as the basis of accountability to data. As I built up theoretical ideas about what counted in problem solving, I formed an empirical rule: Ideas gleaned from the research should be tried out in the classroom, both for inspiration and validation. This, in short, is how one holds theory accountable to data, and vice versa. (The problem-solving research and development I conducted from 1975 to 1985 was, in effect, a decade-long design experiment [Cobb, Confrey, diSessa, Lehrer, and Schauble, 2003; Schoenfeld, 2006]. Theoretical ideas were tested in practice, and both theory and instructional design were modified in the light of performance data.)

What I found first was that Pólya's heuristic problem-solving strategies were much more complex (and therefore more difficult to learn) than he had suggested. For example, a "simple" strategy such as "solve an easier related problem" is not really one strategy; rather, it is a collection of more than a dozen strategies. Note that the theoretical lens of information processing was essential here. I asked this question, "If I start by assuming that the problem solver has typical (human) information-processing capacities, how can I specify any particular strategy so that the problem solver can implement it?" What seemed to be reasonably specified strategies turned out to be vague and ill-defined; I had to go to a finer level of detail for them to be implementable. The fact that the observations were theory-based is what led to the new findings.

Observations revealed the true complexity of implementing Pólya's heuristic strategies. For example, to use the strategy "Make sense of the problem by looking at examples," one must (a) think to use the strategy, (b) know which version of the strategy to use, (c) generate the appropriate examples, (d) gain the insight needed from the examples, and (e) use that insight to solve the original problem. In light of these findings, it was no surprise that Pólya's ideas had been so difficult to implement. Heuristic problem-solving strategies are difficult to learn and to use, and students need detailed training. However, once one had become aware of the relevant level of detail, the strategies could be taught. My college students were amazingly successful (see Schoenfeld, 1985, for details).

Solving the grain-size problem raised a previously unseen problem.[1] Each of the two dozen or so powerful strategies in Pólya's *How to Solve It* was, in itself, 10–20 strategies, meaning that students had to learn hundreds of strategies. Controlling all these strategies was a challenge—one had to have a strategy for figuring out which strategies to try and when to try them. This led me to the study of monitoring and self-regulation, an aspect of metacognition. I discovered how serious this issue was when I asked students to solve problems before they took my problem-solving course. More than 60% of their attempts consisted of reading the problem, picking an approach, and pursuing that approach until they ran out of time. Absent the reconsideration of unsuccessful approaches—and a very large percentage of the initial approaches chosen by students were unsuccessful—the students were guaranteed to fail.

Once I was aware of this problem, it could be addressed. As my students worked on problems in class, I pestered them repeatedly with these questions: What are you doing? Why are you doing it? and How will it help you solve the problem? Over time, the students internalized these questions and improved at monitoring and self-regulation. But this was not enough.

Of fundamental interest to me was the question of what caused students difficulty. Of course, it is not very interesting if students fail to solve a problem because they do not have adequate domain knowledge. Thus, when I had students work problems in my lab, I chose problems they should have been able to solve. At the time, plane geometry was a required course for 10th-grade, college-intending students. Thus, my students should have known enough mathematics to be able to solve straightforward plane geometry problems.

I gave them a simple straightedge-and-compass construction problem. To my great surprise, the students approached the problem in purely empirical fashion. They made conjectures and then tried the constructions they had conjectured might work, in order to see whether they did. In clinical interviews, I discovered that the students actually knew a substantial amount of geometry—they were able to derive the properties of the desired circle. But then, when asked to do the construction, they ignored the result that they had just proved! This observation of anomalous behavior led to the study of beliefs. By now this story is familiar, so I will not repeat it (see Schoenfeld, 1985, 1992). Once again, the observation of unexpected behavior led to a new set of studies addressing the question, Where did such counterproductive behavior come from? To explore this issue seriously required that I spend a fair amount of time observing instruction in local schools, where I discovered that student beliefs originated in their experiences with school mathematics. That is, the practices in which they engaged were the source of their beliefs (Schoenfeld, 1988, 1992).

By the late 1980s, I had found that the following were major determinants of problem-solving success or failure: the knowledge base, heuristic strategies, metacognition (specifically, monitoring and self-regulation), beliefs, and practices. However, I did not have a theoretical description of how and why people made the choices they did while solving problems. That was the next, and most fundamental, question—and the major goal of my research agenda.

On Pushing the Boundaries of the Problem Space

The first decade of my research in education was devoted to the study of people solving (mathematical) problems in isolation, in the research laboratory. This obviously artificial setting was, to put things simply, a reflection of the state of the art: Given the research tools and techniques available in the 1970s and 1980s, it was all we could do to study thinking in isolation. The major goal, of course, was to understand thinking and problem solving in general—in any problem-solving context in any domain. Understanding this calls for developing a theory of human decision making. (Members of my research group know that my long-term research goal—sometimes glimpsed over the horizon, sometimes not—has been the "theory of everything," or TOE.) Changing directions somewhat and moving to the study of tutoring allowed me to delay a head-on attack on the vexing problem I had been unable to solve (how and why people make the decisions they do). However, at the same time it moved my overall agenda forward. Tutoring is a complex form of problem solving in which mathematical decision making and social interactions are involved. Thus, my work expanded into the social domain. In addition, I had the longer term goal of understanding teaching. Tutoring is a stepping-stone toward that larger goal, in that it is less complex but involves some of the same complexities of decision making.

We brought a student into my lab for some tutorial sessions. The social interactions in the session tapes were messy, rendering these sessions unsuitable for extended analysis, I thought, but Abraham Arcavi said, "Alan, there are really interesting things going on in these tapes. We should take a closer look" (personal communication, December 7, 1986). Abraham, Jack Smith, and I did. The student's errors, which were resistant to straightforward tutoring, turned out to be rooted in some very deep misconceptions. Her incorrect, but robust, understandings shaped what she saw, what she remembered, and what she forgot. Specifically, new pieces of information that did not fit with what she knew tended to fade away, even though the new things were correct and what she knew was wrong. When our analyses were completed, we had a new article (Schoenfeld, Smith, and Arcavi, 1993), a new way of thinking about people's knowledge structures and how they changed, and some new methods (microgenetic analyses) for charting that growth and change.

I want to emphasize that this work embodied all three of the core principles discussed in the introduction. First, the work was theory driven. The cognitive-science approach to representing knowledge as networks of nodes and connections allowed us to conceptualize and chart the changes in the student's knowledge structures—and, ultimately, to see the limitations of the perspective with which we had started. (Specifically, if freshly learned material did not connect to established learning structures, it faded away—and, because it did fit, old mal-knowledge that had been temporarily replaced by the correct knowledge could regenerate itself. Because her knowledge structures, including incorrect ones, were robust, they resisted change and it took major work to undo the mal-knowledge.) Second, what was interesting—the robustness of the student's mal-perceptions—was interesting in part because we had theoretical expectations, and her learning trajectory violated them. Third, as noted, we were expanding the space of inquiry into the social (although still in the lab. One step at a time . . .).

At the time, there were two bodies of research on tutoring. One focused on subject matter. It looked closely at student understanding and how to move it forward, but it ignored "human factors." A second focused on human factors, examining issues such as intrinsic versus extrinsic motivation. However, there were no connections between them, and that made no sense. Sometimes, something a student says or does requires an immediate mathematical response, for example, when the student says "$(a + b)^2 = a^2 + b^2$." Of course, how the tutor responds to a statement like this depends very much on what the tutor knows ("What options do I have to address this misconception?") and what he or she believes ("Do I need to work carefully through this, or just give the student the correct formula?"). Sometimes, something a student says or does requires an immediate personal response, for example, when the student looks weary or disheartened. Here, too, how the tutor acts depends very much on what the tutor knows ("What can I do to restore equilibrium?") and believes ("How important is it to pursue the content? How important is it to be sympathetic and back off for the moment?"). But in both cases, whether the event is content-related or affect-related, something has happened that causes the tutor to consider/reconsider how things are going. On the basis of his or her knowledge and beliefs, the tutor either puts new goals in place or continues to pursue the current high-priority goals. That is, the tutor's evolving top-level goals determine the course of action. Such a *goal-directed architecture* allowed us to model tutoring decisions and to unify the two literatures. And it led to the question (and the next expansion of the space), Might this architecture be the correct one to model teaching?

Although classroom environments are typically much more chaotic than tutoring environments, the basic question is the same: Why do teachers make the choices they do? We hypothesized that the answer is the same: A teacher enters with a plan, and then makes adjustments on the basis of (a) what happens, (b) beliefs about what is important to pursue, and (c) the knowledge that she or he can bring to bear.

Once again, we had good luck. Mark Nelson, a student teacher in our teacher preparation program, said to the head of the program,

Dan Zimmerlin, "I didn't like the way today's lesson went. Can you help me understand why?" Zimmerlin said, "Bring the tape to Alan's research group. He wants to study teaching." Four months later, we understood the problem, and we were able to explain why Nelson had done what he did (see Schoenfeld, 2000; Zimmerlin and Nelson, 2000). It turned out that Nelson's pedagogical choices were a function of his knowledge, goals, and beliefs. (His beliefs determined what he would and would not do in the classroom. In this case, they kept him from using some of his knowledge.) We hypothesized that this was the case in general.

It was time to choose a new tape for analysis. In line with Core Principle 2, "explore different dimensions of the problem space," I needed to choose an example that was significantly different from the tape we had analyzed. Nelson was a new teacher, teaching a traditional lesson. So I needed a tape of an experienced teacher. Yet again, there was good luck: Emily van Zee, who was doing a postdoc at Berkeley and attending my research group, brought in for discussion a tape of Jim Minstrell's physics teaching. Minstrell is a very well known teacher-researcher, and van Zee and Minstrell had written an article about his teaching style.

I asked Minstrell if I could use his data for an independent analysis. He said yes, and a year later we had modeled the full hour of instruction. There is a huge amount of detail in the analysis (see Schoenfeld, 1998). Here I will simply point to the main aspects of the analysis. The formal content of Minstrell's lesson involves the use of mean, median, and mode. But the main point of the lesson is that he wants his students to see that such formulas need to be used sensibly. The previous day, eight students had measured the width of a table, obtaining the values 106.8, 107.0, 107.0, 107.5, 107.0, 107.0, 106.5, and 106.0 cm. Minstrell wanted the students to discuss the "best number" for the width of the table: Which numbers—all or some—should they use? How should they combine them? With what precision should they report the answer? He had a flexible script for each part of the lesson: (a) raise the issue; (b) ask for a student suggestion; (c) clarify and pursue the suggestion by asking questions, inserting some content if necessary; (d) once this suggestion has been worked through, ask for more suggestions; and (e) when students run out of ideas, either inject more ideas or move to the next part of the lesson. We analyzed the lesson in fine detail—decomposing the lesson into smaller and smaller episodes, noting which goals were present and at what levels of activation, and observing how transitions corresponded to changes in goals. When things went "according to plan," the lesson was easy to model.

But what can one say when things do not go according to plan? This happened when Minstrell was reviewing various ways to compute the "best value" to represent the eight measurements given above. The class had discussed mean and mode when a student raised her hand and said,

> This is a little complicated, but I mean it might work. If you see that 107 shows up four times, you give it a coefficient of 4, and then 107.5 only shows up one time, you give it a coefficient of 1, you add all those up, and then you divide by the number of coefficients you have.

There is a wide range of possible responses, ranging from "That's a very interesting question. I'll talk to you about it after class" to "Let's make sure we all understand what you've suggested and then explore it." Each has different entailments for how the class will play out. The research challenge: Is it possible to say how Minstrell will respond?

According to our model, Minstrell's fundamental belief about his physics teaching is that physics is a sense-making activity and that students should experience it as such. One of his major goals is to support inquiry and to honor student attempts at figuring things out. Minstrell's knowledge base includes favored techniques such as "reflective tosses," in which one asks questions that get students to explain or elaborate on what they have said. Thus he will choose to pursue the student's suggestion, using reflective tosses (for details, see Schoenfeld, 1998). We modeled Minstrell's decision using a form of *subjective expected utility* (or cost-benefit analysis).[2] This form of modeling has worked consistently in a variety of situations in which we have tried to capture nonroutine decision making.

What next? At a meeting I saw a video of Deborah Ball teaching a third-grade class (the "Shea number" tape). The lesson was amazing. In it, the third graders argued on solid mathematical grounds; the discussion agenda evolved as a function of classroom conversations; the teacher seemed at times to play a negligible role; and sometimes she made decisions that people have said did not make sense. In addition, I had little or no intuition about what happened. Given Core Principle 2, this was ideal. There were major differences from previous tapes we had studied in grade level, content, psychological (developmental) issues, classroom dynamics, the "control structure" for the classroom, and the teacher's role. What a challenge!

Three years later, our analyses showed that Ball employs a "debriefing routine" that consists of asking questions and fleshing out answers in a particular way—and that she used that routine five times in the first six minutes of class. What seems somewhat unstructured on casual viewing turns out to be very highly structured. Moreover, Ball's controversial decision (in which she led a student on a mathematical excursion that ran the risk of derailing her own announced agenda) can be modeled as a principled move entirely consistent with her larger agenda—once you know that the success of the next part of her lesson hinged on students' understanding of the issue she discussed with the student.

Ball's lesson segment has been modeled on a line-by-line basis (see Schoenfeld, 2008). In addition, once the modeling was done, some very interesting consistencies between Ball's routine for getting students to clarify their understandings and Minstrell's interactive routine became apparent. After completing the analysis, I realized that I use a variant of the same routine in teaching my problem-solving courses. It may well turn out to be a general, learnable routine for supporting highly interactive, student-centered classrooms (Schoenfeld, 2002). Moreover, given that the theory of teaching-in-context (the claim that a teacher's in-the-moment actions can be modeled as a function of the teacher's attributed knowledge, goals, beliefs/orientations, and a particular kind of decision making) had proved successful in allowing us to model three radically different cases of teaching (Nelson, Minstrell, and Ball), the theory was demonstrably robust.

What next? There are a number of possible directions, some of which I am pursuing now and some of which I hope to pursue in the future. The one on which I am working at present is an abstraction of the theory of teaching-in-context, a general theory of human in-the-moment decision making. Teaching is somewhat special in the complexity of its interactions, but in many ways it typifies knowledge-intensive domains in which practitioners engage in a substantial amount of well-practiced behavior, punctuated by episodes of nonstandard decision making. Other domains that can be characterized in this way include cooking, for example, preparing and cooking a meal; crafts such as automobile mechanics and electronic troubleshooting; and routine medical diagnosis and practice (see the extensive Artificial Intelligence literature on how doctors' orientations to patients and disease shape their diagnoses, how routine is followed, etc.). Assuming that human brains work the same way across similar domains, then the architecture of the theory of teaching-in-context should apply in those arenas as well.[3] In short, I claim that goal-oriented "acting in the moment"—including problem solving, tutoring, teaching, cooking, and brain surgery—can be explained and modeled using a theoretical architecture in which the following are represented: an individual's knowledge, goals, "orientations" (an abstraction of beliefs that includes values, preferences, etc.), and decision making (captured in an "internal calculus" that can be modeled as a form of cost-benefit analysis). I hypothesize that things work as described in Box 1.

Given this claim, in what domains beyond teaching might I test it? Medicine was one of the domains on my list. I like my doctor, and I know that she is intellectually curious, so I asked her whether I could tape one of our routine visits and analyze it. She said yes. As it happens, it was easy to model her actions. Like most general practitioners, she has a family of "disease-related scripts" that govern her interactions with patients who have known diagnoses. I have adult-onset diabetes, and it is easy to see how her actions conform to a "Type II diabetes script," in which she works through the numbers on my most recent lab tests with me and exhorts me to be a better patient than I am inclined to be. Modeling her actions (which are much less complex in this situation than those of a teacher handling a whole class) provided some confirmation of the generality of the approach. (See Schoenfeld, 2010b, for the general argument and the model of the diagnostic interaction. Note as well that there are extensive psychological and Artificial Intelligence literatures on medical diagnosis.)

Beyond that, the interaction with my doctor had been very productive. This raised some interesting questions. Could I understand why it had been so productive? Because there were only two participants in the conversation, it seemed reasonable to model both. When I modeled doctor and patient (with regard to the categories of analysis—knowledge, goals, orientations, and decision making), it turned out that there was an excellent match between our goals. In particular, when one participant acted in a way that made a goal clear, the other participant picked up on the goal and made it his or her own. Thus, *goal synchronization* appears to be a major factor in making the conversation productive!

Box 1. How Things Work in Outline

- An individual enters into a particular context with a specific body of knowledge, goals, and orientations (beliefs, dispositions, values, preferences, etc.).
- The individual orients to the situation. Certain pieces of information and knowledge become salient and are activated.
- Goals are established (or reinforced, if they pre-existed).
- Decisions are made, consciously or unconsciously, in pursuit of these goals.
  (a) If the situation is familiar, then the process may be relatively automatic, in which case the action(s) taken is (are) in essence the access and implementation of scripts, frames, routines, or schemata.
  (b) If the situation is not familiar or there is something nonroutine about it, then decision making is made via an internal calculus that can be modeled by (i.e., is consistent with the results of) the subjective expected values of available options, given the orientations of the individual.
- Implementation begins.
- Monitoring (whether effective or not) takes place on an ongoing basis.
- This process is iterative, down to the level of individual utterances or actions.
  (a) Routines aimed at particular goals have subroutines, which have their own subgoals.
  (b) If a subgoal is satisfied, the individual proceeds to another goal or subgoal.
  (c) If a goal is achieved, new goals kick in via decision making.
  (d) If the process is interrupted or things do not seem to be going well, decision making is activated once again. This may or may not result in a change of goals and/or the pathways used to try to achieve them.

Source: From Schoenfeld 2010a.

What next? There are a number of possibilities. First, this commentary and my recent book (Schoenfeld, 2010b) make a general claim regarding the architecture of people's in-the-moment acting and decision making. A significant amount of empirical work needs to be done to test and refine that claim. Second, I think that the idea of goal synchronization discussed in the previous paragraph (see also chapter 7 of Schoenfeld, 2010b) has great promise as a theoretical and empirical tool. What makes for a highly productive classroom? I strongly suspect that goal synchronization (between students, and between students and teacher) is a significant part of that story. The analytical tools developed in the chapter that analyzes my conversation with my doctor are general, and I think they can be used productively for the study of classroom interactions. This could provide a powerful way to integrate cognitive and social analyses. Finally, there is the issue of integrating a theory of learning into the current theory. At present, the theory is about acting in the moment: The individual makes decisions on the basis of his or her current knowledge, goals, and orientations. A logical next step is to build models of acting-in-the-moment that incorporate ongoing changes into their descriptions of current knowledge, goals, and orientations. For example, as a result of a classroom dialogue, a teacher might know or believe more than before about a particular student. The teacher might think that a certain approach to a topic is less useful than she or he had previously thought, and have in mind modifications or variations to try the next time. In reality, a teacher's knowledge, goals, and orientations are being continuously updated. It would be interesting and potentially useful to build models that take this kind of learning into account.

Discussion

I return to the three core principles outlined at the beginning of this article and briefly address two issues raised by reviewers. Core Principle 1, the importance of taking theory seriously, is absolutely central. That theme has permeated these reflections, so I will merely summarize some of my main assertions here. First, it is when issues are couched in theoretical terms that one can hypothesize and test their generality. Second, one's theoretical commitments, whether tacit or explicit, shape what one sees and deems important. Thus, there is much to be gained (and many pitfalls to be avoided!) by being explicit about one's assumptions.

Third, making sure to cover all aspects in the theory-model-empirical-data triad is an extremely powerful way of improving and refining one's ideas. From my perspective, taking theory seriously means holding oneself accountable to empirical data. (Armchair educational theory is about as useful as armchair philosophy.) Moreover, how one holds oneself accountable to data is critically important. It is easy to provide ad hoc explanations of individual events, and this is dangerous. Theoretical commitments (along with their instantiations as context-specific models) guard against ad hocism. It is one thing to claim to have a general theory of teaching, problem solving, or decision making. It is quite another to build a model, or even the outline of a model, that uses only constructs sanctioned by the theory and that "captures" the behavior that is being modeled in some significant way. Taking the theory-model-empirical-data triad seriously is thus a way of keeping oneself intellectually honest and also making progress.

Core Principles 2 and 3 lie at the heart of a productive empirical research program. Much of this article's narrative has been organized along those two lines (see also the following), and I hope their import is clear. It is worth noting, once again, that both principles are deeply theory-laden. One's theoretical perspective structures the dimensions of the problem space, so a systematic exploration of those dimensions (Principle 2) is de facto theoretically driven. And, as noted, the most interesting and potentially productive observations (Principle 3) are the ones that do not quite fit with our theoretical expectations and compel us to take a closer look at what is happening.

Reviewers asked whether I would comment on steps one can take to build a productive career, and to discuss some of the ways in which my research group works. For the first, I would reformulate some of the statements above as recommendations. Personally, I think it is essential as a researcher to work on a big problem that one thinks is truly important and about which one cares deeply. There are a sufficient number of big problems on which to work, and the choice is a matter of taste: One could concern oneself with teachers' professional development, helping middle school students (especially disenfranchised students) come to grips with middle school mathematics in meaningful ways, or understanding students' mathematical learning disabilities, to name just three. Next, one has to find a toehold—a part of the problem on which one can make tangible and meaningful progress. One should then focus one's attention on the manageable subproblem while keeping the larger problem in the back of one's mind. My experience has been that once the subproblem has been solved, one sees more clearly and is in a better position than before to make interesting observations, produce generalizations, or explore the problem space. In consequence, addressing the big problem becomes an iterative process: Each solution raises new questions or makes previously intractable questions potentially approachable. This kind of approach, I believe, guarantees that one has plenty to work on while maintaining a sense of direction.[4]

That approach, along with one major addition, shape both the raison d'être and modus operandi of my research group. I see my primary role as a mentor, helping talented young scholars learn to harness their passions in the ways described in the previous paragraph. (The topics I mentioned in that paragraph are each the foci of current members of the group, and they are rich enough to keep those group members productively engaged for many decades.) The missing ingredient for young scholars is one that comes with experience: learning how to take a complex problem and find the right toehold. An important problem is complex and messy; the challenge is to figure out how to address a part of it that is meaningful and manageable. Thus, much of my advising consists of discussing such issues—most often in the whole group rather than in individual sessions. Group members bring their work—at every stage of the work from initial conception through multiple reconceptualizations, to initial data gathering and interpretation (and yet more reconceptualizations) through final paper or dissertation writing—and all of us collectively discuss that work-in-process (including mine) (see Schoenfeld, 1999, for some detail). The collective discussions provide a mechanism for initially watching, and then, over time, becoming increasingly engaged. The result is an apprenticeship into the habits of mind that I hope and believe will serve the students well as researchers. I can only wish for them as much fun in pursuing the issues about which they care as I have had doing the same.

Acknowledgments

I am grateful to the reviewers and to Ed Silver, whose suggestions induced me to turn what was in essence a chronological narrative into a somewhat more traditionally structured research commentary. I also thank Cathy Kessel, Noreen Greeno, Yoshi Shimizu, Marty Simon, and the extended functions group for their comments and suggestions.

Notes

1. This is a general and critically important issue. Almost always, coming to a deeper understanding of some phenomenon allows one to see things that were hitherto invisible. Thus, if one is working on a large and significant problem, it is likely to unfold gradually as aspects of the problem are addressed successfully.

2. I hasten to add that modeling in this way does not presuppose that all teaching decisions are "rational." Indeed, subjective expected utility turns out to be an ironically "rational" way of capturing the consistent irrationalities in people's decision making!

3. Note that the expression *knowledge-intensive domains*, in which practitioners engage in a substantial amount of well-practiced behavior, punctuated by episodes of nonstandard decision making, is theoretically laden, as is the assumption that cognitive architecture will be the same for cooking, automobile, and medical practice. Once again, it is a set of underlying theoretical assumptions that provide the basis for generalization and abstraction.

4. I say this upon reflection; I certainly cannot say that I followed this rule explicitly, although I seem to have been true to it. In that sense, this "Research Commentary" is also the reflection of an accidental metatheorist.

References

Cobb, P., Confrey, J., diSessa, A., Lehrer, R., and Schauble, L. (2003). Design experiments in educational research. *Educational Researcher*, *32*(1), 9–13.

Newell, A., and Simon, H. A. (1972). *Human problem solving.* Englewood Cliffs, N.J.: Prentice-Hall.

Pólya, G. (1945). *How to solve it: A new aspect of mathematical method.* Princeton, N.J.: Princeton University Press.

Schoenfeld, A. H. (1985). *Mathematical problem solving.* Orlando, Fla.: Academic Press.

Schoenfeld, A. H. (1987). Confessions of an accidental theorist. *For the Learning of Mathematics*, 7(1), 30–8.

Schoenfeld, A. H. (1988). When good teaching leads to bad results: The disasters of "well taught" mathematics classes. *Educational Psychologist, 23*, 145–66.

Schoenfeld, A. H. (1992). Learning to think mathematically: Problem solving, metacognition, and sense-making in mathematics. In D. A. Grouws (Ed.), *Handbook of research on mathematics teaching and learning* (pp. 334–70). New York: Macmillan.

Schoenfeld, A. H. (1998). Toward a theory of teaching-in-context. *Issues in Education, 4*(1), 1–94. Retrieved October 21, 2009, from http://www-gse.berkeley.edu/faculry/AHSchoenfeld/AHSchoenfeld.html.

Schoenfeld, A. H. (1999). The core, the canon, and the development of research skills: Issues in the preparation of education researchers. In E. C. Lagemann and L. S. Shulman (Eds.), *Issues in education research: Problems and possibilities* (pp. 166–202). San Francisco: Jossey-Bass.

Schoenfeld, A. H. (2000). Models of the teaching process. *Journal of Mathematical Behavior, 18*, 243–62.

Schoenfeld, A. H. (2002). A highly interactive discourse structure. In J. Brophy (Ed.), *Advances in research on teaching: Vol. 9. Social constructivist teaching: Its affordances and constraints* (pp. 131–70). New York: Elsevier.
Schoenfeld, A. H. (2006). Design experiments. In P. B. Elmore, G. Camilli, and J. Green (Eds.), *Handbook of complementary methods in education research* (pp. 193–206). Mahwah, N.J.: Erlbaum.
Schoenfeld, A. H. (2007). Method. In F. K. Lester Jr. (Ed.), *Second handbook of research on mathematics teaching and learning* (pp. 69–107). Charlotte, N.C.: Information Age.
Schoenfeld, A. H. (2008). On modeling teachers' in-the-moment decision-making. In A. H. Schoenfeld (Ed.), *Journal for Research in Mathematics Education monograph series: Vol. 17. A study of teaching: Multiple lenses, multiple views* (pp. 45–96). Reston, Va.: National Council of Teachers of Mathematics.
Schoenfeld, A. H. (2010a). How and why do teachers explain things the way they do? In M. K. Stein and L. Kucan (Eds.), *Instructional explanations in the disciplines.* New York: Springer.
Schoenfeld, A. H. (2010b). *How we think: A theory of goal-oriented decision-making and its educational applications, New York, Routledge.*
Schoenfeld, A. H., Smith, J. P., III, and Arcavi, A. A. (1993). Learning: The microgenetic analysis of one student's evolving understanding of a complex subject matter domain. In R. Glaser (Ed.), *Advances in instructional psychology* (Vol. 4, pp. 55–175). Hillsdale, N.J.: Erlbaum.
Zimmerlin, D., and Nelson, M. (2000). The detailed analysis of a beginning teacher carrying out a traditional lesson. *Journal of Mathematical Behavior, 18*, 263–79.

The Conjoint Origin of Proof and Theoretical Physics

Hans Niels Jahnke

The Origins of Proof

Historians of science and mathematics have proposed three different answers to the question of why the Greeks invented proof and the axiomatic-deductive organization of mathematics (see Szabó 1960, 356 ff.).*

(1) The *socio-political thesis* claims a connection between the origin of mathematical proof and the freedom of speech provided by Greek democracy, a political and social system in which different parties fought for their interests by way of argument. According to this thesis, everyday political argumentation constituted a model for mathematical proof.
(2) The *internalist thesis* holds that mathematical proof emerged from the necessity to identify and eliminate incorrect statements from the corpus of accepted mathematics with which the Greeks were confronted when studying Babylonian and Egyptian mathematics.
(3) The *thesis of an influence of philosophy* says that the origin of proof in mathematics goes back to requirements made by philosophers.

Obviously, thesis (1) can claim some plausibility, though there is no direct evidence in its favor and it is hard to imagine what such evidence might look like.

Thesis (2) is stated by van der Waerden. He pointed out that the Greeks had learned different formulae for the area of a circle from Egypt and Babylonia. The contradictory results might have provided a strong motivation for a critical reexamination of the mathematical rules in use at

*I would like to thank Gila Hanna and Helmut Pulte for their valuable advice.

the time the Greeks entered the scene. Hence, at the time of Thales the Greeks started to investigate such problems by themselves in order to arrive at correct results (van der Waerden 1988, 89 ff.).

Thesis (3) is supported by the fact that standards of mathematical reasoning were broadly discussed by Greek philosophers, as the works of Plato and Aristotle show. Some authors even use the term "Platonic reform of mathematics."

This paper considers in detail a fourth thesis which in a certain sense constitutes a combination of theses (1) and (3). It is based on a study by the historian of mathematics Árpád Szabó[1] (1960), who investigated the etymology of the terms used by Euclid to designate the different types of statements functioning as starting points of argumentation in the "Elements."

Euclid divided the foundations of the "Elements" into three groups of statements: (1) Definitions, (2) Postulates, and (3) Common Notions (Heath 1956). Definitions determine the objects with which the Elements are going to deal, whereas Postulates and Common Notions entail statements about these objects from which further statements can be derived. The distinction between postulates and common notions reflects the idea that the postulates are statements specific to geometry whereas the common notions provide propositions true for all of mathematics. Some historians emphasize that the postulates can be considered as statements of existence.

In the Greek text of Euclid handed down to us (Heiberg's edition of 1883–1888) the definitions are called ὅροι, the postulates ἀιτήματα and the common notions κοιναί ἔννοιαι. In his analysis, Szabó starts with the observation that Proclus (fifth century AD), in his famous commentary on Euclid's elements, used a different terminology (for an English translation, see Proclus 1970). Instead of ὅροσ (definition) Proclus applied the concept of ὑπόθεσις (hypothesis) and instead of κοιναί ἔννοιαι (common notions) he used ἀξιώματα (axiomata). He maintained the concept of ἀιτήματα (postulates) as contained in Euclid. Szabó explains the differing terminology by the hypothesis that Proclus referred to older manuscripts of Euclid than the one which has led to our modern edition of Euclid.

Szabó shows that ὑπόθεσις (hypothesis), αἴτημα (postulate), and ἀξίωμα (axiom) were common terms of pre-Euclidean and pre-Platonic dialectics, which is related both to philosophy and rhetoric. The classical Greek philosophers understood dialectics as the art of exchanging arguments and counter-arguments in a dialog debating a controversial proposition. The outcome of such an exercise might be not simply the refutation of one of the relevant points of view but rather a synthesis or a combination of the opposing assertions, or at least a qualitative transformation (see Ayer and O'Grady 1992, 484).

The use of the concept of hypothesis as synonymous with definition was common in pre-Euclidean and pre-Platonic dialectics. In this usage, hypothesis designated the fact that the participants in a dialog had to agree initially on a joint definition of the topic before they could enter the argumentative discourse about it. The Greeks, including Proclus, also used hypothesis in a more general sense, close to its meaning today. A hypothesis is that which is underlying and consequently can be used as a foundation of something else. Proclus, for example, said: "Since this science of geometry is based, we say, on hypothesis (ἔξ ὑποθέσεως εἶναι), and proves its later propositions from determinate first principles . . . he who prepares an introduction to geometry should present separately the principles of the science and the conclusions that follow from the principles, . . . " (Proclus 1970, 62).

According to Szabó, the three concepts of hypothesis, aitema (postulate), and axioma had a similar meaning in the pre-Platonic and pre-Aristotelian dialectics. They all designated those initial propositions on which the participants in a dialectic debate must agree. An initial proposition which was agreed upon was then called a "hypothesis." However, if participants did not agree or if one declared no decision, the proposition was called aitema (postulate) or axioma (Szabó 1960, 399).

As a rule, participants will introduce into a dialectic debate hypotheses that they consider especially strong and expect to be accepted by the other participants: numerous examples of this type can be found in the Platonic dialogs. However, it is also possible to propose a hypothesis with the intention of critically examining it. In a philosophical discourse, one could derive consequences from such a hypothesis that are *desired* (plausible) or *not desired* (implausible). The former case leads to a strengthening of the hypothesis, the latter to its weakening. The extreme case of an undesired consequence would be a logical contradiction, which would necessarily lead to the rejection of the hypothesis. Therefore, the procedure of indirect proof in mathematics can be considered as directly related to common customs in philosophy. According to Szabó (1960) this constitutes an explanation for the frequent occurrence of indirect proofs in the mathematics of the early Greek period.

The concept of common notions as a name for the third group of introductory statements needs special attention. As mentioned above, this term is a direct translation of the Greek κοιναί ἔννοιαι and designates "the ideas common to all human beings." According to Szabó, the term stems from Stoic philosophy (since 300 BC) and connotes a proposition that cannot be doubted justifiably. Proclus also attributes the same meaning to the concept of ἀξίωμα, which he used instead of κοιναί ἔννοιαι. For example, he wrote at one point: "These are what are generally called indemonstrable axioms, inasmuch as they are deemed by everybody to be true and no one disputes them" (Proclus 1970, 152). At another point he even wrote, with an allusion to Aristotle: ". . . whereas the axiom is as such indemonstrable and everyone would be disposed to accept it, even though some might dispute it for the sake of argument" (Proclus 1970, 143). Thus, only quarrelsome people would doubt the validity of the Euclidean axioms; since Aristotle, this has been the dominant view.

Szabó (1960) shows that the pre-Aristotelean use of the term axioma was quite similar to that of the term aitema, so that axioma meant a statement upon which the participants of a debate agreed or whose acceptance they left undecided. Furthermore, he makes it clear that the propositions designated in Euclid's "Elements" as axioms or common notions had been doubted in the early period of Greek philosophy, namely by Zenon and the Eleatic School (fifth century BC). The explicit compilation of the statements headed by the term axioms (or common notions) in the early period of constructing the elements of mathematics was motivated by the intention of rejecting Zeno's criticism. Only later, when the philosophy of the Eleates had been weakened, did the respective statements appear as unquestionable for a healthy mind.

In this way, the concept of an axiom gained currency in Greek philosophy and in mathematics. Its starting point lay in the art of philosophical discourse; later it played a role in both philosophy and mathematics. More important for this paper, it underwent a concomitant change in its epistemological status. In the early context of dialectics, the term axiom designated a proposition that in the beginning of a debate could be accepted or not. However, axiom's later meaning in mathematics was clearly that of a statement which itself cannot be proved but is absolutely certain and therefore can serve as a fundament of a deductively organized theory. This later meaning became the still-dominant view in Western science and philosophy.

Aristotle expounded the newer meaning of axiom at length in his "Analytica posteriora":

> I call "first principles" in each genus those facts which cannot be proved. Thus the meaning both of the primary truths and the attributes demonstrated from them is assumed; as for their existence, that of the principles must be assumed, but that of the attributes must be proved. E. g., we assume the meaning of "unit", "straight" and "triangular"; but while we assume the existence of the unit and geometrical magnitude, that of the rest must be proved. (Aristotle 1966, I, 10)

Aristotle also knew the distinction between postulate (aitema) and axiom (common notions) as used in Euclid:

> Of the first principles used in the demonstrative sciences some are special to particular sciences, and some are common; . . . Special principles are such as that a line, or straightness, is of such-and-such a nature; common principles are such as that when equals are taken from equals the remainders are equals. (Aristotle 1966, I, 10)

Thus, Szabó's study leads to the following overall picture of the emergence of mathematical proof. In early Greek philosophy, reaching back to the times of the Eleates (ca. 540 to 450 BC), the terms axioma and aitema designated propositions which were accepted in the beginning of a dialog as a basis of argumentation. In the course of the dialog, consequences were drawn from these propositions in order to examine them critically and to investigate whether the consequences were desired. In a case where the proposition referred to physical reality, "desired" could mean that the consequences agreed with experience. If the proposition referred to ethics, "desired" could mean that the consequences agreed with accepted norms of behavior. Desired consequences constituted a strong argument in favor of a proposition. The most extreme case of undesired consequence, a logical contradiction, led necessarily to rejecting the proposition. Most important, in the beginning of a dialog the epistemic status of an axioma or aitema was left indefinite. An axiom could be true or probable or perhaps even wrong.

In a second period, starting with Plato and Aristotle (since ca. 400 BC) the terms axioma and aitema changed their meaning dramatically; they now designated propositions considered absolutely true. Hence, the epistemic status of an axiom was no longer indefinite but definitely fixed. This change in epistemic status followed quite naturally because at that time mathematicians had started building theories. Axioms were supposed true once and for all, and mathematicians were interested in deriving as many consequences from them as possible. Thus, the emergence of the classical view that the axioms of mathematics are absolutely true was inseparably linked to the fact that mathematics became a "normal science" to use T. Kuhn's term. After Plato and Aristotle, the classical view remained dominant until well into the nineteenth century.

Natural as it might have been, in the eyes of modern philosophy and modern mathematics this change of the epistemic status of axioms was nevertheless an unjustified dogmatization. The decision to build on a fixed set of axioms and not to change them any further is epistemologically quite different from the decision to declare them absolutely true.

On a more general level, we can draw two consequences: First, Szabó's (1960) considerations suggest the thesis that the *practice of a rational discourse* provided a model for the organization of a mathematical theory according to the axiomatic-deductive method; in sum, proof is rooted in communication. However, this does not simply support the sociopolitical thesis, according to which proof was an outcome of Greek democracy. Rather, it shows a connection between proof and dialectics as an *art of leading a dialog*. This art aimed at a methodically ruled discourse in which the participants accept and obey certain rules of behavior. These rules are crystallized in the terms hypothesis, aitema, and axiom, which entail the participants' obligation to exhibit their assumptions.

The second important consequence refers to the *universality* of dialectics. Any problem can become the subject of a dialectical discourse, regardless of which discipline or even aspect of life it involves. From a problem of ethics to the question of whether the side and diagonal of a square have a common measure, all problems could be treated in a debate. Different persons can talk about the respective topic as long as they are ready to reveal their suppositions. Analogously, the possibility of an axiomatic-deductive organization of a group of propositions is not confined to arithmetic and geometry, but can in principle be applied to any field of human knowledge. The Greeks realized this principle at the time of Euclid, and it led to the birth of theoretical physics.

Saving the Phenomena

During the Hellenistic era, within a short interval of time Greek scientists applied the axiomatic-deductive organization of a theory to a number of areas in natural science. Euclid himself wrote a deductively organized optics, whereas Archimedes provided axiomatic-deductive accounts of statics and hydrostatics.

In astronomy, too, it became common procedure to state hypotheses from which a group of phenomena could be derived and which provided a basis for calculating astronomical data. Propositions of quite a different nature could function as hypotheses. For example, Aristarchos of Samos (third century BC) began his paper "On the magnitudes and distances of the sun and the moon" with a hypothesis about how light rays travel in the system earth-sun-moon, a hypothesis about possible positions of the moon in regard to the earth, a hypothesis giving an explanation of the phases of the moon and a hypothesis about the angular distance of moon and sun at the time of half-moon (a measured value). These were the ingredients Aristarchos used for his deductions.

In the domain of astronomy, the Greeks discussed, in an exemplary manner, philosophical questions about the relation of theory and empirical evidence. This discussion started at the time of Plato and concerned the paths of the planets. In general, the planets apparently travel across the sky of fixed stars in circular arcs. At certain times, however, they perform a retrograde (and thus irregular) motion. This caused a severe problem; since the Pythagoreans, the Greeks had held a deeply rooted conviction that the heavenly bodies perform circular movements with constant velocity. But this could not account for the irregular retrograde movement of the planets.

Greek astronomers invented sophisticated hypotheses to solve this problem. The first scientist who proposed a solution was Eudoxos, the best mathematician of his time and a close friend of Plato's. Though the phenomenon of the retrograde movement of the planets was well known, it did not figure in the dialogs of Plato's early and middle period. Only in his late dialog "Nomoi" ("Laws") did Plato mention the problem. In this dialog, a stranger from Athens (presumably Eudoxos) appeared, who explained to Clinias (presumably Plato) that it only seems that the planets "wander" (i.e., perform an irregular movement), whereas in reality precisely the opposite is true: "Actually, each of them describes just one fixed orbit, although it is true that to all appearances its path is always changing" (Plato 1997, 1488). Thus, in his late period Plato acknowledged that we have to adjust our basic ideas in order to make them agree with empirical observations.

I will illustrate this principle by a case simpler than the paths of the planets but equally important in Greek astronomy. In the second century BC, the great astronomer and mathematician Hipparchos investigated an astronomical phenomenon probably already known before his time, the "anomaly of the sun." Roughly speaking, the term referred to the observation that the half-year of summer is about 1 week longer than the half-year of winter. Astronomically, the half-year of summer was then defined as the period that the sun on its yearly path around the earth (in terms of the geocentric system) needs to travel from the vernal equinox to the autumnal equinox. Analogously, the half-year of winter is the duration of the travel from the autumnal equinox to the vernal equinox. Vernal equinox and autumnal equinox are the two positions of the sun on the ecliptic at which day and night are equally long for beings living on the earth. The two points, observed from the earth, are exactly opposite to each other (vernal equinox, autumnal equinox, and center of the earth form a straight line). Since the Greek astronomers supposed that all heavenly bodies move with constant velocity in circles around the center of the earth it necessarily followed that the half-years of summer and winter would be equal.

The Greek astronomers needed to develop a hypothesis to explain this phenomenon. Hipparchos proposed a hypothesis placing the center of the sun's circular orbit not in the center of the earth but a bit outside it (See Figure 1). If this new center is properly placed, then the arc through which the sun travels during summer, observed from the earth, is greater than a half-circle; the anomaly of the sun is explained. Later, Hipparchos' hypothesis was called by Ptolemaios the "eccentric hypothesis" (Toomer 1984, 144 pp).

Another hypothesis competing with that of Hipparchos was the "epicyclic hypothesis" of Appolonios of Perge (third century BC; see Figure 2). It said that the sun moves on a circle concentric to the center of the universe, however "not actually on that circle but on another circle,

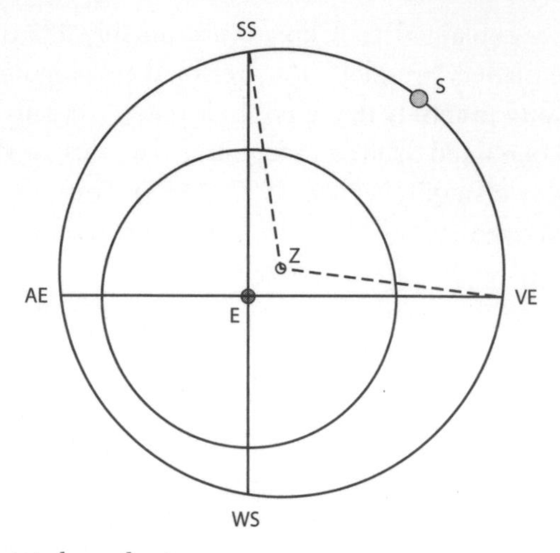

Figure 1. Eccentric hypothesis.

which is carried by the first circle, and hence is known as the epicycle" (Toomer 1984, 141). Hence, the case of the anomaly of the sun confronts us with the remarkable phenomenon of a *competition of hypotheses*. Both hypotheses allow the derivation of consequences which agree with the astronomical phenomena. Since there was no further reason in favor of either one, it didn't matter which one was applied. Ptolemaios showed that, given an adequate choice of parameters, both hypotheses are mathematically equivalent and lead to the same data for the orbit of the sun. Of course, physically they are quite different; nevertheless, Ptolemaios did not take the side of one or the other.

Hence the following situation: The Greeks believed that the heavenly bodies moved with constant velocity on circles around the earth. These two assumptions (constancy of velocity and circularity of path) were so fundamental that the Greeks were by no means ready to give them up. The retrograde movement of the planets and the anomaly of the sun seemed to contradict these convictions. Consequently, Greek astronomers had to invent additional hypotheses which brought the theory into accordance with the phenomena observed. The Greeks called the task of inventing such hypotheses "saving the phenomena" ("σῴζειν τά φαινόμενα").

The history of this phrase is interesting and reflects Greek ideas about how to bring theoretical thinking in agreement with observed phenomena (see Lloyd 1991; Mittelstrass 1962). In written sources the term

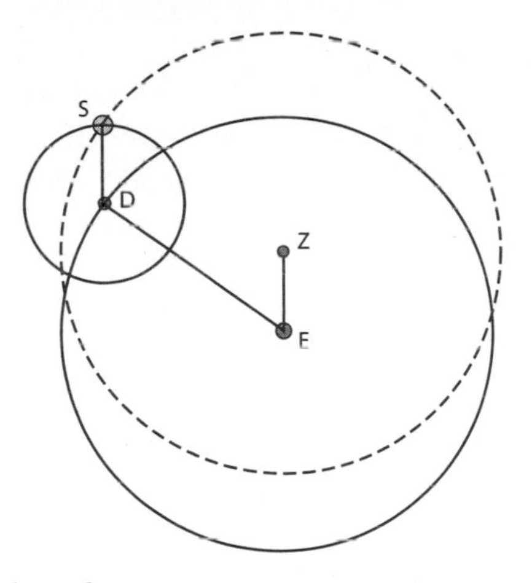

Figure 2. Epicyclic hypothesis.

"saving the phenomena" first appears in the writings of Simplikios, a Neo-Platonist commentator of the sixth century AD, a rather late source. However, the phrase probably goes back to the time of Plato. Simplikios wrote that Plato made "the saving of phenomena" a task for the astronomers. But we have seen that Plato hit upon this problem only late in his life; it is much more probable that he learned about it from the astronomers (e.g., Eudoxos) than vice versa. It seems likely that the phrase had been a terminus technicus among astronomers since the fourth century BC.

A number of philosophers of science, the most prominent being Pierre Duhem (1908/1994), have defended the thesis that the Greeks held a purely conventionalist view and did not attribute any claim of truth to astronomical hypotheses. They counted different hypotheses, like the excentric or epicyclic hypotheses as equally acceptable if the consequences derived from them agreed with the observed phenomena. However, the Greeks in fact never questioned certain astronomical assumptions, namely the circularity of the paths and the constancy of velocities of the heavenly bodies, attributing to them (absolute) truth.

Mittelstrass (1962), giving a detailed analysis of its history, shows that the phrase "saving the phenomena" was a terminus technicus in ancient astronomy and stresses that it was used only by astronomers and in the context of astronomy (Mittelstrass 1962, 140 ff). He questions Simplikios' statement that Plato posed the problem of "saving the phenomena" and contradicts modern philosophers, such as Natorp (1921,

p. 161, 382, 383), who have claimed that the idea of "saving the phenomena" was essential to the ancient Greek philosophy of science. According to Mittelstrass, only Galileo first transferred this principle to other disciplines and made it the basis of a general scientific methodology, which by the end of the nineteenth century was named the "hypothetico-deductive method."

Mittelstrass is surely right in denying that "saving the phenomena" was a general principle of Greek scientific thinking. He can also prove that the phrase was used explicitly only in astronomy. Greek scientific and philosophic thinking was a mixture of different ideas and approaches; there was no unified "scientific method." Nevertheless, Mittelstrass goes too far in strictly limiting to astronomy the idea that a hypothesis is evaluated through the adequacy of its consequences. As Szabó (1960) has shown, such trial was common practice in Greek dialectics and was reflected in early meanings of the terms aitema, axioma, and hypothesis, meanings that the terms kept until the times of Plato and Aristotle. The procedure of supposing a hypothesis as given and investigating whether its consequences are desired abounds in Plato's dialogs. Thus, the idea underlying the phrase "saving the phenomena" had a broader presence in Greek scientific and philosophical thinking than Mittelstrass supposes. Besides, Mittelstrass did not take into account Szabó's (1960) study, though it had already been published.

Hence, I formulate the following thesis: The extension of the axiomatic procedure from geometry to physics and other disciplines cannot be imagined without the idea that an axiom is a hypothesis which may be justified not by direct intuition but by the adequacy of its consequences, in line with the original dialectical meaning of the terms aitema, axioma, and hypothesis.

The Greeks set up a range of possible hypotheses in geometry and physics with a variety of epistemological justifications. For example, Euclid's geometrical postulates were considered from antiquity up to the nineteenth century as evident in themselves and absolutely true. Only the parallel postulate couldn't claim a similar epistemological status of direct evidence; this was already seen in antiquity. A possible response to this lack would have been to give up the epistemological claim that the axioms of geometry are evident in themselves, but the Greeks didn't do that. Another way out would have been to deny the parallel postulate the status of an axiom. The Neo-Platonist commentator Proclus did exactly that, declaring the parallel postulate a theorem whose proof had not yet been found. However, this tactic was motivated by philosophical not mathematical reasons, though the problem was a mathematical one. Besides, Proclus lived 700 years after Euclid; we do not even know how Euclid himself thought about his parallel postulate. Perhaps Euclid as a mathematician was more down-to-earth and was less concerned about his postulate's non-evidency.

A second example concerns statics. In the beginning of "On the equilibrium of planes or the centers of gravity of planes," Archimedes set up seven postulates; the first reads as

> I postulate the following: 1. Equal weights at equal distances are in equilibrium, and equal weights at unequal distances are not in equilibrium but incline towards the weight which is at the greater distance. (Heath, 1953, 189)

This postulate shows a high degree of simplicity and evidence and, in this regard, is like Euclid's geometric postulates. The other postulates on which Archimedes based his statics are similar; they appear as unquestionable. Statics therefore seemed to have the same epistemological status as geometry and from early times up to the nineteenth century was considered a part of mathematics. However, during the nineteenth century statics became definitively classified as a subdiscipline of physics. Meanwhile, the view that the natural sciences are founded without exception on experiment became dominant. Hence, arose a problem: Statics had the appearance of a science that made statements about empirical reality, but was founded on propositions apparently true without empirical evidence. Only at the end of the nineteenth century did E. Mach (1976) expose, in an astute philosophical analysis, the (hidden) empirical assumptions in Archimedes' statics, thus clarifying that statics is an empirical, experimental science like any other.

As a final example, consider the (only) hypothesis which Archimedes stated at the beginning of his hydrostatics ("On floating bodies"):

> Let it be supposed that a fluid is of such a character that, its parts lying evenly and being continuous, that part which is thrust the less is driven along by that which is thrust the more; and that each of its parts is thrust by the fluid which is above it in a perpendicular direction if the fluid be sunk in anything and compressed by anything else. (Heath, 1953, 253)

Archimedes derived from this hypothesis his famous "law of upthrust" ("principle of buoyancy") and developed a mathematically sophisticated theory about the balance of swimming bodies. Obviously, the hypothesis does not appear simple or beyond doubt. To a historically open-minded reader, it looks like a typical assumption set up in a modern situation of developing a mathematical model for some specific aim—in other words, a typical hypothesis whose truth cannot be directly judged. It is accepted as true insofar as the consequences that can be derived from it are desirable and supported by empirical evidence. No known source has discussed the epistemological status of the axiom and its justification in this way. Rather, Archimedes' hydrostatics was considered as a theory that as a whole made sense and agreed with (technical) experience.

Considered in their entirety, the axiomatic-deductive theories that the Greeks set up during the third century BC clearly rest on hypotheses that vary greatly in regard to the justification of their respective claims of being true. Some of these hypotheses seem so intuitively safe that a "healthy mind" cannot doubt them; others have been accepted as true because the theory founded on them made sense and agreed with experience.

In sum, ancient Greek thinking had two ways of justifying a hypothesis. First, an axiom or a hypothesis might be accepted as true because it agrees with intuition. Second, hypotheses inaccessible to direct intuition and untestable by direct inspection were justified by drawing consequences from them and comparing these with the data to see whether the consequences were desired; that is, they agreed with experience or with other statements taken for granted. Desired consequences led to strengthening the hypothesis, undesired consequences to its weakening. Mittelstrass (1962) wants to limit this second procedure to the narrow context of ancient astronomy. I follow Szabó (1960) in seeing it also inherent in the broader philosophical and scientific discourse of the Pre-Platonic and Pre-Aristotelean period.

Intrinsic and Extrinsic Justification in Mathematics

From the times of Plato and Aristotle to the nineteenth century, mathematics was considered as a body of absolute truths resting on intuitively safe foundations. Following Lakatos (1978), we may call this the *Euclidean* view of mathematics. In contrast, modern mathematics and its philosophy would consider the axioms of mathematics simply as statements on which mathematicians agree; the epistemological qualification of the axioms as true or safe is ignored. At the end of the nineteenth century, C. S. Peirce nicely expressed this view: " . . . all modern mathematicians agree . . . that mathematics deals exclusively with hypothetical states of things, and asserts no matter of fact whatever" (Peirce 1935, 191). We call this modern view the *Hypothetical* view.

Mathematical proof underwent a foundational crisis at the beginning of the twentieth century. In 1907, Bertrand Russell stated that the fundamental axioms of mathematics can only be justified not by an absolute intuition but by the insight that one can derive the desired consequences from them (Russell 1924; see Mancosu 2001, 104). In discussing his own realistic (or in his words "Platonistic") view of the nature of mathematical objects, Gödel (1944) supported this view:

> The analogy between mathematics and a natural science is enlarged upon by Russell also in another respect . . . He compares the axioms of logic and mathematics with the laws of nature and logical evidence with sense perception, so that the axioms need not necessarily be evident in themselves, but rather their justification lies (exactly as in physics) in the fact that they make it possible for these "sense perceptions" to be deduced: . . . I think that . . . this view has been largely justified by subsequent developments. (Gödel 1944, 210)

On the basis of Gödel's "Platonistic" (realistic) philosophy, the American philosopher Penelope Maddy has designated justification of an axiom by direct intuition as "intrinsic," and justification by reference to plausible or desired consequences as "extrinsic" (Maddy 1990). According to Maddy, Gödel posits a faculty of mathematical intuition that plays a role in mathematics analogous to that of sense perception in the physical sciences:

> . . . presumably the axioms [of set theory: Au] force themselves upon us as explanations of the intuitive data much as the assumption of medium-sized physical objects forces itself upon us as an explanation of our sensory experiences. (Maddy, 1990, 31)

For Gödel the assumption of sets is as legitimate as the assumption of physical bodies, Maddy argues. Gödel posited an analogy of intuition with perception and of mathematical realism with common-sense realism. If a statement is justified by referring to intuition Maddy calls the justification *intrinsic*. But this is not the whole story. As Maddy puts it:

> Just as there are facts about physical objects that aren't perceivable, there are facts about mathematical objects that aren't intuitable. In both cases, our belief in such 'unobservable' facts is justified by their role in our theory, by their explanatory power, their predictive success, their fruitful interconnections with other well-confirmed theories, and so on. (Maddy, 1990, 32)

In other words, in mathematics as in physics, one can justify some axioms by direct intuition (intrinsic), but others only by referring to their consequences. The acceptance of the latter axioms depends on evaluating their fruitfulness, predictive success, and explanatory power. Maddy calls this type of justification *extrinsic justification*.

Maddy enlarged on the distinction between intrinsic and extrinsic justification in two ways. First, she discussed perception and intuition (1990, 36–81), trying to sketch a cognitive theory that explains how human beings arrive at the basic intuitions of set theory. There she attempted to give concrete substance to Gödel's rather abstract arguments. Second, she elaborated on the interplay of intrinsic and extrinsic justifications in modern developments of set theory (1990, 107–150). Mathematical topics treated are measurable sets, Borel sets, the Continuum hypothesis, the Zermelo-Fraenkel axioms, the axiom of choice, and the axiom of constructability. She found that as a rule there is a mixture of intrinsic and extrinsic arguments in favor of an axiom. Some axioms are justified almost exclusively by extrinsic reasons. This raises the question of which modifications of axioms would make a statement like the continuum hypothesis provable and what consequences such modifications would have in other parts of mathematics. Here questions of weighing advantages and disadvantages come into play; these suggest that in the last resort extrinsic justification is uppermost.

Maddy succinctly stated the overall picture which emerges from her distinction:

> . . . the higher, less intuitive, levels are justified by their consequences at lower, more intuitive, levels, just as physical unobservables are justified by their ability to systematize our experience of

observables. At its more theoretical reaches, then, Gödel's mathematical realism is analogous to scientific realism.

Thus Gödel's Platonistic epistemology is two tiered: the simpler concepts and axioms are justified intrinsically by their intuitiveness; more theoretical hypotheses are justified extrinsically, by their consequences. (Maddy 1990, 33)

In conclusion, until the end of the nineteenth century, mathematicians were convinced that mathematics rested on intuitively secure intrinsic hypotheses which determined the inner identity of mathematics. Extrinsic hypotheses could occur and were necessary only outside the narrower domain of mathematics. This view dominated by and large the philosophy of mathematics. Then, non-Euclidian geometries were discovered. The subsequent discussions about the foundations of mathematics at the beginning of the twentieth century resulted in the decisive insight that pure mathematics cannot exist without hypotheses (axioms) which can only be justified extrinsically. Developments in mechanics from Newton to the nineteenth century enforced this process (see Pulte 2005).

Today, there is a general consensus that the axioms of mathematics are not absolute truths that can be sanctioned by intuition: rather, they are propositions on which people have agreed. A formalist philosophy of mathematics would be satisfied with this statement: however, modern realistic or naturalistic philosophies go further, trying to analyze scientific practice inside and outside of mathematics in order to understand how such agreements come about.

Implications for the Teaching of Proof

As we have seen, the "Hypothetical" view of modern post-Euclidean mathematics has a high affinity with the origins of proof in pre-Euclidean Greek dialectics. In dialectics, one may suppose axioms or hypotheses without assigning them epistemological qualification as evident or true. Nevertheless, at present the teaching of proof in schools is more or less ruled by an implicit, strictly Euclidean view. When proof is mentioned in the classroom, the message is above all that proof makes a proposition safe beyond doubt. The message that mathematics is an edifice of absolute truths is implicitly enforced, because the hypotheses underlying mathematics (the axioms) are not explicitly explained as such.

Therefore, the hypothetical nature of mathematics remains hidden from most pupils.

This paper pleads for a different educational approach to proof *based on the modern Hypothetical view* while taking into account its affinity to the early beginnings in Greek dialectics and Greek theoretical science. This approach stresses the *relation between a deduction and the hypotheses* on which it rests (cf. Bartolini Bussi et al. 1997 and Bartolini Bussi 2009). It confronts pupils with situations in which they can *invent* hypotheses and *experiment* with them in order to understand a certain problem. The problems may come from within or from outside mathematics, from combinatorics, arithmetic, geometry, statics, kinematics, optics, or real life situations. Any problem can become the subject of a dialog or of a procedure in which hypotheses are formed and consequences are drawn from them. Hence, from the outset pupils see proof in the context of the hypothetico-deductive method.

There are mathematical and pedagogical reasons for this approach. The *mathematical* reasons refer to the demand that instruction should convey to the pupils an *authentic and adequate image of mathematics* and its role in human cognition. In particular, it is important that the pupils understand the differences and the connections between mathematics and the empirical sciences, because frequently proofs are motivated by the claim that one cannot trust empirical measurements. For example, students are frequently asked to measure the angles of a triangle, and they nearly always find that the sum of the angles is equal to 180°. However, they are then told that measurements are not precise and can establish that figure only in these individual cases. If they want to be sure that the sum of 180° is true for all triangles they have to prove it mathematically. However, for the students (and their teachers) that theorem is a statement about real (physical) space and used in numerous exercises. As such, the theorem is true when corroborated by measurement. Only if taking into account the fundamental role of measurement in the empirical sciences, can the teacher give an intellectually honest answer to the question of why a mathematical proof for the angle sum theorem is urgently desirable. Such an answer would stress that in the empirical sciences proofs do not replace measurements but are a means for building a network of statements (laws) and measurements.

The *pedagogical* reasons are derived from the consideration that the teaching of proof should explicitly address two questions: (1) What is a proof? (2) Where do the axioms of mathematics come from?

Question (1) is not easy and cannot be answered in one or two sentences. I shall sketch a genetic approach to proof which aims at explicitly answering this question (see Jahnke 2005, 2007). The overall frame of this approach is the notion of the hypothetico-deductive method which is basic for all sciences: by way of a deduction, pupils derive consequences from a theory and check these against the facts. The approach consists of three phases, a first phase of *informal thought experiments* (Grade 1+); a second phase of *hypothetico-deductive thinking* (Grade 7+); and a third phase of *autonomous mathematical theories* (upper high school and university). Students of the third phase would work with closed theories and only then would "proof" mean what an educated mathematician would understand by "proof."

The first phase would be characterized by informal argumentations and would comprise what has been called "preformal proofs" (Kirsch 1979), "inhaltlichanschauliche Beweise" (Wittmann and Müller 1988) and "proofs that explain" in contrast to proofs that only prove (Hanna 1989). These ideas are well-implemented in primary and lower secondary teaching in English-speaking countries as well as in Germany.

In the second phase the instruction should make the concept of proof an explicit theme—a major difficulty and the main reason why teachers and textbook authors mostly prefer to leave the notion of proof implicit. There is no easy definition of the very term "proof" because this concept is dependent on the concept of a theory. If one speaks about proof, one has to speak about theories, and most teachers are reluctant to speak with seventh-graders about what a theory is.

The idea in the second phase is to build local theories; that is small networks of theorems. This corresponds to Freudenthal's notion of "local organization" (Freudenthal 1973, p. 458) but with a decisive modification. The idea of measuring should not be dispersed into general talk about intuition; rather we should build small networks of theorems based on empirical evidence. The networks should be manageable for the pupils, and the deductions and measurements should be organically integrated. The "small theories" comprise hypotheses which the students take for granted and deductions from these hypotheses.

For example, consider a teaching unit about the angle sum in triangles exemplifying the idea of a network combining deductions and measurements (for details, see Jahnke 2007). In this unit the alternate angle theorem is introduced as a hypothesis suggested by measurements. Then a series of consequences about the angle sums in polygons is derived from this hypothesis. Because these consequences agree with further measurements the hypothesis is strengthened. Pupils learn that a proof of the angle sum theorem makes this theorem not absolutely safe, but dependent on a hypothesis. Because we draw a lot of further consequences from this hypothesis which also can be checked by measurements, the security of the angle sum theorem is considerably enhanced by the proof.

Hence, the answer to question (1) consists in showing to the pupils by way of concrete examples the relation between hypotheses and deductions; exactly this interplay is meant by proof.

Question (2) is answered at the same time. The students will meet a large variety of hypotheses with different degrees of intuitiveness, plausibility, and acceptability. They will meet basic statements in arithmetic which in fact cannot be doubted. They will set up by themselves ad hoc-hypotheses which might explain a certain situation. They will also hit upon hypotheses which are confirmed by the fact that the consequences agree with the phenomena. This basic approach is common to all sciences be they physics, sociology, linguistics, or mathematics. We have seen above that the Greeks already had this idea and called it "saving the phenomena." The students' experience with it will lead them to a realistic image of how people have set up axioms which organize the different fields of mathematics and science. These axioms are neither given by a higher being nor expressions of eternal ideas; they are simply man made.

Note

1. For a discussion of the personal and scientific relations between Szabó and Lakatos, see Maté (2006). I would like to thank Brendan Larvor for drawing my attention to this paper.

References

Aristotle. (1966). *Posterior Analytics*. English edition by Hugh Tredennick. London/Cambridge: William Heinemann/Harvard University Press.

Ayer, A. J., and O'Grady, J. (1992). *A dictionary of philosophical quotations*. Oxford: Blackwell Publishers.

Bartolini Bussi, M. G. (2009). Experimental mathematics and the teaching and learning of proof. In V. Durand-Guerrier, S. Soury-Lavergne, and F. Arzarello (Eds.) *CERME 6 Proceedings*, Jan. 28–Feb. 1, 2009, Lyon, France, pp. 221–30.

Bartolini Bussi, M. G., Boero, P., Mariotti, M. A., Ferri, F., and Garuti, R. (1997). Approaching geometry theorems in contexts: from history and epistemology to cognition. In *Proceedings of the 21st conference of the international group for the psychology of mathematics education (Lahti)* (Vol. 1, pp. 180–95).

Duhem, P. (1994). *Sozein ta Phainomena. Essai sur la notion de théorie physique de Platon à Galilée.* Paris: Vrin (Original publication 1908).

Freudenthal, H. (1973). *Mathematics as an educational task.* Dordrecht: Reidel.

Gödel, K. (1944). Russell's Mathematical Logic. In P. A. Schilpp (Hrsg.) *The Philosophy of Bertrand Russell* (pp. 125–53). New York: Tudor. Quoted according to: P. Benacerraf, and H. Putnam (Eds.) *Philosophy of mathematics*, Englewood Cliffs 1964, Prentice-Hall.

Hanna, G. (1989). Proofs that prove and proofs that explain. In *Proceedings of the thirteenth conference of the international group for the psychology of mathematics education* (Vol. II, pp. 45–51). Paris: PME.

Heath, T. L. (Ed. and Transl.) (1953). *The works of Archimedes.* New York: Dover.

Heath, T. L. (Ed. and Transl.) (1956). *The thirteen books of Euclid's elements.* Second edition revised with additions. New York: Dover.

Jahnke, H. N. (2005). A genetic approach to proof. In M. Bosch (Ed.) *Proceedings of the Fourth Congress of the European Society for Research in Mathematics Education, Sant Feliu de Guíxols 2005*, 428–37.

Jahnke, H. N. (2007). Proofs and Hypotheses. *Zentralblatt für Didaktik der Mathematik (ZDM), 39*(2007), 1–2, 79–86.

Kirsch, A. (1979). Beispiele für prämathematische Beweise. In W. Dörfler, and R. Fischer (Eds.) *Beweisen im Mathematikunterricht* (pp. 261–74). Klagenfurt: Hölder-Pichler-Temspsky.

Lakatos, I. (1978). A renaissance of empiricism in the recent philosophy of mathematics. In I. Lakatos (Ed.) *Philosophical papers* (Vol. 2, pp. 24–42). Cambridge: Cambridge University Press.

Lloyd, G.E.R. (1991). Saving the appearances. In G.E.R. Lloyd (Ed.), *Methods and problems in Greek science* (pp. 248–77). Cambridge: Cambridge University Press.

Mach, E. (1976). *Die Mechanik. Historisch-kritisch dargestellt* (Unveränderter Nachdruck der 9. Auflage, Leipzig 1933, 1. Auflage 1883). Darmstadt: Wissenschaftliche Buchgesellschaft.

Maddy, P. (1990). *Realism in mathematics.* Oxford: Clarendon Press.

Mancosu, P. (2001). Mathematical Explanation: Problems and Prospects. *Topoi, 20*, 97–117.

Máté, A. (2006). Árpád Szabó and Imre Lakatos, or the relation between history and philosophy of mathematics. *Perspectives on Science, 14*(3), 282–301.

Mittelstrass, J. (1962). *Die Rettung der Phänomene. Ursprung und Geschichte eines antiken Forschungsprinzips.* Berlin: Walter de Gruyter & Co.

Natorp, P. (1921). *Platos Ideenlehre. Eine Einführung in den Idealismus*, 2nd ed. Leipzig: Meiner. 1921.

Peirce, C. S. (1935). The essence of mathematics. In C. Hartshorne, P. Weiss (Eds.) *Collected Papers of Charles Sanders Peirce* (Vol. III, pp. 189–204). Cambridge: Harvard University Press.

Plato. (1997). *Plato: Complete works.* J. M. Cooper (Ed.) Cambridge/Indianapolis, Ind.: Hackett Publishing Company.

Proclus (1970). *A Commentary on the First Book of Euclid's Elements.* Translated, with Introduction and Notes, by Glenn R. Morrow. Princeton, N.J.: Princeton University Press.

Pulte, H. (2005). *Axiomatik und Empirie. Eine wissenschaftstheoriegeschichtliche Untersuchung zur Mathematischen Naturphilosohie von Newton bis Neumann.* Darmstadt: Wissenschaftliche Buchgesellschaft.

Russell, B. (1924). Logical Atomism. In J. M. Muirhead (Ed.), *Contemporary British Philosophy, first series* (pp. 357–83). London: George Allen & Unwin.

Szabó, Á. (1960). *Anfänge des Euklidischen Axiomensystems*. Archive for History of Exact Sciences 1, 38–106. Page numbers refer to the reprint in O. Becker (Ed.) (1965) Zur Geschichte der griechischen Mathematik (pp. 355–461). Darmstadt: Wissenschaftliche Buchgesellschaft.

Toomer, G. J. (Ed. and Transl.) (1984). *Ptolemy's Almagest*. London: Duckworth.

van der Waerden, B. L. (1988). *Science awakening I, Paperback edition*. Dordrecht: Kluwer.

Wittmann, E. C., and Müller, G. (1988). Wann ist ein Beweis ein Beweis. In P. Bender (Ed.) *Mathematikdidaktik: Theorie und Praxis. Festschrift für Heinrich Winter* (pp. 237–57). Berlin: Cornelsen.

What Makes Mathematics Mathematics?

Ian Hacking

We have seldom paused to ask what counts as mathematics. Certainly we have thought a good deal about the nature of mathematics—in my case, for example, about the logicist claim that mathematics is logic.[1] But we took "mathematics" for granted, and seldom reflected on why we so readily recognize a conjecture, a fact, a proof idea, a piece of reasoning, or a sub-discipline, as mathematical. We asked sophisticated questions about which parts of mathematics are constructive, or about set theory. But we shied away from the naïve question of why so many diverse topics addressed by real-life mathematicians are immediately recognized as "mathematics."

> **Mathematics.** Originally, the collective name for geometry, arithmetic, and certain physical sciences (as astronomy and optics) involving geometrical reasoning. In modern use applied, (a) in a strict sense, to the abstract science which investigates deductively the conclusions implicit in the elementary conceptions of spatial and numerical relations, and which includes as its main divisions geometry, arithmetic, and algebra; and (b) in a wider sense, so as to include those branches of physical or other research which consist in the application of this abstract science to concrete data. When the word is used in its wider sense, the abstract science is distinguished as *pure mathematics*, and its concrete applications (e.g. in astronomy, various branches of physics, the theory of probabilities) as *applied* or *mixed mathematics*.

The *Oxford English Dictionary (OED)* definition is an excellent one. Why not stop right here, and answer our question by quoting the dictionary? Because the kinds of things we call mathematics are, in a word, so curiously miscellaneous. The dictionary already hints at that, in part by its implication that the concept of mathematics itself has a history, with the name applying in different ways to different categories over the course of time.

A Mathematician's Miscellany[2]

The arithmetic that all of us learned when we were children is very different from the proof of Pythagoras' theorem that many of us learned as adolescents. When we began to read Plato, we saw in the *Meno* how to construct a square double the size of a given square, and realized that the argument is connected to "Pythagoras." But that is totally unlike the rote skill of doubling a small integer at sight, or a large one by pencil.

Both types of examples are unlike the idea that Fermat had when he wrote down what came to be called his last theorem. We nevertheless seem immediately to understand his question about the integers. The situation is very different from the proof ideas that lie behind Andrew Wiles' discovery of a way to prove the theorem. Few of us have mastered even a sketch of that argument. Is it "the same sort of thing" as the proof that there is no greatest prime? I am not at all sure.

The mathematics of theoretical physics will seem a different type of thing again, but we should not restrict ourselves to theory. Papers in experimental physics are rich in mathematical reasoning. Take a recent very successful field, very cold atoms, virtually at absolute zero (*cf.* Hacking 2006). There are two fundamentally different types of entity, bosons and fermions. All elements but one have isotopes that are bosons. Ions of any chosen species of boson go into the same ground state when they lose enough energy. Then they all have the same wave function, which leads some experimenters to speak of macroscopic wave functions, which would once have seemed to be a contradiction in terms. This all started in 1924 when Einstein read a letter about photons from S. N. Bose, and saw from the equations that something very strange should happen near 0° K: there would be a new kind of matter. Only in 1995 was it possible to get a few thousand trapped ions of rubidium-47 cold enough to create what we now call Bose-Einstein condensate.

Although the experimental results in this field are brand new, the mathematics that Einstein took for granted was mostly old-fashioned, taken from what has been compared to a physicist's toolkit (Krieger 1987, 1992). Much of it had been around for more than a century when Einstein made use of it. The mathematics in the toolbox—and the way it is used—is very different from that of the geometer or number theorist.

The mathematical part of the physicist's toolbox is mostly old, but something entirely new has been added. We have powerful computational techniques to make approximate solutions to complex equations that cannot be solved exactly. They enable practitioners to construct simulations that establish intimate relations between theory and experiment. Today, most experimental work in physics is run alongside simulations. Is the simulation of nature by powerful computers (applied) mathematics, in the same way that modeling nature using Lagrangians or Hamiltonians is called applied mathematics?

Modern condensed matter physics, of which the theory and practice of cold atoms is a part, employs sophisticated mathematical models of physical situations. Economists also construct complicated models. They run computer simulations of gigantic structures they call "the economy" to try to guess what will happen next. The economists are as incapable of understanding the reasoning of the physicist as most physicists are of making sense of modern econometrics. Are they both using mathematics?

We are not really sure whether to say that programmers writing hundreds of meters of code are doing mathematics or not. We need the programmers to design the programs on which we solve, by simulation and approximation, the problems in physics or economics. What part is mathematics and what part not? We do not dignify as mathematics the solving of chess problems, white to mate in three. Few people will call programming a computer to play chess an instance of mathematics. Arithmetic for carpentry or commerce seems very different from the theory of numbers. What, then, makes mathematics mathematics?

Only Wittgenstein Seems to Have Been Troubled

It is curious that Wittgenstein seems to have been the first notable philosopher ever to emphasize the differences between the miscellaneous activities that we file away as mathematics. "I should like to say," he wrote in his *Remarks on the Foundations of Mathematics*, that "mathematics is a MOTLEY of techniques of proof" (Wittgenstein 1978: III § 46, 176). When he repeated this idea, saying that he wanted "to give an account of the motley of mathematics," he went out of his way to emphasize it. Where the translators use a single capitalized word, Wittgenstein's

German has a capitalized adjective followed by an italicized noun: "ein BUNTES *Gemisch* von Beweistechniken."

In Luther's Bible, *bunt* is the word for Jacob's coat of many colors, and the word in general means parti-colored, and, by metaphor, miscellaneous. "Motley" is an apt translation of Wittgenstein's double-barreled phrase, "buntes Gemisch," for it implies a disorderly variety within a group. The German noun "Treiben" denotes bustling activity; "ein buntes Treiben" is emphatic, meaning a real hustle and bustle with all sorts of different things going on. Likewise, "ein buntes Gemisch" is not just a mixture, but rather a mixture of all sorts of different kinds of things. When the adjective is in capital letters, and the noun in italics, BUNTES *Gemisch*, Wow!

Thus the metaphor captures an aspect of grouping different from Wittgenstein's well-known family resemblances. This is not to deny that there are family resemblances between the motley examples of mathematics that I have just given; it is instead to suggest that mere family resemblance is not enough to collect them together.

There are of course disagreements about how to use Wittgenstein's notion of family resemblance, and about what he himself intended. If, as writers as diverse as Renford Bambrough (1961) and Eleanor Rosch (1973) have suggested, virtually all general terms are family resemblance terms, then the notion gives no help with the identity of mathematics in particular. I favor the "minimalist" interpretation firmly advanced by Baker and Hacker (1980: 320–43). They argue that Wittgenstein's use of the notion is critical rather than constructive, and that he used it chiefly for what they call psychological and formal concepts. The latter includes language, number, *Satz*. He also used the idea implicitly in his discussion of names. Baker and Hacker also mentioned passages where the notion of a family resemblance occurs in accounts of "the concepts of proof, mathematics, and applications of a calculus, and of the ways in which mathematics forms concepts" (Baker and Hacker 1980: 340; Hacker deleted this observation in his revised edition, 2005). None of the citations bears on the motley of mathematics, or the question of what makes mathematics mathematics. That phrase, "BUNTES *Gemisch*," was making a positive point different from Wittgenstein's remarks about family resemblance.

In an essay that Timothy Smiley invited for a British Academy symposium on *Mathematics and Philosophy*, I connected Wittgenstein's "motley of mathematics" with some of his other thoughts about mathematics

(Hacking 2000). I shall not pursue those ramifications here. I wanted only to notice that one philosopher had taken my question seriously. I shall not mention him again.

We do take for granted that the answer to the title question of what makes mathematics mathematics, will not be a set of necessary and sufficient conditions for being mathematics.

The Philosophy of Mathematics

An innocent abroad, who consulted the online *Stanford Encyclopedia of Philosophy*, and then the *Routledge Encyclopedia of Philosophy*, would conclude that there is no question about what counts as mathematics. She might be left wondering about what counts as the philosophy of mathematics itself. *Stanford* has a heading, "Philosophy of Mathematics" (Horsten 2007). *Routledge* does not. Instead it has "Mathematics, Foundations of," which covers something of the same waterfront (Detlefsen 1998). This is not about what cognoscenti would call Foundations of Mathematics, as represented say by the excellent and very active online *FOM*: "a closed, moderated, e-mail list for discussing *Foundations of Mathematics*." It is about the philosophy of mathematics.

Many of the topics discussed in the one article are discussed in the other. Yet although between them they list some 170 articles and books in their bibliographies of classic contributions, only 13 of these items occur in both lists. Only eight items cited by *Stanford* were published after *Routledge* appeared, so that does not explain the discrepancy.

A complete accounting of all related entries in the two encyclopedias generates a little more overlap, but the initial contrast is striking. The prudent innocent will judge that the philosophy of mathematics covers a lot of topics, and that different philosophers have different opinions of what is most central or interesting. Perhaps the discrepancy between the two encyclopedias is due in part to different ideas about what mathematics "is." Only in part, for sure, but that is food for thought.

A Different Type of Demarcation Problem: What Counts as a Proof?

The title question, "What makes mathematics mathematics?" looks like a demarcation problem. Our explorer will not find the boundaries of mathematics discussed in the two encyclopedias, but she will encounter some other boundary questions. This is because there are different opinions about what mathematical arguments count as sound proofs.

Only some mathematics is constructive; for brevity I group intuitionist criticism as constructivist, although the motivations are not identical. Constructivists of various stripes can tell a classically sound proof when they see one. They seem not to deny that classical arguments are the product of mathematical insight, or that they are produced by mathematicians. Perhaps there have been iconoclasts, somewhere, sometime, arrogant enough to say, "That is not mathematics *at all*," but that does not seem to be a common reaction. We may call the constructivist critiques of classical proofs *retroactive*, because they apply to proofs, inter alia, that had been on the books long before the criticisms were made. It does not seem that these retroactive critiques address the larger question of what makes mathematics mathematics.

There are also what I shall call *conservative* criticisms of *new* mathematical techniques. One is described in the final paragraphs of the Stanford essay, which in turn refers to its first citation, to Appel, Haken, and Koch (1977). That was a first publication of the proof of the conjecture that any map can be colored using only four colors. The proof relies on a computer to check that some 1,936 graphs have a certain property, and relies on dozens of pages of written argument to show that any map falls into one of these 1,936 categories. "The proof of the four-color theorem gave rise to a debate about the question to what extent computer-assisted proofs count as proofs in the true sense of the word" (Horsten 2007: end of third-to-last paragraph). The need for computers focused a general debate. Topologists, struggling with the map problem itself, seem to have been more vexed by the length and imperspicuity of the hand-checking that was also needed. The problem is not about using a computer. Many topologists now use computers to generate counterexamples and eliminate red herrings. Conservatives insist only that in the end, no lemma in a proof should call for a computer confirmation.

Perhaps some conservatives say that computer-assisted proofs that never mature into perspicuous proofs are not mathematics. The issue is, nevertheless, not about the boundaries of mathematics, but about the boundaries of admissible proofs. Indeed we count mistaken proofs as mathematics. Thus Sir Alfred Kempe's 1879 and P. G. Tait's 1880 non-proofs of the four-color conjecture stood for a decade until flaws were found. Even today they are on any common understanding, "mathematics." There are live current questions, about what makes a proof a proof, but they are *prima facie* distinct from my question, of what makes mathematics mathematics.

Three Kinds of Answer

There are three inviting answers to my question. They represent different attitudes, perhaps three different casts of mind:

(1) Mathematics has a peculiar subject matter, which people versed in the discipline simply recognize.
(2) Mathematics is a cognitive field ultimately determined by a domain-specific faculty or faculties of the human mind. It is a task of cognitive science and of neurology to investigate the faculty(ies) or "module(s)" in question.
(3) Mathematics is constituted less by its content than by disciplinary boundaries that have emerged in the course of contingent historical practices.

These three answers are compatible. It will occur to any unsophisticated person that mathematics obviously has a peculiar subject matter, which is investigated by means of one or more mental faculties. Perhaps, as Kant thought, there is a distinct faculty for arithmetical reasoning, and another for geometrical reasoning. The disciplinary boundaries in our teaching and our professions mark our present grasp of that subject matter.

There seems to remain, in this terminus of temporary good-will, a chicken-and-egg question about (1) and (2) that tacitly ignores (3). Is the peculiar subject matter of mathematics a consequence of our mental faculties, so that in some curious sense there is no mathematics without the human brain to process it? Or are the mental faculties simply honed to accord with a human-independent body of fact? Here we get two fundamentally opposed attitudes to mathematics, both of which take present disciplinary boundaries as irrelevant.

The Two Attitudes or Viewpoints

These attitudes are better represented not by the philosophers' distinctions between "realism" and "antirealism" but by a classic debate between a mathematician and a neurologist. One author, Alain Connes, won a Fields Medal for mathematics. The other, Jean-Pierre Changeux, directed the research laboratory in Molecular Neurobiology at the Institut Pasteur. Both were colleagues at the Collège de France, and so were able to be frank and direct without the rancor that sometimes attends such discussions. Connes mentions one phenomenon that impresses him deeply:

> Here we come upon a characteristic peculiar to mathematics that is very difficult to explain. Often it's possible, though only after considerable effort, to compile a list of mathematical objects defined by very simple conditions. Intuitively one believes that the list is complete, and searches for the general proof of its exhaustiveness. New objects are frequently discovered in just this way, as a result of trying to show that the list is exhausted. Take the example of finite groups. (Connes and Changeux 1995: 19)

Mathematicians first thought that there are just six types of finite groups, a list complete by the end of the nineteenth century. During a period quite late in the twentieth century, exactly 20 more types were discovered—the sporadic groups. And that is all there are: end of the story. The last finite group to be found is called the *monster*, for such it is, with more than $(8) \times (10!)$ elements. Assuming that there is no deep underlying mistake in the proof, it feels as if this last idiotic group was just there all the time, laying in wait for us, with a monstrous grin on its face. And if human beings had not been smart enough to figure this out, the monster would still have been there, happy as a clam, indifferent to our stupidity.

The neurobiologist agrees that phenomena like this are astonishing. But they are the consequence of cognitive procedures that are formed within the human genetic envelope of possibilities. It is a contingent fact that human beings devised group theory, within which certain structures would form, ultimately based on combinatorial practices to which our minds are given.[3]

Realism and Antirealism

The *Routledge Encyclopedia* addresses closely related issues in its entries for "Antirealism in the philosophy of mathematics" (Moore 1998), and for "Realism in the philosophy of mathematics" (Blanchette 1998). The difference between these two philosophies is represented in terms of a question: Are there mathematical truths that we could never know? Realists are defined as those who answer "Yes," and antirealists as those who answer "No."

Among philosophers, the classic realist stance is platonism. I use a lowercase "p," because the name, which was invented by Bernays in 1936, derives more from folk-knowledge of Plato than from historical texts. Given these two limited options, Plato would prefer realist platonism to antirealism. Connes is a platonist, but not a Platonist.

Among British philosophers, the antirealist analysis due to Michael Dummett has become classic, although he is not even mentioned in the Stanford article. The "antirealist" neurologist finds Dummett's version of antirealism uninteresting and perhaps unintelligible. Even to most cognitive scientists it is just so much conversation about language. It does not, they think, get to the heart of the matter, namely the fact that mathematics is a by-product of our brain and its "genetic envelope," namely the field of possible structures that it can construct. To which the platonist makes the obvious retort: "Exactly so, 'structures'! Structures that we reconstruct, perhaps, but the form of the structure was there for us to discover."

A third quiet voice might be heard here, that of the under-represented "applied" mathematician. These structures were mostly discovered when investigating nature. They are not merely useful representations of nature, says an heir to Galileo: they are found in nature.

At any rate, even though (1) and (2) appear to be compatible, they stand for very different ways of thinking about mathematics. Neither has much truck with (3), the idea that what counts as mathematics is the product of a contingent history of human endeavors and the emergence of disciplinary boundaries. The three answers betoken different interests and, in the case of (3), a research program that befits the new discipline of Science Studies. I shall connect this third perspective with the first two, not because I profess Science Studies, but because it widens our understanding of the title question. I shall do this by a route that is anathema to most sociologists of science, for I proceed through Immanuel Kant.

Kant

An unexpected paragraph comes right at the start of Detlefsen's survey in the *Routledge Encyclopedia*. It follows his assertion that Greek and medieval thinkers "continue to influence foundational thinking to the present day":

> During the nineteenth and twentieth centuries, however, the most influential ideas [in the philosophy of mathematics] have been those of Kant. In one way or another and to a greater or lesser extent, the main currents of foundational thinking during this period—the most active and fertile period in the entire history of the subject—are nearly all attempts to reconcile Kant's foundational ideas with various later developments in mathematics and logic. (Detlefsen 1998: 181)

Kant does not loom so large in most other introductions to the subject. He is not even mentioned in Horsten's "Philosophy of Mathematics" in the *Stanford Encyclopedia*.

I accept that there is something absolutely right in Detlefsen's stage-setting. For among Kant's innumerable legacies was the conviction that there is a specific body of knowledge, mathematics, of striking importance to any metaphysics and epistemology. In Bertrand Russell's words of 1912: "The question which Kant put at the beginning of his philosophy, namely 'How is pure mathematics possible?' is an interesting and difficult one, to which every philosophy which is not purely sceptical must find some answer" (Russell 1912: 85). We know what troubled him. "This apparent power of anticipating facts about things of which we have no experience is certainly surprising."

Detlefsen singles out two other problems, namely mathematics' "richness of content and its necessity." These are among the mathematical phenomena that have made mathematics loom so large in the work of some, but only some, figures in the canon of Western philosophy. In Hacking (2000) I emphasized the alleged a priori character of mathematical knowledge, and the alleged necessity of mathematical truths. Richness of content should, of course, be added. Alain Connes's reaction to the finite group called "the monster" is a fine example. In an axiomatic account of mathematics, we start with what seem to be rather trivial assertions, of the sort that John Stuart Mill called "merely verbal," and proceed by leaps and bounds to all sorts of astounding discoveries. Richness of content beyond belief. Kant seldom mentions this, but he surely had it in mind (see e.g. Kant 1787: B 16f).

On this occasion I shall hardly touch on these three topics of richness, necessity, and the a priori character of mathematics. They are not why I pick up on Kant. Instead I focus on the clause to which we never pay attention, the conviction that *there is a specific body of knowledge*, mathematics, of striking importance. Perhaps Kant helped to lodge that proposition in our heads, so that mathematics is just a given, a domain that makes some philosophers curious. Of course mathematics has mattered to philosophers all the way back at least to Plato, but, as we shall see, Plato's own demarcation of mathematics is different from our own.

Kant's Vision of the Ur-History of Mathematics

Kant had already published the *Critique of Pure Reason* when he sat back, and reflected, that even reason has a history. That pivotal moment, between the first and second editions of the *Critique*, took place when Europe turned from the timeless reason of the Enlightenment to the historicist world that we still to some extent inhabit. In his new Introduction for the second edition, Kant betrayed a wonderful enthusiasm for a defining moment in the history of human reason (as he saw it). Kant, of all people, has become a historicist. He used such purple prose that I quote it in full:

> In the earliest times to which the history of human reason extends, *mathematics*, among that wonderful people, the Greeks, had already entered upon the sure path of science. But it must not be supposed that it was as easy for mathematics as it was for logic—in which reason has to deal with itself alone—to light upon, or rather to construct for itself, that royal road. On the contrary, I believe that it long remained, especially among the Egyptians, in the groping stage, and that the transformation must have been due to a *revolution* brought about by the happy thought of a single man, the experiment which he devised marking out the path upon which the science must enter, and by following which, secure progress throughout all time and in endless expansion is infallibly secured. The history of this intellectual revolution—far more important than the discovery of the passage round the celebrated Cape of Good Hope—and of its fortunate author, has not been preserved.

> But the fact that Diogenes Laertius, in handing down an account of these matters, names the reputed author of even the least important among the geometrical demonstrations, even of those which, for ordinary consciousness, stand in need of no such proof, does at least show that the memory of the revolution, brought about by the first glimpse of this new path, must have seemed to mathematicians of such outstanding importance as to cause it to survive the tide of oblivion. A new light flashed upon the mind of the first man (be he Thales or some other) who demonstrated the properties of the isosceles triangle. The true method, so he found, was not to inspect what he discerned either in the figure, or in the bare concept of it, and from this, as it were, to read off its properties; but to bring out what was necessarily implied in the concepts that he had himself formed *a priori*, and had put into the figure in the construction by which he presented it to himself. If he is to know anything with *a priori* certainty he must not ascribe to the figure anything save what necessarily follows from what he has himself set into it in accordance with his concept. (Kant 1787: B x–xii, 19)

We no longer countenance the hero in history, "be he Thales or some other." Even if there was a historical Thales in whose head the legendary penny dropped, there had to be uptake, there had to be people to talk to, to correspond with, to turn a transient thought into knowledge that endured. And thanks to the labors of generations of scholars, we can no longer dismiss the Egyptians and the peoples of Mesopotamia as in "the groping stage."

We can now turn Kant's prose into something closer to the historical facts, thanks to Reviel Netz (1999). He would prefer Eudoxus to Kant's Thales, but the important point is that there was a moment of radical change in the human mastery of mathematics. Kant got that right, by present lights. Using the metaphor recently favored in paleontology, Netz suggests "that the early history of Greek mathematics was catastrophic"—a sudden change in the very "feel" of mathematical thinking. In a lower key: "A relatively large number of interesting results would have been discovered practically simultaneously" (Netz 1999: 273). Netz suggests a period of at most eighty years. We have no need to dismiss the Babylonians, Egyptians, and others who taught mathematics to the Greeks, in order to see that at the time of "Thales or some other" a revolution in reason was wrought.

What Was So Revolutionary?

In the modern spirit of iconoclasm called post-modern, post-colonial, and so forth, the Greeks are no longer "that wonderful people." Their elevation to a central role in Western history, as the only begetters of all things wise and beautiful, was (it is now said) an act of European imperialism. It was all the more important in German thought in the epoch before the German-speaking lands had become sufficiently united to exert their power outside of central Europe. Let us not argue the point.

Greeks may have been glorified in the era of European triumphalism, but there really was something revolutionary, "catastrophic," that happened in those eighty years of Mediterranean history. Some Greeks, including Ionians, discovered new mathematical facts and structures, but that, from Kant's point of view, was not what counted most. The revolutionary discovery that made it all possible was proof. Greeks uncovered what may well be an innate capacity of all human beings, the ability to make demonstrative proofs.

A New Metaphor: Crystallization

Kant was absolutely right. The discovery of proof was revolutionary. But because the idea of scientific revolution has been so over-worked since the days of Thomas Kuhn, it is wise to choose another metaphor, saying that a *crystallization* of mathematics occurred at the time of Eudoxus ("or some other").[4]

This metaphor is intended to capture several aspects of what happened. First, the new method of demonstrative proof did not lack precursors or anticipations. As in the case of literal crystallization, there was a great deal going on before the new structure appeared. Second, this event inaugurated a whole new way of doing things, stabilized within its own local and historically contingent practices, and yet capable of transfer to new civilizations, where it would be stabilized in very different social environments. Like many a crystal of purest ray serene, it could be thrust into darkness before being uncovered again in a period of reopening and rebirth.

A Style of Scientific Thinking

Demonstrative proof is a distinctive style of scientific thinking. Because the very word "style" is so evocative, the expression, "style of scientific thinking" can be used in many ways. I use it in an artificially narrow sense that I acquired from the historian of science Alistair Crombie (1994). He proposed some six distinct fundamental "styles of scientific thinking in the European tradition" that emerged and matured at distinct times and places. Brief tags for his six styles are: mathematical, hypothetical modeling, experimental exploration, taxonomic, statistical, and genetic. I believe he taught an important insight, but I shall not argue the case here. I mention it because my argument below is part of a larger analysis of the history of scientific reason. It can be seen as a continuation of Kant's original "historicist" thought stated in the long paragraph quoted above.

Indeed in the very next paragraph after that, Kant turns to natural science (*Naturwissenschaft*) which, he tells us, "was very much longer in entering upon the highway of science (*Wissenschaft*)." He takes us to the world of Galileo and of Torricelli, and speaks of a "discovery [that] can be explained as being the sudden outcome of an intellectual revolution," another crystallization (Kant 1787: B xii.). Following Crombie, I file Galileo under the style of hypothetical modeling and Torricelli under the style of experimental exploration. Then I proceed to the synthesis that Kant probably intended, what I call the laboratory style of scientific thinking. That is all a matter of footnotes to Kant's recognition of these "intellectual revolutions." They were not only intellectual; they were also a matter of new things we could do with hand and eye. They briefly changed the role of Europe in world history, and permanently changed the role of our species on our planet.

The metaphor of "crystallization" may suggest something too rigid, too mineralogical, and too fixed. Styles of scientific thinking *evolve*, do they not? Perhaps we should extend the metaphor to the science of near-life. It is often observed that viruses are equivocal: some of the time they are inanimate, but when they find a host they are alive. They become alive by exploiting molecules in the host cell to create a kind of metabolism that serves them well. They are not parasites, which are autonomous organisms, but individuals that thrive by incorporating themselves into the lives of their hosts. When alive, they evolve far more rapidly than their hosts do. When inanimate, they are said to be like crystals. I seize upon the analogy. A style of scientific thinking is like a virus, a crystallization that can evolve in a host, a community, a network of human beings.

Enough of metaphor. To return to real people, I emphasize that what Eudoxus and company did, was not only to establish some new mathematical facts, techniques, and proof ideas. They also discovered a new way to find things out, namely by reasoning and proof. This was not a mathematical discovery, but the discovery of a human capacity of which our species had, in earlier times, only glimmerings here and there. It was the discovery and then exploitation of a mental faculty or faculties, of precisely the sort that cognitive science and neurobiology is now investigating.

Netz's book is widely admired for its reconstruction of the diagrams that are notoriously missing from surviving ancient texts. Few readers attend to his subtitle, *A Study in Cognitive History*, yet that aspect is exactly what is fundamental and wholly original. It is the first detailed analysis of the cultural history of the discovery of a cognitive capacity.

Some cognitive scientists conjecture that mental modules—one or more—enable us to engage in mathematical reasoning. Netz himself argues against excessive modularity, and sides with Jerry Fodor in favor of more general processing devices (Netz 1999: 4–7). He urges never to forget Fodor's "First Law of the Nonexistence of Cognitive Science." Fodor's first law preaches humility. "The more global . . . a cognitive process is, the less anybody understands it" (Fodor 1983: 107). The cognitive processes needed for mathematical demonstration are pretty global.

Mention of cognition returns us to the second answer to our title question: (2) Mathematics is a cognitive domain ultimately determined by a domain-specific faculty or faculties of the human mind. The above discussion enriches answer (2), but is wholly compatible with the "platonist" answer (1). Now we pass to reflections that depend on a wholly fortuitous aspect of cultural history, thus directing us towards answer (3).

Why Should Greeks Have Cared?

Why did some Greek thinkers think that the newly discovered capacity for demonstrative proof was so important? (This is different from the question, of why future Europeans such as Kant and Bertrand Russell thought that it was important.) Netz (1999: 209ff), following Geoffrey Lloyd (1990), suggests an answer. City-states were organized in many ways, but Athens is of central importance. It was a democracy of citizens, all of whom were male and none of whom were slaves. It was a democracy for the few; but within those few, there was no ruler. Argument ruled. If you could make the weaker argument appear the stronger, you won.

Athenians were the most consistently argumentative bunch of self-governors of whom we have any knowledge. We read Aristotle for his logic and not for his rhetoric. Greeks read him for his rhetoric; his logic was strictly for the Academy. The trouble with arguments about how to administer the city and fight its battles is that no arguments are decisive. Or they are decisive only thanks to the skill of the orator, or the cupidity of the audience. But there was one kind of argument to which oratory seemed irrelevant. Any citizen, and indeed any young slave who was encouraged to take the time, and to think under critical guidance, could follow an argument in geometry. He could come to see for himself, perhaps with a little instruction, that an argument was sound. He could even create the argument, find it out for himself. In geometry, arguments speak for themselves to the inquisitive mind.

Cynics will say that this is a lie from the start. A "*little*" instruction? The instruction is just a kind of rhetoric, mere oratory. Look at the classic, the demonstration to be found in the *Meno*, of how to double a square. The slave boy is said to discover the technique by himself, unaided. He is coached by leading questions from Socrates.

But in addition to prompting there is something else: the extraordinary phenomenon, accessible to almost any thoughtful reader of the *Meno*, of *seeing* that the square on the diagonal is twice that of the given square. In company with the diagram, and talk about the diagram, there is a new kind of experience, of conviction based solely on the perception of a new truth. Geoffrey Lloyd remarked that this phenomenon is truly impressive to members of an argumentative society that has no recourse to a ruler, and whose final criterion is nothing more than talk and persuasion.

Plato, the Kidnapper

"So what?" asks the politician in the public arena. You can prove only recondite or useless facts. Quite aside from the uselessness of proof in political debate, it is not even useful to the architect. It is no good saying that geometrical theorems could be useful to a surveyor. (a) The surveyors already knew most of the practical facts required, for they had been acquired from empirical Egyptian mathematicians. (b) Netz reports that not a single surviving text suggests a connection between the problems and solutions of the geometers, and the practical interests of architects. Only later did mathematics become "useful." Maybe Archimedes used it in the famous problem of burning mirrors. Military mathematics came into its own only in the age of Napoleon. The great mathematicians such as Laplace solved problems connected with artillery, but they were interested not in proofs but in solutions to problems of motion.

Nobody debating military strategy or the tax on corn in the Agora was able to use geometrical proof. So why should Greeks have cared about proof? An answer may be that hardly anyone did. Netz's book is about an epistolary tradition involving a small band of mathematicians exchanging letters around the Mediterranean Sea. They cared about new discoveries and new proofs, but not about the very idea of proof. Enter Plato, kidnapper.

I take the label from Bruno Latour's brilliant critical exposition of Netz's book. His opening sentence reads:

> This is, without contest, the most important book of science studies to appear since Shapin and Schaffer's *Leviathan and the Air-Pump*. (Latour 2008: 441)

Alongside Netz, Latour is referring to Shapin and Schaffer (1985): the book subtitled *Hobbes, Boyle, and the Experimental Life*. I completely agree with Latour's judgment, but for reasons quite opposite to his. Latour sees both books as magnificent illustrations of his network theory of knowledge—which they certainly are. But I also see Shapin and Schaffer as having presented a decisive crystallization of the laboratory style of scientific thinking, and Netz as presenting a decisive crystallization of the mathematical style of scientific thinking which we call demonstrative proof. In both cases a new style of scientific thinking became established in the practices of discovery, of creating knowledge, or, to be more colloquial but more exact, in the human repertoire of *finding out*.

Latour rightly takes Netz's analysis as a compelling example of knowledge sustained by a network of creators and distributors of that knowledge. Nowhere is that better illustrated than by Archimedes, who, working out of Syracuse in Sicily, created and maintained an unparalleled body of new understanding, and yet had only a handful of disciples and correspondents around the Mediterranean. But what specially fascinates Latour is the *isolation* of this network from the rest of the ancient world, be it learned, political, or vernacular.

> To the great surprise of those who believe in the Greek Miracle, the striking feature of Greek mathematics, according to Netz, is that it was completely peripheral to the culture, even to the highly literate one. Medicine, law, rhetoric, political sciences, ethics, history, yes; mathematics, no. (Latour 2008: 445)

The Greek and Hellenistic mathematicians were a handful of specialists talking with and writing to each other around the Mediterranean basin, and no one else cared:

> —with one exception: the Plato-Aristotelian tradition. But what did this tradition (itself very small at the time) take from mathematicians? . . . Only one crucial feature: that there might exist one way to convince which is apodictic and not rhetoric or sophistic. The philosophy extracted from mathematicians was not a fully fledged practice. It was only a way to radically differentiate itself through the right manner of achieving persuasion. (Latour 2008: 445)

Latour overstates his case. The philosophical tradition took a good deal from mathematics: what about the golden mean, for example, or the profound role of proportion in ethical theory? That is irrelevant to Latour's case. He proposes that the philosophers focused on proof in order to differentiate themselves from the common herd. Thus their use of proof as above rhetoric was nothing more than a rhetorical trick.

Latour pays little heed to the ways in which the philosophers were profoundly impressed by the human capacity to prove. They were, as I like to put it, bowled over by demonstrative proofs. In consequence they vastly exaggerated the potential of proof. It is easy to argue that the ensuing theory of knowledge impeded the growth of scientific knowledge from the time of Archimedes to the time of Galileo. To defeat the lust for demonstrative proof, we needed another crystallization. Or in Kant's phrase, we needed the "intellectual revolution" that he associated with Galileo and Torricelli. That was the discovery of other human talents—not purely intellectual ones—and it led to the laboratory style of scientific thinking. The definitive history of that crystallization was *Leviathan and the Air-Pump*, the very book that Latour rightly pairs with Netz's *Shaping of Deduction*.

Let us here agree with Latour: Plato kidnapped a certain idea of proof and made it a dominant theme in Western philosophical thought. But let us not grant to Latour the idea that proof is unimportant. Let us not allow Latour to kidnap Netz, that is, to allow us to forget Netz's own fundamental concern, cognitive history (which Latour barely mentions).

On the other hand, let us extend Latour's insight. Kant codified, for the modern world, Plato's kidnapping of mathematics. He made the a priori, the apodictic, and the necessary the hallmarks of mathematics, even though they are noticeable only here and there in the motley of mathematical activity. This leads us, for a final observation about ancient times, back to Plato's own demarcation of mathematics.

Plato on the Difference between Philosophical and Practical Mathematics

An important tradition in reading Plato on mathematics derives from Jacob Klein (1968). He argued that Plato made a fundamental distinction between the theory of numbers and calculating procedures. Here is a brief summary of the idea, due to one of Klein's students:

> Plato is important in the history of mathematics largely for his role as inspirer and director of others, and perhaps to him is due the sharp distinction in ancient Greece between arithmetic (in the sense of the theory of numbers) and logistic (the technique of computation). Plato regarded logistic as appropriate for the businessman and for the man of war, who 'must learn the art of numbers or he will not know how to array his troops.' The philosopher, on the other hand, must be an arithmetician 'because he has to arise out of the sea of change and lay hold of true being'. (Boyer 1991: 86)

We need not subscribe either to the terminology or the details of the interpretation to propose that ("real") mathematics, for Plato, did not include the arithmetic we learned in school, and later applied in business transactions, or ordering supplies for the troops. A redescription, owing more to Netz than to Klein, would be that Euclid's *Elements* made a decision, to emphasize diagrammatic proofs rather than numerical examples. Hence despite the depth of some work on numbers to be found in Apollonius and Archimedes, there was no tradition of "advanced arithmetic" in antiquity, in the way in which there was "advanced geometry."[5]

Plato, then, put to one side the daily uses of arithmetic in technologically and commercially advanced societies such as those of Greece or Persia. Those uses are what Klein for his own reasons called "logistic." In my opinion we should avoid the notion that computation is for practical affairs in "the sea of change." That is the philosophical gloss of appearance and reality all over again. A primary point, closer to the experience of doing or using mathematics, is that computation is algorithmic. It proceeds by set rules. One does not understand a calculation: one checks that one has not made a slip. There is no experience of proof as in the theory of numbers or geometry.

Quite possibly there were manuals that taught how to calculate, complete with shortcuts. They would be comparable to what we sophisticates dismiss as "cookbooks," mere "how to do it" instructions that do not convey insight or understanding of "how it works." We may conjecture that that sort of text has not been preserved, partly because on Platonic and then Aristotelian authority, it did not present "science," *scientia*. Just possibly, if there had been a classical text of advanced arithmetic in antiquity, the questions of a priori knowledge, apodictic certainty, and necessity would have been posed in that context. I like to imagine that those questions might, instead, have been *exposed* as pseudo-questions, at least in the context of the theory of numbers.

These cursory remarks suggest that Plato (or his heirs) created a disciplinary boundary between mathematics, the science that every philosopher must master, and computation, the technique of commerce and the military. This bears some relationship to the recent distinction between pure and applied mathematics, but the fundamental difference is that the one involves perspicuous proof, insight, and understanding, while the other involves routine.

It is important to add that, from the perspective of twentieth-century British analytic philosophy, with its talk of "puzzlement" in philosophy, the results of calculation are not "puzzling," in the way that proof can be experienced as puzzling (*cf.* Hacking 2000: §4.3). Notice also that in Plato's vision (version Klein), there is far less of a motley of mathematics than in ours—because the routine computational side of mathematics is not (real) mathematics at all.

Pure and Mixed Mathematics

Francis Bacon was his usual prescient self when he devised that now abandoned term, "mixed mathematics," which appears at the end of the *OED* definition.[6] He was captured by the powerful image of the Tree of Knowledge. Every "branch" (as we still say) of knowledge had to have its place on a tree. He needed a branch of the main limb of mathematics on which optics and mathematical astronomy could flourish, for they were mathematics in the older sense of the term, as noticed by the *OED*. These he called mixed, which also included music, architecture, and engineering.

They were not mixed, I think, because they mixed deduction and observation. It was rather a matter of the sphere to which they applied. Mixed mathematics was not pure mathematics "applied" to nature, but an investigation of the sphere in which the ideal and the mundane were intermingled. Both the mixed and the pure were that part of natural philosophy that fell under metaphysics, viz. the study of fixed and unchanging relations.

The Enlightenment was an era of classification, where Natural History had a proud place as the science of nature observed. Hierarchical classifications, which we now conceive as branching trees, were the model for all knowledge, be it of minerals or diseases; the tree was also the model for the presentation of knowledge itself. Bacon's Tree of Knowledge was hardly new; Raymond Lull's was more graphic. We now think of tree-diagrams as one of the most efficient ways to represent certain kinds of information, but preserved tree diagrams begin to appear surprisingly late in human history (Hacking 2007). By Bacon's time, however, they flourished in genealogy, logic, and many other fields, and were cast in wonderful glass on cathedral windows. That was the past: Bacon's own Tree of Knowledge was a benchmark for the future.

The image of a branching Tree of Knowledge was to persist for centuries after Lull and Bacon. It is most notably incarnate in D'Alembert's preface to *La grande encyclopédie*. It is a prominent pull-out page of Auguste Comte's *Cours*, the massive 12-year production that continued the encyclopedic project of systematizing all knowledge and representing its growth in both historical and conceptual terms. The Tree of Knowledge that was planted so firmly in the early modern world by Francis Bacon has long been institutionalized in the structure of our universities with their departments and faculties.

Probability—Swinging from Branch to Branch

Many a new inquiry had to be forced on to the tree. Where would the Doctrine of Chances, a.k.a. the Art of Conjecturing, fit? It was by definition not about the actual world, nor about an ideal world. It was about action and conjecture; it was the successor to a non-theory of luck. There was no branch on a Tree of Knowledge on which to hang it. Probability was uneasily declared a branch of mixed mathematics, less because of its content than because of its practitioners, such as the Bernoullis, who were mathematicians par excellence. The mixed, as we shall see, morphed into the applied. Hence the residual place for the "theory of probabilities" as "mixed or applied" mathematics alongside astronomy and physics in the *OED* entry.[7]

The Tree of Knowledge became the tree of disciplines. This may have a somewhat rational underlying structure, of the sort at which Bacon or D'Alembert aimed, but it is largely the product of a series of contingent decisions. This can be nicely illustrated by the location of probability theory in various sorts of institutions around the world. It was once a paradigm of the mixed, so you would expect it to continue as applied mathematics. That is certainly not what happened in Cambridge, where the Faculty of Mathematics is divided into two primary departments. One is Applied Mathematics and Theoretical Physics, the home of Newton's Lucasian chair. The other is the Department of Pure Mathematics and Mathematical Statistics. Probability appears to have jumped from branch to branch of the Tree of Knowledge. In truth, to continue the ancient arboreal metaphor, it is an epiphyte. It can lodge and prosper anywhere in a tree of knowledge, but is not part of its organic structure at all.

Pure and Applied

There is no space, here, to adumbrate the transition of nomenclature from "mixed" mathematics to "applied" mathematics. *Perhaps* the switch was from an idea of mixing mathematics and the study of nature, to one of applying mathematics to nature. That picture may well be too anachronistic or at least too simple a vision.

Galileo's own famous image is a compelling alternative. The Book of Nature is written in the Language of Mathematics. Galileo did not apply abstract structures to nature. He found the structures in nature, and articulated their properties, thereby reading the Book of Nature itself. Husserl (1936) rightly seized upon what Galileo was doing as radically new, and said that Galileo mathematized nature. Galileo might have retorted that Husserl had things upside down: "I did not mathematize Nature, for she is already mathematical, and waiting to be read."

Galileo's contribution may have been, as Netz puts it, a footnote to Archimedes (Netz and Noel 2007: *26*). It was certainly not a footnote to Plato. Galileo had no truck with Plato's conception of mathematics as outside this world. I realize that this statement flies in the face of a received tradition established by Alexandre Koyré, according to which Galileo was permeated by Platonism. Let us compromise, and say that in the world of Galileo, mathematics had an entirely new role.

The situation looks more straightforward half a century later. Newton distinguished practical from rational mechanics. He took geometry to be a limiting case of practical mechanics, important to builders and architects. Geometry, the very possibility of which so astonished Plato, was placed alongside the practical arts, which Plato did not count as mathematics at all.

There is little reason to think that Newton, the greatest mathematician of his age, cared much about the phenomenon or experience of proof which Plato had made central to his fetishism of mathematics. To continue Latour's metaphor, slightly tongue-in-cheek, we may venture that Galileo and Newton liberated mathematics from the philosophical bonds in which kidnapper Plato had enslaved it.

Newton's rational mechanics was among other things the general theory of motion, and hence of what is constant underneath ever-changing Nature. That can be presented as conforming to Plato's imperative, to discover the reality behind appearance. Whatever it was, it was mathematics, set out in a book with an unambiguous title: *Philosophiæ Naturalis Principia Mathematica*, the mathematical principles of natural philosophy.

Pure Kant

"Pure"—*rein*—evidently plays an immense role in Kant's first *Critique*, starting with its title. The primary contrast for both the English and the German adjectives is "mixed."[8] Hence Bacon's branching of mathematics into pure and mixed. The next, moralistic sense of being free from corruption or defilement, especially of a sexual sort, comes a close second. At the start of his rewritten Introduction for the second edition of the first *Critique*, Kant emphasizes what, for him, was the primary contrast: "The Distinction Between Pure and Empirical Knowledge" (Kant 1787: B 1).

Kant's question, which Russell repeated with such enthusiasm, was "How is pure mathematics possible?" What contrasts with "pure" on this occasion? We hear "applied." Galileo and Newton did not speak of applied mathematics. Kant's opposite of pure mathematics was empirical.[9] In fact Kant asked a pair of questions, one after the other:

> How is pure mathematics possible?
> How is pure science of nature (*Naturwissenschaft*) possible? (Kant 1787: B 20)

Today we are puzzled, and some are baffled, by the idea of a pure science of nature. In his footnote Kant clearly contrasts it to "(empirical) physics."[10] Kant cites, as an example of pure science of nature, primary propositions "relating to the permanence in the quantity of matter, to inertia, to the equality of action and reaction, etc." On one reading, Kant is talking about Newtonian mechanics.

Kant embedded pure mathematics in the Transcendental Aesthetic, the launching pad for his entire theory of knowledge. Plato had made mathematics a matter of Ideas in a realm other than that of appearance. Kant made it part of transcendental idealism, and arithmetic and geometry conditions of all possible experience. This was a radical innovation, and yet a continuation of Plato's leitmotif. Kant was restoring a Platonic vision of pure mathematics as something utterly separate and absolutely fundamental to the nature of knowledge. Kant was a kidnapper too. He kidnapped mathematics from the mathematicians by insisting that some of what they did was pure.

The great mathematicians of the generation that flourished in the era of the first *Critique*, men such as Lagrange, Legendre, and Laplace, did not see things that way. They were mathematicians. In general the scientists of that era made no difference between pure and applied. It was the tidy Kant who put a category of pure mathematics up front.

That is the core truth behind Detlefsen's starting point in the *Routledge Encyclopedia*, the statement that during the nineteenth and twentieth centuries, the most influential ideas in the philosophy of mathematics have been those of Kant. The philosophy of mathematics, as many of us understand it, starts from an unquestioned assumption that there *is* pure mathematics. We then proceed with Plato's and Kant's vision as something to accept, to modify, to explain, or to reject. The philosophy of mathematics is implicitly about the philosophy of pure mathematics, with a coda, asking how some of it is so applicable to nature.

The separation of the pure from the applied could not happen on Kant's say-so alone. It also called for some highly contingent events in disciplinary organization.

Applied Mathematics

Our idea of applying pure mathematics to nature should not be read back into Kant or Newton. We do have a convenient benchmark for the distinction between pure and applied. In 1810 Joseph Gergonne (1771–1859) founded what is usually regarded as the first mathematics journal, *Annales de mathématiques pures et appliquées*. Most of the articles were contributions to geometry, Gergonne's own field of expertise. Germany followed suit in 1826, when A. L. Crelle (1780–1855) founded the *Journal für die reine und angewandte Mathematik*. The focus was quite different. Crelle published most of Niels Henrik Abel's (1802–1829) papers that transformed analysis.

The nominal distinction between pure and applied did not take hold for some time. I very much doubt that readers of Gergonne's and Crelle's journals knew which articles were pure and which were applied, if, indeed, they asked the question at all. Lagrange inventing Lagrangians, and even Hamilton inventing Hamiltonians, did not think they were applying mathematics to nature. They were investigating nature mathematically. Only after the fact do we abstract the mathematics from nature and "purify" it, and then, retroactively, speak of applying the pure mathematics to scientific problems.

Pure Mathematics

Here I should like to be not just insular but local. The analytic tradition in the philosophy of mathematics is properly traced back to Frege, but a lot of the stage-setting is grace of Whitehead and Russell. Their opus is a third benchmark. Compare their book with Newton's. Two great works are titled *Principia Mathematica*. They are entirely different in content and project.

Perhaps nobody really believed in Whitehead and Russell's great book. Possibly only the authors read all three volumes. But even the great German set-theorists set themselves up with that work as a monument, even if it turned out, to everyone's surprise, to be essentially incomplete. So it is worth the time to consider the mathematical milieu that Whitehead and Russell took for granted—and their conception of pure mathematics for which they hoped to lay the foundations.

At least in British curricula we can locate the point at which "Pure Mathematics" became a specific institutionalized discipline. In 1701, Lady Sadleir had founded several college lectureships for the teaching of Algebra at Cambridge University. In 1863, the endowment was transformed into the Sadleirian Chair of Pure Mathematics, whose first tenant was Arthur Cayley. This coincided with an important shift in the teaching of mathematics. The old Smith's Prize, founded in 1768, was the way in which a young Cambridge mathematician could establish his genius. In the old days, up until 1885, it was awarded after a stiff examination in what would now be called applied mathematics. One Wrangler who went on to tie for the Smith's Prize became the greatest British mathematician of the nineteenth century. But what we call mathematics has changed. We do not call him a mathematician but a physicist. I mean James Clerk Maxwell. Many names hallowed in the annals of physics, such as Stokes, Kelvin, Tait, Rayleigh, Larmor, J. J. Thomson, and Eddington won the Smith's Prize for mathematics.[11]

Yet after 1863, what was called mathematics at Cambridge was increasingly pure mathematics rather than Natural Philosophy. It was within this conception of mathematics that Russell came of age. Likewise it was in this milieu that G. H. Hardy became the preeminent local mathematician, whose text, *Pure Mathematics*, became a sort of official handbook of what mathematics is, or how it should be studied, taught, examined, and professed at Cambridge. Russell's vision of mathematics was not determined by Hardy's, or vice versa, but the two visions are coeval, a product of a disciplinary accident in the conception of mathematics.

Contingency, Necessity, and Neurology

I have sketched only the beginning of an argument, that what is counted as mathematics depends in part on a complex and very contingent history. I do not mean to imply that the history could have gone any way whatsoever. It was constrained by its content, and by human capacities. They no longer constrain in the same way. The advance of fast computation is changing the entire landscape of human knowledge, including that of mathematics. That is a topic for the future. Here I have been concerned with the past.

Our picture of the philosophy of mathematics is of philosophical reflection on a definite and predetermined subject matter. I suggest that the subject matter itself is much less determinate than we have imagined. This does not undercut the debate between the two attitudes I have mentioned, (1) platonic and (2) neurobiological, or the traditional philosophical debate between "realists" and "antirealists" of mathematics. It will surely go on as before. I urge only that the more difficult but perhaps more answerable question should now become: how have the platonic and neurobiological constraints jointly interacted with the contingent history of mathematics from "Thales" to now?

Notes

1. This essay appeared originally in a festschrift for T. J. Smiley of Cambridge University. The "we" of this first paragraph refers to we, his students. My own exploration of logicism is Hacking (1979), a long-delayed by-product of a research paper (Hacking 1963) that profited from Smiley's constant advice.

2. This phrase is the title of J. E. Littlewood's (1953) charming potpourri of mathematical anecdote and examples.

3. I got the phrase "genetic envelope" from a conversation with Changeux; I do not think he has used it in print.

4. The idea of a crystallization of a style of scientific thinking was introduced in Hacking (2009), and will be developed in a continuation of that book. My Crombian theme of styles of scientific thinking was launched long ago (Hacking 1982, 1992).

5. These remarks are my version of a short discussion with Netz in March 2009.

6. The assertion that the term "mixed mathematics" is original to Bacon is due to Brown (1991).

7. Two late quotations from the *OED* speak of applying, but differ in their meaning. In 1706: "*Mixt Mathematicks*, are those Arts and Sciences which treat of the Properties of Quantity, apply'd to material Beings, or sensible Objects; as Astronomy, Geography, Navigation, Dialling [sundials], Surveying, Gauging &c." In 1834, in Coleridge: "We call those [sciences] *mixed* in which certain ideas of the mind are applied to the general properties of bodies."

8. *OED*: Without foreign or extraneous admixture: free from anything not properly pertaining to it; simple, homogeneous, unmixed, unalloyed. Grimm's *Deutsches Wörterbuch*: frei von fremdartigem, das entweder auf der Oberfläche haftet oder dem Stoffe beigemischt ist, die eigenart trübend.

9. I know of only one occasion where he spoke of applied mathematics, namely in his *Lectures on Metaphysics* (delivered from the 1760s to the 1790s). "Philosophy, like mathematics as well, can be divided into two parts, namely into the *pure* and into the *applied*" (Kant 1790–1: 307).

10. Kant wrote "eigentilichen (empirischen) Physik," which Kemp Smith renders "(empirical) physics, properly so called." Like the English noun "physics," *Physik* in Kant's time still meant natural science in general. Kant might have meant something more like "real (empirical) physics."

11. Partly thanks to T. J. Smiley, by 1961 an essay in modal logic could win a Smith's Prize. Robert Smith, Plumian Professor in Astronomy, doubtless turned over in his grave. He intended the encouragement of Mathematics and Natural Philosophy. It may amuse students of today's economy to know that Smith endowed the prize with profits from the South Sea Bubble. The structure of the competition for the Smith's Prize and related prizes was reorganized in 1998.

References

Baker, G. P. and Hacker, P.M.S. (1980) *Wittgenstein: Understanding and Meaning, An Analytical Commentary of the Philosophical Investigations*, vol. 1, Oxford: Blackwell. 2nd ed., extensively revised by P.M.S. Hacker (2005).

Bambrough, R. (1961) "Universals and family resemblances," *Proceedings of the Aristotelian Society* 61: 207–22.

Blanchette, P. A. (1998) "Realism in the philosophy of mathematics" in E. Craig (Ed.) *Routledge Encyclopedia of Philosophy*, vol. 8, London and New York: Routledge, pp. 119–24.

Boyer, C. B. (1991) *A History of Mathematics*, 2nd ed., revised by U. C. Merzbach, New York: Wiley.

Brown, G. I. (1991) "The evolution of the term 'mixed mathematics,'" *Journal of the History of Ideas* 52: 81–102.

Connes, A., and Changeux, J. P. (1995) *Conversations on Mind, Matter and Mathematics*, Princeton, N.J.: Princeton University Press. Originally published as *Matière à pensée*, Paris: Odile Jacob, 1989.

Crombie, A. C. (1994) *Styles of Scientific Thinking in the European Tradition: The history of argument and explanation especially in the mathematical and biomedical sciences and arts*, 3 vols, London: Duckworth.

Detlefsen, M. (1998) "Mathematics, Foundations of" in E. Craig (Ed.) *Routledge Encyclopedia of Philosophy*, vol. 6, London and New York: Routledge, pp. 181–92.

Fodor, J. (1983) *The Modularity of Mind: An Essay on Faculty Psychology*, Cambridge, Mass.: MIT Press.

Hacking, I. (1963) "What is strict implication?," *Journal of Symbolic Logic* 28: 51–71.

———. (1979) "What is logic?," *The Journal of Philosophy* 76: 285–319.

———. (1982) "Language, truth and reason" in M. Hollis and S. Lukes (Eds.) *Rationality and Relativism*, Oxford: Blackwell, pp. 48–66.

———. (1992) "'Style' for historians and philosophers," *Studies in History and Philosophy of Science* 23: 1–20.

———. (2000) "What mathematics has done to some and only some philosophers," in T. J. Smiley (Ed.) *Mathematics and Necessity*, Oxford: Oxford University Press for the British Academy, pp. 83–138.

———. (2006) *Another New World Is Being Constructed Right Now: The Ultracold*, preprint 316, Berlin: Max-Planck Institut für Wissenschaftsgeschichte.

———. (2007) "Trees of logic, trees of porphyry," in J. Heilbron (Ed.) *Advancements of Learning: Essays in Honour of Paolo Rossi*, Florence: Olshki, pp. 146–97.

———. (2009) *Scientific Reason*, Taipei: National Taiwan University Press.

Horsten, L. (2007) "Philosophy of Mathematics," in E. N. Zalta (Ed.) *The Stanford Encyclopedia of Philosophy*, http://plato.stanford.edu/archives/win2007/entries/philosophy mathematics/.

Husserl, E. (1936) "Der Ursprung der Geometrie als intentional-historisches Problem," an appendix in D. Carr (Trans.) *The Crisis of European Sciences and Transcendental Phenomenology: An Introduction to Phenomenological Philosophy*, Evanston, Ill.: Northwestern University Press, 1970, pp. 353–78.

Kant, I. (1787) *Critique of Pure Reason*, page references are to the translation by N. Kemp Smith, London: Macmillan, 1929.

———. (1790–1) "Metaphysic L_2" in his *Lectures on Metaphysics*, K. Ameriks and S. Naragon (Trans.), Cambridge: Cambridge University Press, 1997, pp. 299–354.

Klein, J. (1968) *Greek Mathematical Thought and the Origin of Algebra*, E. Brann (trans.), Cambridge, Mass.: MIT Press.

Krieger, M. H. (1987) "The physicist's toolkit," *American Journal of Physics* 55: 1033–8.

———. (1992) *Doing Physics: How Physicists Take Hold of the World*, Bloomington and Indianapolis, Ind.: Indiana University Press.

Latour, B. (2008) "The Netz works of Greek deductions," *Social Studies of Science* 38: 441–59.

Littlewood, J. E. (1953) *A Mathematician's Miscellany*, London: Methuen.

Lloyd, G. (1990) *Demystifying Mentalities*, Cambridge: Cambridge University Press.

Moore, A. W. (1998) "Antirealism in the philosophy of mathematics" in E. Craig (Ed.) *Routledge Encyclopedia of Philosophy*, vol. 1, London and New York: Routledge, pp. 307–11.

Netz, R. (1999) *The Shaping of Deduction in Greek Mathematics: A Study in Cognitive History*, Cambridge: Cambridge University Press.

Netz, R., and Noel, W. (2007) *The Archimedes Codex: Revealing the Secrets of the World's Greatest Palimpsest*, London: Wiedenfeld and Nicolson.

Russell, B. (1912) *The Problems of Philosophy*, page references are to the 1946 ed., London: The Home University Library.

Rosch, E. (1973) "Natural categories," *Cognitive Psychology* 4: 328–50.

Shapin, S., and Schaffer, S. (1985) *Leviathan and the Air-Pump: Hobbes, Boyle and the Experimental Life*, Princeton, N.J.: Princeton University Press.

Wittgenstein, L. (1978) *Remarks on the Foundations of Mathematics*, 3rd ed., G. H. von Wright, R. Rhees, and G.E.M. Anscombe (Eds.), G.E.M. Anscombe (Trans.), Oxford: Blackwell.

What Anti-realism in Philosophy of Mathematics Must Offer

Feng Ye

This article proposes a new approach to anti-realism in the philosophy of mathematics. The realism versus anti-realism debate in the philosophy of mathematics comes from some conflicting intuitions regarding mathematics. The basic intuition favoring realism (or anti-anti-realism) is (Burgess 2004; Rosen and Burgess 2005; Colyvan 1999, 2002; Baker 2001, 2005):[1] as long as mathematicians and scientists attempt to refer to mathematical entities and assert mathematical theorems in their *best* theories, we already have our best reason to believe that mathematical entities exist and mathematical theorems are true, because we should respect scientists' understanding of their own theories and take their words literally, and there are no other stronger or more superior standards for justifying existence and truth. Note that this does not require mathematical entities to be strictly indispensable in the sciences. Burgess and Rosen emphasize respecting working mathematicians' literal understanding of their own assertions, against various attempts by anti-realists to paraphrase away references to mathematical entities in the sciences; Colyvan and Baker further emphasize that various pragmatic values of mathematics in the sciences make mathematical entities indispensable for our *best* scientific theories, even if they are not strictly indispensable. This position implies that it is the duty of philosophers to resolve any philosophical problems due to accepting the existence of abstract entities, e.g., the epistemological problem for abstract entities (Benacerraf 1973), and it is not the right of any philosopher to claim, based on any philosophical principle (or actually philosophical prejudice), that mathematical entities do not exist and mathematicians and scientists are systematically wrong.

However, some philosophers are not convinced by these. They hold the opposite intuition that mathematical entities are unlike robustly real physical objects, and that scientists seem to treat mathematical entities differently. Therefore, for instance, some anti-realistic philosophers contend that scientific confirmation cannot reach mathematics (Maddy 1997, 2005a,b; Sober 1993; Leng 2002); some argue that we are entitled to "take back" the assertions on the existence of mathematical entities in the sciences (Melia 2000), or to treat mathematical entities as fictions (Field 1980; Hoffman 2004), or to interpret mathematical assertions non-literally (Yablo 2001, 2002; Chihara 2005; Hellman 2005). The debate sometimes boils down to the question: Who has the burden of proof (Rosen and Burgess 2005)? Realists presume that realism is the prima facie position, because scientists apparently hold the realists' view, and they claim that anti-realists must prove that scientists are systematically wrong. On the other side, anti-realists presume that the epistemological difficulty for abstract entities makes it clear that anti-realism is the prima facie position, and they claim that realists must prove the existence of abstract entities and conclusively refute anti-realism, and that realists must also resolve the epistemological problem for abstract entities.

Now, a philosophical theory should not be like a legal self-defense, where one can presume one's own innocence and wait for the other side to prove that one is guilty. In particular, anti-realists should not just argue that realists have not *conclusively proved* that we have to be committed to abstract entities. We expect a philosophy of mathematics to be able to resolve puzzles due to genuinely conflicting intuitions. For that, one has to analyze intuitions from both sides impartially and carefully. In particular, anti-realists have to pay attention to realists' points carefully and *sympathetically*. They should positively account for all the *intuitive* observations that realists cite to support realism. They should not just try to find faults in realists' well-constructed arguments aiming to refute anti-realism and prove that mathematical entities really exist (such as the indispensability argument).

Therefore, in this article, I will try to explore the strongest challenges to anti-realism by interpreting the intuitions that appear to support realism sympathetically. A fundamental idea underlying these challenges is that, after denying that abstract mathematical entities exist, anti-realists should explain what then really exists in mathematics and should provide a *literally truthful* account of every aspect of mathematical practices by referring to what really exists. Anti-realists should especially account for the features of mathematical practices that are taken by realists as evidence supporting realism, such as the objectivity of mathematics, the apparent apriority and necessity of mathematics, and the applicability of mathematics. In other words, wherever realists have (or appear to have) an explanation of some feature of mathematical practices by assuming the existence of abstract mathematical entities, anti-realists should not simply reject drawing the realistic conclusion about the existence of abstract entities without providing another equally "realistic" (i.e., literally truthful) explanation of that feature of mathematical practices. Only then can anti-realists really resolve the genuine puzzles about mathematics and can they hope to convince realists or working mathematicians and scientists.

Moreover, I will argue that, for anti-realists to be coherent, they must not assume the objectivity of infinity in any format (not even the objectivity of potential infinity), and they must work under the assumption that there are only strictly finitely many concrete objects *in total*. That is, anti-realists must show that assumptions implying infinitely many entities are *in principle* dispensable in mathematical applications, which means showing that some sort of strict finitism is *in principle* sufficient for the applications.

These may be the strongest challenges to anti-realism and the strictest constraints on anti-realism. No current anti-realistic philosophies can meet all these challenges and constraints. I will argue that this is why they cannot convince realists, and why they are subject to serious objections from realists, if one reads realists' intuitions sympathetically.

I must make it very clear that I am not trying to refute anti-realism here, although some of my comments on current anti-realistic philosophies may sound like attempts to refute anti-realism. I am only trying to say that current anti-realistic philosophies have not offered (and some of them have not even tried to offer) the most essential thing that an anti-realistic philosophy of mathematics must offer. For instance, sometimes they simply label the phenomena to be explained by names such as "empirical adequacy" (Hoffman 2004) or "nominalistic adequacy" (Melia 2000), without really offering (or trying to offer) a "realistic" explanation for the phenomena and resolving the genuine puzzles (i.e., explaining why mathematics is empirically or nominalistically adequate if not true). Sometimes, they do try to offer an explanation (e.g., Field 1980), but in those explanations, they are committed to something that is equally suspicious from the nominalistic point of view (i.e., infinity and continuity of space-time, etc.). I will elaborate on these issues in due course.

In commenting on these anti-realistic accounts, I am only raising issues to motivate a new anti-realistic philosophy of mathematics that is perhaps better. I am not trying to refute the basic tenets of anti-realism.

Then, at the end of the article, I will introduce an on-going research project for a new and positive anti-realistic account for mathematical practices, to meet all the challenges and requirements for anti-realism raised in this article. The philosophical bases of this approach are naturalism and physicalism. It pursues a completely scientific study of human mathematical practices, viewing human mathematical practices as human brains' (versus non-physical minds') cognitive activities. It tries to answer philosophical questions about mathematics on a truly naturalistic and completely scientific basis. These will include questions regarding the meanings of mathematical language, the objectivity of mathematics, the apriority and necessity of mathematics, and the applicability of mathematics.

Challenges for Anti-realism

> **Challenge 1**: *Anti-realism must explain what really exists in mathematics (if not mathematical entities), and must show how these real things can account for the meanings of mathematical statements and mathematicians' knowledge, intuitions, and experiences.*

Mathematical statements are certainly meaningful for mathematicians (in a broad sense) and mathematicians certainly have knowledge, intuition, and experiences regarding mathematics. After denying that mathematical entities exist, anti-realists must then say what really exists in mathematics and must provide a literally truthful account for meaning, knowledge, and intuition, and so on in human mathematical practices by referring to these real things. For instance, mathematical statements as concrete syntactical entities (realized as ink marks on papers, for instance) are certainly real things involved in human mathematical practices, but it seems that there must be something else in order for those statements to be meaningful. Recall that formalism is sometimes understood as the claim that a mathematical theory is a formal system of *meaningless* symbols. However, ordinary mathematical statements are certainly meaningful for working mathematicians. An anti-realist philosopher's job should be to describe, very *realistically* (i.e., using literally truthful assertions), how a mathematical sentence can be meaningful without assuming an abstract reality independent of our minds.

Verificationism is therefore in a better position to claim that the meanings of symbols consist in their uses and that some of our knowledge is knowledge about the uses of language, not knowledge about external entities. However, it is still unclear if this is sufficient to account for meaning, knowledge, and intuition in mathematical practices. For instance, it is unclear if this can account for mathematicians' intuitions "about mathematical structures," or their apparent knowledge that "a Riemann space is approximately isomorphic to real space-time." These intuitions and knowledge do not seem to consist of intuitions or knowledge about how to manipulate symbols. At least much more work needs to be done by verificationists to show that verificationism can indeed account for meaning, knowledge, and intuition in mathematical practices. Moreover, very importantly but unfortunately, instead of exploring what the meanings of the language of *classical* mathematics consist of and what scientists' actual knowledge consists of when they use the language of *classical* mathematics, some verificationists (e.g., Dummett 1973) declare scientists' *successful* uses of the language of classical mathematics as *illegitimate* uses and suggest that only intuitionistic uses, which are almost never practiced by scientists, are legitimate.

Fictionalism (e.g., Field 1980) accepts the literal interpretation of the meanings of mathematical statements. That is, according to fictionalists, mathematical statements purport to refer to abstract mathematical entities. However, fictionalism claims that existential mathematical statements are thus "literally false" (and universal statements may be "vacuously true") since mathematical entities do not exist. This is also misleading. If this "literal meaning" is what working scientists assign to mathematical statements and it accounts for scientists' knowledge, experiences, and intuitions regarding mathematics and its applications, then, as realists like Burgess (2004) contend, we should respect scientists and admit that mathematical entities exist; if this "literal meaning" is not what working scientists mean and it cannot account for scientists' knowledge, experiences, and intuitions regarding mathematics and its applications, then the true meanings of mathematical statements should be more properly construed in other ways, and calling mathematical theorems "literally false" is unhelpful and confusing. It hints that scientists are just making false assertions about nothing pointlessly, and that scientists have neither mathematical knowledge nor valid mathematical intuitions. In other words, if mathematical entities do not exist and mathematical theorems are false (or vacuously true), then what does mathematicians' grasp of the meanings of mathematical statements consist of, and what are mathematicians' knowledge and intuitions about? Fictionalism actually leaves the true meaning, knowledge, and intuition, and so on in mathematical practices unaccounted for. However, I am not saying that fictionalism cannot account for meaning, knowledge, and intuition in mathematical practices. Rather, I am suggesting that if they do try, then they may realize that they need to adopt the new approach that I will introduce at the end of this article, and then fictionalism as a distinctive philosophical position will be unnecessary.

Some fictionalists might think that accounting for meaning, knowledge, and intuition is unimportant because what makes mathematics applicable has nothing to do with what mathematicians grasp as meaning or what they have as knowledge and intuitions. However, this amounts to saying that working mathematicians and scientists come across the right mathematical theory for a type of applications by chance, or that scientists just arbitrarily choose statements that may be false and apply them in the sciences. This is incredible. More importantly, this clearly puts some philosophical principle (i.e., the nominalistic intuition) over scientists' judgments. For scientists, there are scientifically valid reasons why, for instance, the Riemann space theory is applicable for describing real space-time, for instance, the reason that Riemann spaces *are* (approximately) isomorphic to space-time. Such reasons are based on scientists' understanding of the meanings of statements in the Riemann space theory, based on their geometrical intuitions "about Riemann spaces," and based on their knowledge "about Riemann spaces." The realists' claim is that scientists' understanding, knowledge, and intuitions imply the literal existence of Riemann spaces. Anti-realists must account for scientists' understanding of meaning, their knowledge, and intuitions positively (without assuming that Riemann spaces really exist, of course). Otherwise, anti-realists will not be able to meet the realists' challenge.

Moreover, accounts for meaning, knowledge, and intuition, and so on must be realistic accounts, or accounts consisting of literally true assertions. For instance, anti-realists should not say, "scientists use Riemann spaces as models for simulating real space-time," because this assertion is "literally false" for anti-realists, since Riemann spaces do not exist. It amounts to saying that scientists use nothing for simulating real space-time. Similarly, one should not say, "mathematicians have geometrical intuition about Riemann spaces as fictional entities," because fictional entities do not exist and one cannot have intuitions about nothing. I will not venture into the metaphysics of fictional entities here. From the nominalistic and naturalistic point of view, the real problem in these "literally false" explanations seems to be that they have not really explained what it means to talk about "fictional things," or to have knowledge or intuitions "about fictional things." Until we have a literally truthful account without referring to alleged fictional things, realists can always come back and claim that the indispensability of alleged fictional entities in the sciences shows that they are *not* mere fictions.

Finally, an anti-realistic account for meaning, knowledge, and so on should focus on the current practices of classical mathematics, and should respect mathematicians' understanding of classical mathematics. It should not invent new mathematics or paraphrase mathematical statements into something unrecognizable by mathematicians. Because, the real issue that is at stake is: Do our actual mathematical practices and working scientists' understanding of them imply realism? For example, if figuralism (Yablo 2001, 2002) is applied to statements about Riemann spaces, it will imply that statements about Riemann spaces have some real content that is not about Riemann spaces and is actually not about any particular things, because, according to figuralism, mathematical theorems are logical truths, and logical truths are true about *everything*. Mathematicians could not have understood that alleged real content of the statements in the Riemann space theory. No geometrical intuition conveyed by the original statements is in that alleged real content. That alleged real content could not be what mathematicians really mean. Again, the realists' challenge is that working mathematicians' actual understanding, knowledge, and intuitions imply the literal existence of mathematical entities such as Riemann spaces (Burgess 2004).

Similarly, mathematicians are obviously not talking about ideal agents, possible concrete inscriptions on papers, etc., (cf. Hoffman 2004; Chihara 2005). Approaches to anti-realism by paraphrasing mathematical statements into statements about ideal agents, possible inscriptions on papers, and so on, have not directly met the realists' challenge. (Besides, ideal agents do not exist. Therefore, the claim that mathematicians are referring to ideal agents is again literally false.)

On the other hand, it seems that the realistic reading of mathematical statements is not the only one assumed by working mathematicians and scientists, as some realists seem to imply. Many authors (e.g., Leng 2005) have pointed out that mathematicians and scientists do not have a unanimous view regarding the nature of mathematics. In particular, physicists sometimes like to call mathematics a language or formalism, and they sometimes speak as if mathematics is just manipulating symbols. Therefore, there are genuine puzzles due to *genuinely* conflicting intuitions about the nature of mathematics. This is essentially different from the issue of the existence of atoms, about which perhaps no working scientists have any doubt today. A philosopher's task, then, is to solve those puzzles for scientists.

The real point is that anti-realists must not start from some obscure philosophical principles alien to scientists, such as the metaphysical intuition of nominalism, or even Ockam's razor principle. For a naturalist, these principles must be less certain than what working scientists *unanimously* hold. Instead, anti-realists must speak in the language of science, and must provide an account that is acceptable by working scientists, according to their scientific methodologies and standards. If anti-realists can provide a literally truthful *scientific* account of scientists' understandings, knowledge, and intuitions regarding mathematics and its applications, without assuming that mathematical entities literally exist (or exist in any mysterious sense that only some philosophers appear to be able to understand), then they may be able to clarify the mystery surrounding the nature of mathematics for working scientists, and they may be able to convince working scientists.[2]

On this view, philosophers do not even have to be so modest as to claim that philosophical analyses never do any good to the sciences, or that philosophers should never suggest anything to working mathematicians (cf. Burgess 2004; Leng 2005). Mach's analysis of relativity of space was alleged to have influenced Einstein. Philosophical analyses certainly cannot substitute for constructive scientific work. What philosophical analyses can offer is to dispel away dogmatic faiths without real empirical supports or illusions due to our thinking habits, which, under some circumstances, may actually hinder new scientific explorations. The realistic *faith* about classical mathematics could be such a dogmatic faith, if it indeed comes solely from our thinking habits, and if it is not really justified by the sciences and is not compatible with our overall scientific worldview.[3] Rejecting that realistic faith will suggest taking a more liberal view on possible mathematical practices and will encourage exploring new ways of doing mathematics.

> ***Challenge 2**: Anti-realism must account for the genuine relationships between some (alleged) mathematical entities (or structures) and some physical things.*

Scientists choose Riemann spaces to model space-time structures for good reasons. Even if Riemann spaces do not literally exist, it is still a matter of fact that *in some sense*, Riemann spaces and real physical space-time structures are structurally similar. Structural similarity appears to be a genuine relationship between some mathematical structures and some physical things. There are also other types of relationships between the mathematical and the physical. For instance, a function may approximately represent the states of a physics system in some way, and a stochastic process may approximately simulate some real random events. Such relationships all seem to be genuine and are the objective reasons why mathematical theories are applicable in the sciences. On the other hand, nothingness certainly could not structurally resemble any real things and could not be related to any real thing in any meaningful way. Therefore, if mathematical entities do not exist, what resembles those real physical things in these genuine relationships between the mathematical and the physical? Anti-realists should not simply deny such relationships, which will again leave working scientists' reasons for the applicability of mathematics unaccountable. Anti-realists must explain what really exists on the mathematical side, and then show that such genuine relationships between the mathematical and the physical are *realistically* (literally truthfully) accountable based on what really exists on the mathematical side. Anti-realists must also show that our knowledge of such relationships is explainable.

Moreover, anti-realists must show how the content of a specific mathematical theory is relevant to the existence of such relationships. For example, the content of the definition of Riemann spaces is certainly relevant to the fact that Riemann spaces resemble real space-time structures, and the content of the theory of finite groups is relevant to the fact that a finite group does not in any way resemble real space-time structures. In other words, it is not enough to say generally that pragmatic consequences decide which mathematical theory is useful to model reality in a particular area. Scientists do not *randomly* pick some literally false statements about nothing and then try to apply them in an arbitrary area in the sciences. They choose (or discover, or define, or invent, or imagine) Riemann spaces to model large-scale space-time structures, because they *really* discern some genuine relationship between these two, *in particular*, based on their understanding of statements "about Riemann spaces" and their knowledge "about Riemann spaces." Anti-realists have to admit scientists' actual intuitions and judgments regarding the relationship between Riemann spaces and physical space-time, and have to explain them realistically and scientifically, by referring to what really exists in mathematical practices (without assuming that Riemann spaces really exist).

There are anti-realistic approaches that resort to some general concepts that apply to mathematics as a whole, such as nominalistic (or empirical) adequacy, to account for the usefulness of mathematics (Melia 2000; Hoffman 2004). These concepts may be of some interest, but they say nothing about such genuine relationships between the mathematical and the physical, and nothing about scientists' reasons for the applicability of a particular mathematical theory to a particular type of natural phenomena based on such relationships. "Nominalistic (or empirical) adequacy" is a name for the observed results in scientists' mathematical practices. It is not an explanation of why those results obtain. It does not explain, for instance, what is special about the Riemann space theory that makes it applicable in modeling space-time, and why scientists did not use the theory of finite groups, or anything else, to model space-time. (Riemann spaces and finite groups do not exist anyway.) Actually, if Riemann spaces did exist and were literally (though approximately) isomorphic to physical space-time, there would be an explanation of their nominalistic or empirical adequacy. The realists' claim is exactly that this justifies the existence of Riemann spaces. Anti-realists cannot meet this challenge without providing an equally literally truthful account for scientists' valid judgments without assuming that Riemann spaces literally exist.

Recall that in answering the same question for empirical adequacy, van Fraassen reminds us that any explanation must stop somewhere anyway, and then he claims that it stops at explaining the empirical adequacy of postulating unobservable (physical) entities in the sciences. Similarly, some anti-realists appear to be claiming that there is no more explanation for the applicability of a mathematical theory for a type of natural phenomena (e.g., the applicability of Riemann space theory to space-time), and that scientists' explanation for it (i.e., by assuming the literal existence of Riemann spaces and a real isomorphism between Riemann spaces and space-time) is just wrong. For instance, Hoffman (2004) takes her fictionalist view to be a completion of van Fraassen's views. I will not try to contest such claims here, but I would like to point out that anti-realists will certainly be in a better position to meet the realist challenge if anti-realists can offer a literally truthful explanation of such genuine relationships between the mathematical and the physical, as well as an explanation of the applicability of mathematics based on such relationships.

> ***Challenge 3:*** *Anti-realism must identify and account for various aspects of objectivity in mathematical practices and applications.*

Even if mathematical entities do not exist, our mathematical knowledge should still have *objective* content. We are not making assertions out of our wishes in doing mathematics. One could wish that Goldbach's conjecture is true, but we know that there is something objective and independent of our wishes there. A natural attempt to explain such objectivity from anti-realists' perspective is to claim that correctness in following the logical rules in a mathematical proof is an objective matter. Then, the challenge for anti-realists is: admitting such objective correctness in rule following appears to be committed to rules as abstract entities and be committed to objective truths about abstract entities, in particular, when rules are understood as mathematical functions that can operate on infinitely many instances of arguments.

Another aspect of objectivity in mathematics has to do with the relationships between the mathematical and the physical. Hoffman's (2004) recent exposition of fictionalism appears to imply that scientists *pretend* that Riemann spaces exist and are (approximately) isomorphic to space-time structures, just like kids pretend that a sofa is a mountain when playing games. However, the (approximate) structural isomorphism between Riemann spaces and real space-time structures seems to be *objective*, not wishful pretending, and it seems to be the *objective* reason for our successes in modeling space-time structures by Riemann spaces. If scientists were indulging in wishfully pretending things as kids do in games, scientists would not be successful in their work. Frege's claim that applications raise mathematics from a game to science is well known. Realists' charge against anti-realism is just that such strong objectivity in the sciences, which is not in kids' games, shows that mathematics is not merely a make-believe. Until anti-realists can clearly explain what this *objective* relation between the mathematical and the physical consists of and how it is the *objective* reason for the applicability of Riemann spaces (without assuming that Riemann spaces exist), they have not yet met the realists' challenge.

Anti-realists who completely deny any objective realistic truths (or seek to account for mathematics and our scientific knowledge, in general, only as socio-cultural constructions or conventions) may not care about this challenge. Criticizing them is usually the realists' job. I propose this as a challenge to anti-realism in mathematics, because I take anti-realism in mathematics as a clarification of and a defense for common-sense realism and scientific realism. It tries to solve puzzles due to alleged mathematical truths about infinity and abstract objects, which appear to be "out of this universe." For this, anti-realism in mathematics must distance itself from views that deny objectivity altogether.

> ***Challenge 4***: *Anti-realism must explain the apparent obviousness, universality, apriority, and necessity of simple arithmetic and set theoretical theorems, and they must also provide a consistent account of logic.*

We have a strong intuition that "5 + 7 = 12" expresses some obvious, universal, necessary, and a priori truth. It does not help to say that "5 + 7 = 12" is "literally false," as some anti-realists seem to be saying, which only adds more puzzles. "5 + 7 = 12" is certainly meaningful to everyone. It has content. Kids do learn something when they learn "5 + 7 = 12." There must be some truth in it even if numbers do not "literally exist" and even if "5 + 7 = 12" is "literally false" in whatever sense. Anti-realism must explain what the content of "5 + 7 = 12" is and why it is obviously true *in some proper sense*. It must also answer questions regarding the universality, apriority, and necessity of "5 + 7 = 12," and give reasonable explanations as to why we strongly believe that "5 + 7 = 12" is a priori, necessary, and universal. Moreover, arithmetic, set theory, and logic are tightly entangled. Some simple theorems in arithmetic and set theory, such as "5 + 7 = 12" or "A ∪ B = B ∪ A," appear to be logical truths in disguise. The common wisdom is that logical truths are universal, a priori, and necessary truths. The universality, apriority, and necessity of arithmetic are obviously closely related to the same characteristics of logic. Anti-realism must provide an account of logic consistent with the anti-realistic ontology and epistemology, and consistent with anti-realists' accounts of arithmetic and simple set theory.

One attempt to explicate the truth of "5 + 7 = 12," adopted by figuralism (Yablo 2002), claims that the real content of the statement is expressed by the following logical truth in the first-order logic

$$\exists_5 xP(x) \wedge \exists_7 xQ(x) \wedge \neg\exists x\,(P(x) \wedge Q(x)) \rightarrow \exists_{12} x\,(P(x) \vee Q(x)).$$

This may be fine; however, for arithmetic statements with quantifiers, Yablo's suggestion is that they are logical truths expressed by infinitely long sentences, namely, infinite conjunctions and disjunctions of logical truths in the above format in the first-order logic. Now, we do not speak infinitely long sentences. Infinitely long sentences are actually mathematical constructions and are thus abstract entities. If infinitely long sentences do not really exist as abstract entities, it is unclear what Yablo has said about the alleged real content of quantified arithmetic statements. Nothingness surely cannot express any meaningful content. If infinitely long sentences simply do not exist, there is nothing there to express that alleged content. Therefore, one naturally suspects that what really express the alleged content are actually still the original quantified statements about numbers.

It seems that what Yablo actually has there is another mathematical theory "about infinitely long sentences" as mathematical entities, which could be defined by using set theory with the axiom of infinity, *and* Yablo has a "true" predicate for those infinitely long sentences, recursively defined in set theory. Therefore, Yablo actually translates real, concrete quantified statements "about numbers," into real, concrete statements in this mathematical theory "about infinitely long sentences" (as abstract entities), and then claims that those infinitely long sentences (as abstract entities) are "true," that is, they satisfy that recursively defined "true" predicate in the mathematical theory "about infinitely long sentences." In the end, the alleged real content is still the content of statements that appears to refer to abstract entities (i.e., infinitely long sentences). This perhaps shows that one has to deal with the content of statements that appears to refer to alleged abstract entities (or fictional entities) directly. There is no magic to get "real content" out of these statements.

> **Challenge 5**: *Anti-realism must be able to account for mathematics under the assumption that there are only finitely many concrete objects in total, but it must also account for our apparent valid intuitions "about infinity."*

Philosophers favoring anti-realism or nominalism may still hold different views regarding infinity. For example, both Field (1998) and Yablo (2002) claim that arithmetic theorems involving infinity are objectively true (when those theorems are properly stated). Field refers to a "cosmological assumption" on infinity of the universe in defending that arithmetic statements implying infinity have objective truth values. However, as of today, physics has not provided a definite account regarding whether the universe is infinite. More importantly, in almost all areas of the sciences so far, the applicability of mathematics is independent of the conjectures of physics about whether space-time is infinite or finite, or discrete or continuous. We apply infinite mathematics also in economics, which is certainly about finite and discrete things. If an account for mathematics and its applications depends on a specific assumption from physics about space-time, it must have missed something essential about the nature of our mathematical knowledge, and must have missed the true reasons for the applicability of mathematics.

Now, if anti-realists accept the objectivity of infinity, but claim that it does not mean that this physical universe is infinite, then they must explain where infinity is if this physical universe is indeed finite and discrete. These anti-realists will also face a similar epistemological problem as realists face, namely, explaining how knowledge about objective infinity is possible, given that we are finite beings with finite experiences. They might try to exploit some concept of scientific confirmation. Then, it is likely to be again some sort of holistic confirmation based on the pragmatic values of assuming infinity in the sciences, and it is likely that realists can simply take over this strategy to explain our knowledge about *any* abstract objects, as Quine does. After all, as long as the door is open to accept one objective truth that appears to be about things essentially beyond this concrete universe, it should not be too difficult to go ahead and accept others for similar reasons.

Similarly, some philosophers resort to an objective "mathematical possibility," according to which infinity is "mathematically possible." Now, if the universe is indeed finite and there are only finitely many concrete objects *in total*, what justifies this claim about the mathematical possibility of infinity? It seems that it will be again some sort of holistic justification based on their usefulness in the sciences. Then, the door to Quinean realism is opened again.

Therefore, rejecting an objective infinity is the only way to be a coherent nominalist. This assertion was already made long ago by Goodman and Quine (1947) and it seems to be ignored by some contemporary philosophers. This assertion implies that for any statement containing quantifiers intended to range over infinite domains, anti-realists must either offer an anti-realistic interpretation of the statement, or admit that the statement can be vacuously false or vacuously true. These include statements about the consistency of formal systems, statements about Turing machines, and statements expressing simple universally quantified arithmetic laws.

We do seem to have a strong intuition that there are objective truths involving infinity, for instance, the commutative law of addition, or the commutative law of addition expressed as an assertion about a Turing machine implementing the addition function, or the consistency of some very simple formal systems, etc. Therefore, anti-realists must show that the literal truth of such assertions involving infinity is neither presumed in the sciences or implied by the sciences, and nor is it needed in the anti-realistic account for mathematical practices and mathematical applications. In other words, similar to the cases for meaning, knowledge, and relationships between the mathematical and the physical, anti-realists must account for our various intuitions "about infinity" without assuming the objectivity of infinity.

> ***Challenge 6****: Anti-realism must explain the applicability of mathematics, and for that purpose, one has to show that the apparent references to infinity and abstract mathematical entities in mathematical applications are in principle dispensable.*

Field (1980) tried to explain how a bunch of "false statements about nothing" could be useful by using the notion of conservativeness. His strategy depends on the assumption that one can nominalize scientific theories by eliminating references to abstract mathematical entities, and thus prove that classical mathematics is conservative over a nominalistic version of science. There were other similar programs for nominalizing mathematics in the 1980s and 1990s (see Burgess and Rosen 1997). However, all these programs assume infinity in one way or another. Therefore, as I have argued above, they are not really nominalistic.

On the other hand, if a nominalization program is successful, it does help anti-realism. There are objections to this claim. First, nominalized scientific theories must be much more complex than the theories stated in classical mathematics. One may argue that simplicity justifies realism about classical mathematics. However, consider this example. We can simulate the population growth on the Earth by a differentiable function and a differential equation. This is fairly simple. Now, if we have a gigantic computer (or just a gigantic brain) that can simulate all people on the Earth (on aspects related to reproduction), then we will have a *literally more accurate* description of the population growth on the Earth and it will give us more accurate predictions on future population growths. This literally more accurate description will refer to finite and discrete entities only and it will not refer to infinity or abstract mathematical entities such as real numbers, differentiable functions, etc. It will be a nominalistic description. On comparing these two descriptions of the population growth on the Earth, we see that classical mathematics can offer a simpler description of discrete and finite things only because it helps us to ignore some details and build a simpler but literally less accurate model. We will have to eliminate infinity, if we want a more accurate description of the phenomena. It is doubtful that such simplicity can justify realism about infinity and abstract entities in classical mathematics. Note that this is not a peculiar example. Considering the fact that our physics has not yet reached things below the Planck scale (about 10^{-35} m, 10^{-45} s etc.), all current applications of continuity and infinity in the sciences (at least except for physics at the Planck scale) gloss over microscopic details. The kind of simplicity that classical mathematics brings in these applications does not seem to confirm the existence of infinity. These applications are similar to the cases where one imagines some non-existent fictional things as simpler substitutes for real but more complex things.

Melia (2000) claims that mathematics brings a simpler theory, not a simpler world, and he argues that such simplicity does not justify mathematics. However, if we agree with realists to put the mathematical world on a par with the physical world, this combined world does become significantly simpler if we postulate infinite mathematical entities, *and* if we ignore the fact that there is a price for it, that is, its part that describes the physical world becomes *literally less accurate*. Ignoring this price, realists seem to be at a good position to claim that positing mathematical entities does bring the same kind of simplicity as positing atoms does. Therefore, a mere distinction between simplicity of the world and simplicity of a theory cannot convince realists, for whom the world includes mathematical entities as well (or, they claim that you should at least have an open mind to allow your world to include abstract mathematical entities).

Colyvan (1999, 2002) and Baker (2001, 2005) imply another objection to the thesis that successful nominalizations of scientific theories will help anti-realism. They claim that mathematics is not only used to build models to represent phenomena and deduce known conclusions. They cite examples to show that mathematics has unification powers, predicting and discovering powers, the ability to apply to new and unexpected areas of natural phenomena, and genuine explanatory powers. They claim that these pragmatic values support realism, by which they seem to imply that those nominalized theories will not have these pragmatic values. However, a nominalized physics theory, if that can be worked out at all, will have the same pragmatic values as the classical theory except for simplicity, because they express the same physics laws, and even their mathematical formats can also be very similar (because the nominalized version typically parodies the classical version, e.g., Field 1980; Ye 2011).

From the logical point of view, the real puzzle about applicability is due to the gap between infinity in mathematics and finitude of the real world (from the Planck scale to the cosmological scale). In order to resolve this puzzle, we must clarify, for instance, the logic of using infinite and continuous models to approximate and simulate finite, discrete things. This puzzle is independent of any philosophical view about the ontological and epistemological status of those alleged models as abstract entities. What anti-realists can hope is that a complete *logical* clarification of the puzzle may turn out to favor anti-realism since it is likely that a complete *logical* clarification will eventually show that infinity and apparent references to abstract entities all in principle can be eliminated in mathematical applications, which will then show exactly how mathematics helps to derive literally true nominalistic conclusions about *finite* concrete things from literally true nominalistic premises about *finite* concrete things *in plain logic*. Moreover, it may show that the literal existence of infinity and abstract entities is irrelevant for explaining applicability, because what really explains applicability is the fact that mathematical proofs used in applications allow eliminating apparent references to infinity and abstract entities, so that the proofs can preserve literal truths about finite concrete things. That is, a real logical explanation of this puzzle of the applicability of infinity to finite things may in the end have to imply that infinity is in principle dispensable.

With doubts about dispensability, some recent anti-realists try to look for some easy arguments to show that even if abstract mathematical entities are indispensable, we still *do not have to be committed to* them. For instance, Leng (2002) claims that mathematical entities are used to build models to represent physical things, and she claims that science confirms the existence of those physical things only, but not the existence of those models.

Now, computers are used to build models to simulate other things. However, computers really exist. Indeed, the *failure* of a computer modeling should not be taken as evidence that the computer does not exist or that our assertions about data and programs in the computer are false, but successes of computer modeling *do* require that computers literally exist and *do* confirm that our assertions about data and programs in computers are *literally* true. Similarly, realists claim that while the failure of mathematical modeling does not imply that the mathematical model does not exist or that our assertions about the model are false, but successful modeling *does* require that the model literally exists and *does* confirm that our assertions about the model are *literally* true. For instance, if one makes an error in doing calculations about a mathematical model, one will not be successful in using the model in applications. Anti-realists may insist that models are fictional entities. However, this cannot convince realists. They claim that scientific applications raise mathematics from a game or fiction to science, and they charge that anti-realists are intellectually dishonest and are placing their principle of nominalism above scientists' judgments.

Besides, remember that according to anti-realists, the claim "scientists are using the fictional model *X* to simulate *Y*" is literally false, because *X* does not exist. Therefore, these anti-realists are making literally false assertions in their philosophical papers, according to themselves.

It clearly implies that these anti-realists have not yet really explained how mathematics is applied, or how mathematical models work. They have not provided a *literally true* explanation of how "fictional models" are applied to derive literal truths about real things. Since fictional things do not exist, in a literally true explanation, one should not refer to "fictional models" again. That is, one has to show that they are in principle dispensable.

Melia (2000) is another instance. Melia claims that classical mathematics is not conservative over nominalistic theories because some assertions about concrete things are not expressible in a nominalistic language. Melia cites some examples to show how assertions about concrete things have to be expressed by referring to abstract entities. Then, Melia proposes the so-called "weaseling strategy," which suggests that we can take back what we asserted previously about the existence of abstract entities in applying mathematics to the sciences. It means that classical mathematics is nominalistically adequate in the following sense: the consequences about concrete things obtained by applying mathematics are true of concrete things.

Now, since all scientifically reliable assertions about concrete things in this universe are accurate only up to some finite precisions (i.e., above the Planck scales 10^{-35} m, 10^{-45} s, etc.), Melia's examples regarding how assertions about concrete things have to be expressed by referring to abstract entities are beside the point for a real nominalist, because these examples all assume infinity. We may need apparent references to infinity and abstract mathematical entities to give a simple description about finite concrete things. An example will be using a differentiable function to represent population growth on the Earth. However, apparent references to infinity and abstract mathematical entities may not be strictly indispensable.

On the other side, sometimes we do take back what we asserted at first in everyday life and in the sciences. However, if we do so and are then confronted with the accusation that we have to take back our conclusions as well, we usually have to show that we do not *really* have to commit to that which was asserted previously. That is, we have to show that what was asserted previously can *in principle* be eliminated. That is the case, for instance, when we refer to a rigid body in elementary mechanics. We believe that such references to fictional entities can in principle be eliminated. Now, what can one offer to *justify* the nominalistic adequacy of one's "weaseling practice" if one admits that abstract entities are strictly indispensable? It seems that the only available strategy is again some sort of holistic confirmation based on the pragmatic values of positing abstract entities. Then, why does not this lead to Quinean realism? Again, we need an explanation of nominalistic adequacy, and it has to be a literally truthful and nominalistic explanation, which means that it should not refer to abstract entities again. If we can explain the applicability of infinite mathematics to finite real things in the universe by showing that apparent references to infinity and abstract entities can in principle be eliminated, then we will have such an explanation of nominalistic adequacy. Otherwise, anti-realists either have to leave nominalistic adequacy unexplained or have to open a door for the Quinean holistic justification for abstract entities.

These are the challenges for anti-realists. In summation, anti-realists must provide a positive account for the practices of classical mathematics including its applications. Especially, anti-realists must account for those aspects that are taken as evidence supporting realism by realists. They should not simply label some phenomena by a name (i.e., "nominalistic adequacy") without giving a real explanation, and should not be negative only (i.e., arguing that they *do not have to be committed* to something). They should not fall into resorting to some holistic confirmation to justify some obscure things, such as infinity, possibility of infinity, "weaseling strategy," and so on. They must explicitly say what are real on the mathematical side in mathematical practices, and then refer to those real things to give a very *realistic* account for mathematical practices. Finally, they must do these without assuming that the universe is infinite or that there are infinitely many concrete objects *in total*. They must realize that the real puzzle of the applicability of mathematics is just the logical puzzle of how infinite mathematics is applied for describing finite real things, which constitute almost all mathematical applications in the current sciences.

Toward a Scientific Account for Mathematical Practices

These challenges and requirements seem to have cornered anti-realism. Is there still a chance for anti-realism in the philosophy of mathematics? The answer is yes. Actually, the analyses in the last section very naturally lead to a completely naturalistic and scientific account for human mathematical practices.

First, the analyses above show that what is missing from current anti-realists is a realistic and literally truthful account for aspects of mathematical practices, by referring to what really exists in mathematical practices. Therefore, what really exists and what is really happening in mathematical practices? Since the alleged "Riemann spaces" do not literally exist, if one asks, "What is that mathematician doing when she talks 'about Riemann spaces'?", the only straightforward answer seems to be, "She is imagining something." This idea is not new. As far as I can trace it, the earliest explicit exposition of it is Renyi (1967). Renyi explicitly suggested that mathematical entities are our imaginations. On the other side, perhaps all contemporary anti-realists more or less have this picture in mind.

Therefore, what we really need is a very realistic and literally truthful explanation of what it is to imagine something. The natural thought is to characterize imagining something as *having relevant mental representations with the same or similar structures as mental representations of real external things, but without any corresponding external things to be directly represented*. I emphasize "directly represented" here, because these mental representations *are* indirectly related to real external things in some way. In other words, our imaginations do not literally create "imaginary entities." Only our hands can create things out of pre-existing materials. When we imagine, our minds create mental representations that reside in our brains.

Then, for mathematics, this means that while there are no mathematical entities, there *are* scientists' mental representations that they create and manipulate in doing and applying mathematics. These are what really exist on the mathematical side in mathematical applications (vs. other physical things as the subject matter of application on the other side), and they are the real things that are used as models for simulating real things. (Scientists use their brains to model other things much like they use computers to model other things.) Anti-realists' task will then be to describe the cognitive functions of these mathematical mental representations in human mathematical practices, and to describe the relationships between these mental representations and other real things in the physical world, to explain various aspects of mathematical practices, including the applicability of mathematics. An account for mathematical practices is thus a continuation and extension of cognitive science, dealing specifically with human mathematical cognitive activities. It is a completely scientific description of a class of natural phenomena. This is what is missing from the current anti-realistic philosophies. With that missing, one cannot answer, *in realistic terms*, what the meanings of mathematical statements are, or what the relationships between the mathematical and the physical consist of, or what exactly are used to model real things in mathematical applications, or how exactly those models work.

This picture of human mathematical practices is consistent with the following nominalistic but completely naturalistic ontological and epistemological assumptions: (1) there is this physical universe, which may be finite and discrete (and we do not know yet), and *only* things in this universe really exist; (2) humans and their brains are parts of this natural world, and their knowledge (including mathematical knowledge) stored in their brains and realized as neural structures comes from their finite brains' interactions with finite concrete things in this universe, either individually or programmed into their genes as a result of evolution; and (3) there is no existence or truth beyond and above this concrete universe. These basic assumptions, which will be simply called *naturalism* here, seem to be clear and coherent. It is not Quinean naturalism. It is consistent with physicalism in contemporary philosophy of mind (e.g., Papineau 1993), and it is consistent with the common-sense realism, scientific realism, and the general scientific and naturalistic worldview, except for the fact that classical mathematics generates a puzzle.

The puzzle is that classical mathematics appears to include knowledge about things essentially out of this concrete universe. Quinean pragmatic mathematical realism actually implies that successful applications of classical mathematics in modern sciences force us to reject this naïve naturalism and force us to accept a more sophisticated view of existence that puts abstract mathematical entities on a par with other concrete things in this universe. I take anti-realism in philosophy of mathematics as an effort to resolve the puzzle and defend this naïve naturalistic worldview.

A research project following these ideas is underway. The project will account for various aspects of human mathematical practices by referring to the cognitive functions of mathematical mental representations in brains and their physical connections with physical entities outside brains. This article actually presents the motivation and sets the goals for the project. Another article (Ye 2010a) elaborates on the kind of naturalism this project relies on. It also argues that the Quinean indispensability argument is actually an argument from the point of view of a Transcendental Subject, and is therefore not a naturalistic argument. It should not disturb a true naturalist. In other words, the article argues that Quine is implicitly inconsistent with himself regarding the basic tenets of naturalism.

Then, a few other articles (Ye 2010b, online-a, -b, -c) and a monograph (Ye 2011) present the positive accounts for various aspects of mathematical practices. They constitute a part of the work done so far in the project. Ye (online-a) discusses some aspects of meaning, knowledge, and intuition in mathematical practices, as well as the relationships between the mathematical and the physical, by referring to the cognitive functions of mathematical concepts and thoughts as mental representations in brains, and by referring to their physical connections with physical entities outside brains. Ye (online-b) identifies various senses of objectivity from the naturalistic point of view and explains why admitting objectivity in mathematical practices does not imply the existence of abstract entities. Ye (online-c) discusses the apriority of logic and arithmetic from the naturalistic point of view. Finally, the monograph Ye (2011) first explains how the question of applicability of mathematics can be formulated as a scientific question and transformed into a logical question. Then, it develops a strategy for explaining the applicability of mathematics, in particular, the applicability of infinite mathematics to this finite physical world. The strategy involves showing first that the applications of classical mathematics are in principle reducible to the applications of strict finitism, a fragment of the quantifier-free primitive recursive arithmetic, and then showing that the applications of strict finitism can be interpreted as sound logical inferences from literally true premises about strictly finite, concrete physical objects, to literally true conclusions about them. These will constitute a logically plain explanation of why infinite mathematics can preserve literal truths about strictly finite things in mathematical applications, by showing that infinity can in principle be eliminated. Two short articles (Ye 2010b, online-d) give a summary of the monograph.

The project is still in progress. It is possible that even if it is successful, it still cannot convince some realists. In particular, since its basis is naturalism, it will not convince those who explicitly reject naturalism (e.g., Gödel). However, this completely naturalistic and scientific description of human mathematical practices as human brains' cognitive activities will show that it is scientifically redundant and meaningless to assume that human mental representations created in human brains in mathematical practices "represent" or "correspond to" the alleged abstract entities. It offers a more coherent scientific and naturalistic picture of human mathematical cognitive activities. Then, this can perhaps convince those whose primary concern is about respecting science versus metaphysical (i.e., nominalistic) intuitions. Moreover, such research should have its own values, independent of any philosophical positions, since it is scientific research into human mathematical cognitive processes and into the exact logic in applying *infinite* mathematics to this *finite* physical world.

Finally, I will briefly compare this research project with other related approaches. Some cognitive scientists have studied the origin and psychological nature of mathematical concepts from the psychological point of view (e.g., Lakoff and Núñez 2000). However, they did not discuss issues that concern philosophers and logicians, such as objectivity in mathematics, the apriority of logic and arithmetic, and the applicability of mathematics and so on. This research project focuses on these philosophical and logical issues, not on the psychological aspects of mathematical practices.

This research also follows the spirit of philosophical naturalism (or physicalism) pursued by Papineau (1993) and is intended to be a substantial improvement on it. In particular, I suggest addressing philosophical issues on meaning and so on, by directly referring to mathematical mental representations from the point of view of cognitive science, while Papineau relies on Field's fictionalism (Field 1980), which is not a truly naturalistic and realistic scientific theory. The strategy for explaining the applicability of mathematics here is reminiscent of Field's notion of conservativeness, but Field assumed infinity in his nominalization program and did not really explain how infinite mathematics is applied to strictly finite physical things. The mathematical tool for explaining applicability here is strict finitism.

Acknowledgements

The research for this article is supported by the Chinese National Social Science Foundation (grant number 05BZX049). My research in philosophy of mathematics began during my graduate study at Princeton many years ago. I am deeply indebted to Princeton University and my advisors

John Burgess and Paul Benacerraf for the scholarship assistance and all the help and encouragement they offered. An earlier version of this article was presented at the Shanghai Conference on Philosophy of Mathematics and the Beijing Conference on Analytic Philosophy, Philosophy of Science, and Logic, both in May 2005. I would like to thank the participants for their comments. Finally, I would like to thank Esther Rosario, the language editor for *Synthese*, for the numerous corrections on my English presentations and for her many suggestions that help make the article clearer and more readable.

Notes

1. See Burgess (2004) for a characterization of the anti-anti-realism position. In this article, I will use the term "realism" in a broad sense. It will apply to both anti-anti-realism and other stronger realistic positions.

2. This is exactly what my truly naturalistic and scientific approach to philosophy of mathematics wants to do. See the last section for more details, including an explanation of what I mean by "true" naturalism.

3. In particular, the view that cognitive subjects can "refer to" or "be committed to" abstract entities appears incompatible with the physicalist view that cognitive subjects are just human brains, that is, some special kind of physical systems, which can have physical interactions with their physical environments only. See the last section of this article and other articles of mine on this subject.

References

Baker, A. (2001). Mathematics, indispensability and scientific progress. *Erkenntnis, 55*, 85–116.

Baker, A. (2005). Are there genuine mathematical explanations of physical phenomena? *Mind, 114*, 223–38.

Benacerraf, P. (1973). Mathematical truth. *Journal of Philosophy, 70*, 661–79.

Burgess, J. P. (2004). Mathematics and bleak house. *Philosophia Mathematica, 12*(3), 18–36.

Burgess, J. P., and Rosen, G. (1997). *A Subject with no object*. Oxford: Clarendon Press.

Chihara, C. (2005). Nominalism. In S. Shapiro (Ed.), *The oxford handbook of philosophy of mathematics and logic* (pp. 483–514). Oxford: Oxford University Press.

Colyvan, M. (1999). Confirmation theory and indispensability. *Philosophical Studies, 96*, 1–19.

Colyvan, M. (2002). Mathematics and aesthetic considerations in science. *Mind, 111*, 69–74.

Dummett, M. (1973). The philosophical basis of intuitionistic logic. In P. Benacerraf and H. Putnam (Eds.), *Philosophy of mathematics: Selected readings* (Reprinted, 1983, pp. 97–129). Cambridge: Cambridge University Press.

Field, H. (1980). *Science without numbers*. Oxford: Basil Blackwell.

Field, H. (1998). Which undecidable mathematical sentences have determinate truth values? In H. G. Dales and G. Oliveri (Ed.), *Truth in mathematics*. Oxford: Oxford University Press.

Goodman, N., and Quine, W. V. (1947). Steps toward a constructive nominalism. *Journal of Symbolic Logic, 12*, 105–22.

Hellman, G. (2005). Structuralism. In S. Shapiro (Ed.), *The oxford handbook of philosophy of mathematics and logic* (pp. 536–62). Oxford: Oxford University Press.

Hoffman, S. (2004). Kitcher, ideal agents, and fictionalism. *Philosophia Mathematica, 12*(3), 3–17.

Lakoff, G., and Núñez, R. (2000). *Where mathematics comes from: How the embodied mind brings mathematics into being*. New York: Basic Books.

Leng, M. (2002). What's wrong with indispensability? *Synthese, 131*, 395–417.

Leng, M. (2005). Revolutionary fictionalism: A call to arms. *Philosophia Mathematica (III), 13*, 277–93.

Maddy, P. (1997). *Naturalism in mathematics*. Oxford: Clarendon Press.

Maddy, P. (2005a). Three forms of naturalism. In S. Shapiro (Ed.), *The oxford handbook of philosophy of mathematics and logic* (pp. 437–514). Oxford: Oxford University Press.

Maddy, P. (2005b). Mathematical existence. *Bulletin of Symbolic Logic, 11*, 351–76.

Melia, J. (2000). Weaseling away the indispensability argument. *Mind, 109*, 455–79.

Papineau, D. (1993). *Philosophical naturalism*. Oxford: Basil Blackwell.

Renyi, A. (1967). *Dialogues on mathematics*. San Francisco: Holden-Day.

Rosen, G., and Burgess, J. P. (2005). Nominalism reconsidered. In S. Shapiro (Ed.), *The oxford handbook of philosophy of mathematics and logic* (pp. 515–35). Oxford: Oxford University Press.

Sober, E. (1993). Mathematics and indispensability, *The Philosophical Review, 102*, 35–57.

Yablo, S. (2001). Go figure: A path through fictionalism. *Midwest Studies in Philosophy, 25*(1), 72–102.

Yablo, S. (2002). Abstract objects: A case study. *Noûs, 36*, 220–40.

Ye, F. (2010a). Naturalism and abstract entities. *International Studies in the Philosophy of Science*, 24, 129–46.

Ye, F. (2010b). The applicability of mathematics as a scientific and a logical problem. *Philosophia Mathematica*, 18, 144–65.

Ye, F. (2011). *Strict finitism and the logic of mathematical applications. Synthese Library, vol. 355.* Dordrecht: Springer.

Ye, F. (online-a). On what really exist in mathematics. http://sites.google.com/site/fengye63/. Accessed June 23, 2011.

Ye, F. (online-b). Naturalism and objectivity in mathematics. *ibid.*

Ye, F. (online-c). Naturalism and the apriority of logic and arithmetic. *ibid.*

Ye, F. (online-d). A strictly finitistic system for applied mathematics. *ibid.*

Seeing Numbers

Ivan M. Havel

The ability to represent time and space and number is a precondition for having any experience whatsoever.
—*Randy Gallistel*

In his influential book *The Principles of Psychology* (1890) the well-known psychologist and philosopher William James listed seven "elementary mental categories" that he postulated as having a *natural* origin [Jam07, p. 629]. In an alleged order of genesis he listed, in the third place, after the ideas of time and space, the idea of number.

Mentioning the idea of number along with the ideas of time and space as something natural is interesting both philosophically and from the viewpoint of cognitive science. In this respect it is worth noting that one of the symptomatic features of mainstream cognitive science is the tendency not to talk so much about ideas as about their *representations*, either in the mind or, even better, in the brain. Just notice the maxim by Randy Gallistel in the above motto (quoted in [Cal09]). The hypothesis that the human sense of number and the capacity for arithmetic finds its ultimate roots in a basic cerebral system has been frequently proposed and elaborated, for instance by the neuroscientist Stanislas Dehaene [Deh97, Deh01, Deh02].

In this essay I am going to take up a somewhat different perspective.[1] Instead of following up with various extraneous concerns offered by brain research and behavioral and/or developmental psychology I will base my considerations on the way things, in our case numbers,[2] are *actually experienced* (a digression towards animal "arithmetic" in a subsequent section is an exception). Hence the nature of my interest is more phenomenological (in the philosophical sense of the word) than scientific and, I dare say, my claims are speculative, rather than conforming to empirical or deductive knowledge. Correspondingly, I use the term "experience" in its philosophical sense, when referring to any mental state associated with one's conscious "living through" a certain event or situation—"conscious" in the sense that there is something "it is like" to be in that state of the mind.

First of all we should make clear what kind of "thing" numbers are, i.e., what is the nature of entities that humans may experience as numbers? The way mathematicians formulate, introduce, or conceptually represent *their* idea of a number is quite a different question. This, however, does not mean that the relation between the mathematicians' concept and human experience should be ignored as irrelevant.

Our first observation is that the very word "number" (in English as well as its equivalents in some other languages) is rather multivalent. This fact may lead to difficulties, especially in the context of our study—but, at the same time, it may hint at certain interesting hypotheses about the origin of this very multivalence. Thus I shall distinguish here at least three different senses of the word: (1) "numbers" as *numerical quantities*, i.e. counts *of* something, (2) "numbers" as *abstract entities* emerging in the human mind and allowing mathematical formalization in the framework of one or another formal theory,[3] (3) "numbers" as natural or conventional *symbols* or *numeral signs* that represent numbers in either one of the previous two senses. In view of the importance of this conceptual distinction in our study, I shall henceforth distinguish the above three senses by using, respectively, the terms (1) *count*, (2) *number*, and (3) *numeral*. However, when there is no danger of confusion, I shall often use the term *number* in the general, more encompassing sense.

Human Sense of Numerical Quantities

Are we, people, endowed with anything like a *sense of counts*? Here the term "sense" (another multivocal word) vaguely refers to human ability or disposition to recognize some (not too large) counts of items that are perceived, remembered, or imagined. The allusion to perceptual "senses" (vision, hearing, touch, etc.) may be intuitively apt but it is worth elaborating a little (cf. [Hav09a, Chapter 5]).

Let us start with the intended meaning of the general notion of having (or being endowed with) *the sense of X* where *X* is a certain predetermined quality (in our case, somewhat oddly, the term "quality" refers to numerical "quantity"). First, such "sense" is something to be attributed to a person who is, so to speak, the possessor or owner of the said sense, and second, "having the sense of" refers to one's *disposition*, rather than to factual *employing* such disposition in a concrete situation (analogically to, say, the sense of humor, sense of responsibility, etc.). This subtle distinction is not always properly taken care of in the scientific literature but it is usually implied by the context. It may turn out to be particularly relevant to cognitive science which attributes various "senses of . . ." or "feelings for . . . " to conscious subjects. I shall call such subjective dispositions, or "senses of something," simply *inner senses* (take it just as a technical term).

Most such inner senses are implicit, pre-reflective features of our everyday experience even if sometimes we may subject them to conscious reflection, especially when they are actually being employed. This holds, mutatis mutandis, for the usual perceptual senses as well as for the sense of counts. All of them are subjectively experienced as well as objectively inferable dispositions. Unlike perceptual senses, the sense of counts does not have its own dedicated physiological organ; rather it indirectly utilizes various perceptual organs (as for its dedicated brain area there exist various conjectures).

We may tentatively put forth the idea of a minimal, pre-reflective inner sense of counts as something innate, already built into the very structure of experience. For this, however, we would have to distinguish various subcategories of counts: depending on whether they correspond to small, moderately large, or very large numbers (prematurely said).

Indeed, we directly perceive, *without* counting, very small counts.[4] For instance, we normally "see at a glance" the triplicity in triangles or tripods, quaternity in squares, quintuplicity in five-point stars—all that without any actual process of counting angles (or legs or vertices or tips)—but when the group becomes larger we gradually become wrong in a direct grasp of the count. This may happen around seven, eight, or more items in the group (depending on the individual and context). For larger groups we cannot but resort to a slower but more reliable actual counting procedure. Let us refer to this transition phase as to the *first horizon* of number apprehension. Then the *second horizon* of number apprehension might vaguely delimit what can be *conceivably* counted in practice (possibly in thought only); finally, numbers that are beyond the second horizon and stretch towards the potential infinity can only be grasped through indirect theoretical tools.

The sense of counts should be distinguished from another, perhaps originally independent *inner sense of numerosity* comprising the ability to notice that a certain group of entities either swells or shrinks, or that it is either larger or smaller than another group of entities. The sense of numerosity *need* not entail, in general, the ability of counting or the idea of a count.

Let us quote W. James' account of (the sense of) counts, namely the ideas of number, of the increasing number-series, and of the emergence of arithmetic [Jam07, pp. 653–4]:

> Number *seems to signify primarily the strokes of our attention in discriminating things. These strokes remain in the memory in groups, large or small, and the groups can be compared. The discrimination is, as we know, psychologically facilitated by the mobility of the thing as a total. But within each thing we discriminate parts; so that the number of things which any one given phenomenon may be depends in the last instance on our way of taking it. [. . .] A sand-heap is one thing, or twenty thousand things, as we may choose to count it. We amuse ourselves by the counting of mere strokes, to form rhythms, and these we compare and name. Little by little in our minds the number-series is formed. This, like all lists of terms in which there is a direction of serial increase, carries with it the sense of those mediate relations between its terms which we expressed by the axiom "the more than the more is more than the less." That axiom seems, in fact, only a way of stating that the terms do form an increasing series. But, in addition to this, we are aware of certain other relations among our strokes of counting. We may interrupt them where we like, and go on again. [. . .] We thus distinguish between our acts of counting and those of interrupting or grouping, as between an unchanged matter and an operation of mere shuffling performed on it. [. . .] The principle of constancy in our meanings, when applied to strokes of counting, also gives rise to the axiom that the same number, operated on (interrupted, grouped) in the same way will always give the same result or be the same.*

Some authors use the term *number sense* for "our ability to quickly understand, approximate, and manipulate numerical quantities" [Deh01]. In this study I dare to claim, probably despite James, that our ability to perform actual counting (above the rudimentary sense of counts) may *not* be a necessary prerequisite for the number sense. In my view, two inner senses—the direct sense of (small) counts and the sense of numerosity—may be more essential.

Can Animals Count?

There is a growing number of studies with animals exhibiting certain limited abilities to count and, as it is often claimed, to perform elementary arithmetic operations [Cal09]. We may be enticed to immediate hypotheses about the evolutionary origin of such abilities. This, under the prevailing Darwinian paradigm, would make us look for one or another survival advantage of such abilities, analogous to the advantages of having, say, the sense of colors, of shapes, of spatial directions, etc.

Perhaps we may conjecture that what appeared relatively early in the animal world might have been abilities that are not based on the process of counting. Such abilities may be primarily two: (1) the ability to identify small counts *at a glance*—a count of eggs, offspring, wolves in a pack, etc.—and (2) the ability to distinguish, without counting, between a smaller amount and a larger amount—of grain, leaves of grass, ants in a colony, etc.—(as James puts it, "the more than the more is more than the less.") Only much later, perhaps among humans, actual counting procedures came to be used, and after that there emerged the abstract concept of a number together with the idea of the "number line" endowed with various arithmetical operations.

Various experiments show that there are certain reasons to attribute the sense of (small) counts to animals. There are reports on primates, elephants, salamanders, chimpanzees, birds, even fish and bees, which can reliably recognize small counts of presented objects [Cal09]. The favorites are four-day-old chicks (thus no training could be assumed) that reportedly are able to correctly determine that $1 + 2$ is greater than $4 - 2$, that $0 + 3$ exceeds $5 - 3$, and that $4 - 1$ is more than $1 + 1$. However, one has to be careful with interpreting such experiments. Researchers frequently use certain appealing phrases like "number recognition" or even "arithmetic skills" when talking about animals. My own small survey of the literature has revealed, first, that such experiments with animals involved counts of some specific objects, usually objects with survival importance to the animal. Second, in most cases the tested abilities could be explained simply by the ability to discriminate between larger and smaller amounts, without any need to do actual counting. (The rash claims about "arithmetic" skills of animals could be enticed by the habitual tendency of us, numerate humans, to do counting, adding, subtracting, etc., even when we deal with relatively small counts.)

There are two suitable ideas that may support various hypotheses about evolutionary origins of the number sense. One idea was already mentioned: looking for obvious survival advantages. The second idea is more logical. For instance, the sense of counts already presumes other inner senses, namely the sense of sameness and difference, or more specifically, the sense of individuality of elements of a group and the sense of that very type of similarity which characterizes the group. (Surprisingly enough, James placed "ideas of difference and resemblance, and of their degrees" into a later, fourth position in his list of natural mental categories.)

Representing Numbers

How did we, the human species, develop the abstract *concept of a number*, as something implicitly related to counts but without reference to particular entities counted? Talking and thinking in a human way about abstract numbers is only possible with their appropriate symbolic representations.[5] Here comes the third meaning of the word "number," viz. that of a *numeral*. Let us use this word in the most general sense, including not only word numerals (like "one," "two," "three", . . .) and their combinations ("thirty-six") but also other kinds of symbol, or better, sets of symbols, that unambiguously represent (abstract) numbers. You can think of the usual graphical signs (1, 2, 3, 4, . . . , or 01, 10, 11, 100, . . . , or I, II, III, IV, . . .). Let us allow, for the purposes of the present study, even more general representations, for instance geometrical shapes, whether drawn, written, or merely imagined.

To avoid misunderstanding: I do not assume that the concept of numeral has to be derived from, or dependent on, a prior concept of (abstract) number. We could equally well associate numerals directly with counts of some entities, real or imagined. The simplest idea is the *analog representation* of counts (or numbers). Think, for instance, of scribbling down or imagining groups of some concrete objects like dots, strokes, tokens, marks, knots on a rope, etc.[6] Each such group can be also viewed as a symbolical numeral directly representing the count of items in the

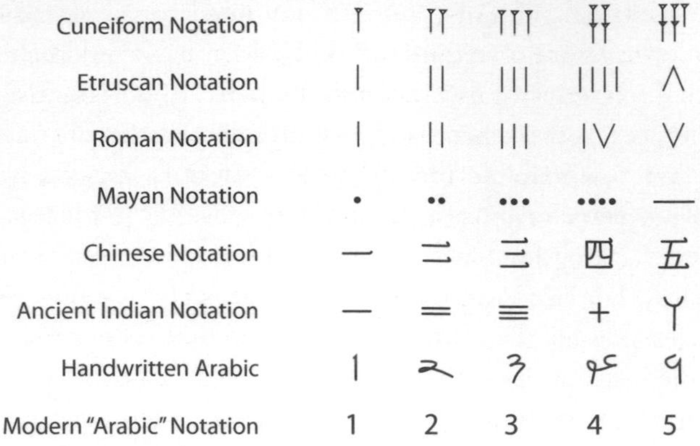

Cuneiform Notation	𒁹	𒈫	𒐈	𒐉	𒐊
Etruscan Notation	I	II	III	IIII	Λ
Roman Notation	I	II	III	IV	V
Mayan Notation	•	••	•••	••••	—
Chinese Notation	一	二	三	四	五
Ancient Indian Notation	—	=	≡	+	Y
Handwritten Arabic	1	2	3	4	5
Modern "Arabic" Notation	1	2	3	4	5

Figure 1. Numeral signs in various cultures (redrawn from [Ifr94]).

same group. In mathematics we speak about the unary numeral system for representing numbers.

As a matter of fact, many cultures use analog notation (signs) for three smallest numbers, sometimes even for four or five; see Fig. 1. Obviously, for slightly larger numbers (near the first horizon of number apprehension) a danger of misinterpretation may increase. Probably because of this and for the sake of compression specific signs have been used instead of analog signs. Moreover, and more importantly, arbitrarily large numbers can be represented by sequential juxtaposition of figures.

It is worth noting that some languages use different words for counts of different categories of object. This fact may suggest that the concept of count may be more natural than the concept of number.

It is interesting to note, in passing, that some languages use *object-specific* numeral systems which depend on the kind of objects counted [BB08]. For instance, on one of the island groups in Polynesia, tools, sugar cane, pandanus, breadfruit, and octopus are counted with different sequences of numerals. There is a current dispute about whether such object-specific counting systems were predecessors of the abstract conception of numbers and number line (Beller and Bender argue in the opposite).

Numbers Dancing in our Heads

We can follow up the above considerations in various directions. We have already observed that there are three different conceptualizations of the prearithmetic idea of number: (1) number as a *count* of some identifiable (maybe visible or tangible) items, (2) (invisible, intangible) *abstract number*, and (3) number as a *numeral* of certain type (visible, speakable, etc.). In theory we can easily point to inherent, indeed even necessary, interrelationships between these three conceptualizations. For instance, we may be interested in comparing three respective roads to infinity: (1′) the intuition of gradual but unbounded swelling the group of items toward larger and larger counts (passing over the first two to the third of the above mentioned horizons of number apprehension), (2′) developing a formal (axiomatic) theory of natural numbers, and (3′) assuming such numerical representations that allow for depiction of arbitrarily large counts (or of arbitrarily large abstract numbers).

I am not going to discuss the theoretical issue of infinity here. Instead I am going to pose a different question, which may be more important if human cognition is at issue. Let us look "inside our minds," so to speak, and ask whether and how far we (humans) could mentally grasp the idea of a number.[7] If we are able to "see at a glance" (i.e., without counting) small counts of things, why not venture into imagining an analogously direct access to much larger counts, or perhaps even to abstract numbers?

True, in our culture we are too captive in the framework of words (numerals), symbols (number signs), arithmetic operations, and indeed, of the whole number line. Speaking for myself, whenever I hear, say, the sound "thirty-six" I immediately happen to hear "six times six" (as a leftover of memorizing the multiplication table in primary school), or alternatively, I could imagine the formula "$36 = 2^2 \cdot 3^2$" (provided I were obsessed with prime number representations). Surprisingly enough, I never imagine a rectangle of six rows and six columns of dots (or something), or a prism made of small cubes, four horizontal, three vertical, and three backwards. Why not? Wouldn't it be easier to form mental images of various geometrical shapes, to remember them, and perhaps to manipulate them in various exciting ways?

Consider for example Fig. 2. In its upper part there are nine groups of dots; each group corresponds to a different count of dots (a single ton, pair, triplet, quartet, etc.). Let us fix the relative position of dots in each group by shaded lines. In the lower part of Fig. 2 only the lines are depicted. Why not treat these line figures as numeral signs, easy to be imagined and remembered? Notice that most of these figures are

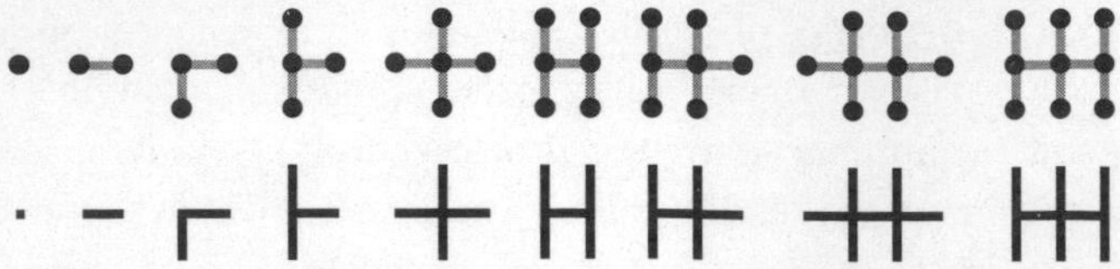

Figure 2. Turning dots into figures.

formed by adjoining previous figures in the sequence; this immediately leads to the tentative idea of using such an adjoining process for pictorially representing much larger numbers. Incidentally, each such numeral can be viewed as both analog and symbolic representation of a certain number.[8] (In one of the following sections this representation will be used to develop a more expressive pictorial representation of numbers.)

There is a plausible hypothesis that autistic savants with extraordinary numerical powers can mentally grasp numbers in some synesthetic way (pictorial, auditory, tactile), or maybe even in a form of some dynamic objects, rather than in the ordinary numeral representation.[9] In this respect it is worth mentioning the case of Daniel Tammet (known for his record of reciting 22,514 digits of π from memory). Let me quote some of Tammet's own reflections (from a recently published conversation with him in *Scientific American* [Leh09]):

> *I have always thought of abstract information—numbers for example—in visual, dynamic form. Numbers assume complex, multi-dimensional shapes in my head that I manipulate to form the solution to sums, or compare when determining whether they are primes or not. [. . .] In my mind, numbers and words are far more than squiggles of ink on a page. They have form, color, texture and so on. They come alive to me, which is why as a young child I thought of them as my "friends." [. . .] I do not crunch numbers (like a computer). Rather, I dance with them. [. . .] What I do find surprising is that other people do not think in the same way. I find it hard to imagine a world where numbers and words are not how I experience them! [. . . My] number shapes are semantically meaningful, which is to say that I am able to visualize their relationship to other numbers. A simple example would be the number 37, which is lumpy like oatmeal, and 111 which is similarly lumpy but also round like the number [numeral figure] three (being 37 × 3). Where you might see an endless string of random digits when looking at the decimals of* π*, my mind is able to "chunk" groups*

> *of these numbers [figures] spontaneously into meaningful visual images that constitute their own hierarchy of associations.*

This is a rare case of a savant able to report on his exceptional inner experience. No wonder such a report generates more questions than answers. As I already mentioned, contemporary discussions mostly seek solutions from brain research, which is expectable in view of the tremendous recent progress in various brain imaging techniques. My opinion is, however, that we could hardly expect from the brain scientist direct and valuable answers to phenomenally formulated questions about subjective experience. When Tammet says, for instance, that numbers "assume complex, multi-dimensional shapes," or that he "dances with them," it is of little help to the neuroscientist in his quest for adequate phenomena in the brain. By no means do I want to say that Tammet's reports are meaningless—quite conversely, I believe that his semi-metaphorical statements say more, in a sense, about the human mind than the empirical scientist could formulate in the language of neuronal dynamics.

The Riddle of Prime Twins

Here "prime twins" is a little pun: I do not mean twin primes (i.e. prime numbers of the form $p, p + 2$), but John and Michel, the famous autistic, severely retarded twins, studied in 1966 by the neurologist Oliver Sacks [Sac85]. I am not going to discuss here their significance for scientific (neurological or psychological) research but merely use them as a valuable source of, or motivation for, wild hypotheses about the way the human mind can deal with numbers in the extreme. Let us quote Sacks' own report on one of his encounters with the twins [Sac85, p. 191]:

> *[They] were seated in a corner together, with a mysterious, secret smile on their faces, a smile I had never seen before, enjoying the strange pleasure and peace they now seemed to have. I crept up quietly, so as not to disturb them. They seemed to be locked in a singular, purely numerical, converse. John would say a number—a six-figure number. Michael would catch the number, nod, smile and seem to savour it. Then he, in turn, would say another six-figure number, and now it was John who received, and appreciated it richly. They looked, at first, like two connoisseurs wine-tasting, sharing rare tastes, rare appreciations. I sat still, unseen by them, mesmerized, bewildered.*

Sacks wisely wrote down their numbers and later, back at home, he consulted a book of numerical tables and what he found was that all the six-figure numbers were *primes*! The next day he dared to surprise the twins and ventured his own eight-figure prime. The twins paused a little time and then both at the same time smiled. The exchange of primes between Sacks and the twins continued during the following days, with gradual increase of the length of the numeral up to one of twenty figures brought out by the twins, for which, however, Sacks had no way to check its primality. For generating and recognizing larger primes the twins needed more time, typically several minutes.

Let us point out four conspicuous aspects of the twins' performance: (i) their striking emotional fondness for the primes, even if (ii) "they [could] not do simple addition or subtraction with any accuracy, and [could] not even comprehend what multiplication or division means" [Sac85, p. 187], (iii) there were delays in their reactions—the larger the longer numerical lengths of the primes processed, and (iv) there were certain indications that visualization was at play, as Sacks writes about another of the twins' abilities (calendar calculation) [Sac85, p. 187]:

> *[Their] eyes move and fix in a peculiar way as they do this—as if they were unrolling, or scrutinizing, an inner landscape, a mental calendar. They have the look of 'seeing', or intense visualization [. . .]*

One of the above listed aspects (the third one, about delays) may suggest that production or recognition of a prime was not an instantaneous act but involved a certain internal "procedure" that consumed a certain time—a nontrivial time compared to the usually instantaneous responses of number prodigies. Well, what may such a procedure be?

It is not my intention to put serious effort into trying to explain the particular extraordinary powers of Sacks' twins[10] (or other number prodigies). I rather take up the twin case as an incentive to deal with numbers in alternative ways, not grounded in ordinary arithmetic.

Primality sans Arithmetic

A simple characterization of primes which is not based on arithmetical operations may be as follows: A number is a prime if and only if it corresponds to the count of items that cannot be arranged into a regular rectangle (except for a row or line). Imagine, for instance, a platoon of soldiers that cannot be drawn up into a rectangular formation (file by three, file by four, etc., except for a single row or line). Then we can easily design an effective finitary procedure *P* which, given a group *X* of individual elements (let us call them tokens), would systematically arrange it into various rectangular formations, always checking whether some of the tokens are left over or not. If the only possible arrangement without a leftover is a single row (or line) then the count of all tokens in *X* is a prime number.

The following four observations about such primality test *P* seem to be particularly relevant to our approach: (1) the procedure *P* may be carried out, in principle, with the help of visual imagery only, (2) feasibility of *P* in imagination could reach a certain (biological or psychological) limit if *X* happens to be too large, (3) the larger *X* the more steps may *P* require (hence more time would be needed for realizing *P*), and, last but not least, (4) in order to execute *P* there is no need to know the precise (numerical) count of tokens in *X*. What is only needed is a presentation of group *X* as a whole and allowing for simple mobility of its tokens. (Incidentally, one can imagine a mechanical apparatus that would realize the procedure).

Now, the oddity of prime number prodigies may be primarily related to point (2). With the current relatively moderate (albeit rapidly growing) scientific knowledge of the structure and dynamics of the brain, and without any knowledge at all about inner experiences of the prodigies, we can only conjecture that, for the prodigies, the feasibility horizon of procedure *P* must be substantially farther than for normal people.

Of course, there may exist (and probably do exist) some unimaginable, entirely different solutions for dealing with primes. We are still far from understanding even the particular case of Sacks' prime twins. How, in addition to some marvelous mental imagery, were they able to communicate about numbers also in the ordinary form of (Arabic) numerals or (English) words? And why did they sense precisely primes as something joyful?

Numbers Turned into Curves

Following up with the first question (about communication), I shall develop here an illustrative example of a relatively simple direct translation between two types of number representation: on the one hand, numerals in the usual form of sequences of decimal digits (or expressions formed from number names), and on the other hand, made-up pictorial "numerals" representing even huge numbers, preferably in a way that would allow "seeing them at a glance."

Let us discuss the latter issue, that of pictorial representation. To be more specific, consider the numeral system introduced in one of the preceding sections (cf. Fig. 2). There I mentioned the possibility of extending such system to arbitrarily large numbers. However, the presumed miraculous imagery of even the most prodigious savants cannot be unbounded. Under the theory of embodied mind, whatever the (yet unknown) nature of pictorial number representation may be, there have to be certain limits to it. Such limits could somehow resemble the first horizon of human direct number apprehension (as mentioned in the second section of this text), except that they may be located considerably far away ("far away" on *our* number line).

My speculative idea for a more powerful savant number representation (assume that it can be still called "representation") is based on the notion that it may have a *hierarchical nature.*[11] I shall illustrate it with a concrete example (admittedly fabricated and most likely without any relation to reality). In the first step choose a certain, perhaps small set N_0 of elementary numerals, for instance the nine figures in Fig. 2. This determines the first level of number representation. In the next step arrange elements of N_0 into a predetermined fixed formation, let us say into a 3×3 regular grid $G(9)$ with one additional point (for zero), as shown in Fig. 3. Obviously, in the Euclidean plane this grid can be uniquely determined by three reference points (a, b, c), also depicted in Fig. 3.

Imagine now a certain number n presented in the form of a decimal numeral (sequence of digits), say, $n = 5\ 950\ 425\ 853$. Draw a smooth curve through the grid, starting in point a and successively passing through all and only those vertices that represent digits 3, 5, 8, 5, 2, 4, 0, 5, 9, 5 in that order (for technical reasons in the reverse order comparing to n). There may be infinitely many such curves but each would represent precisely number n (one of them is the upper left curve in Fig. 4, the other curves in the figure represent other numbers; I included them for the readers who like pictures).

For the inverse procedure, consider a given planary curve (for instance one of those in Fig. 4) together with three reference points, a, b,

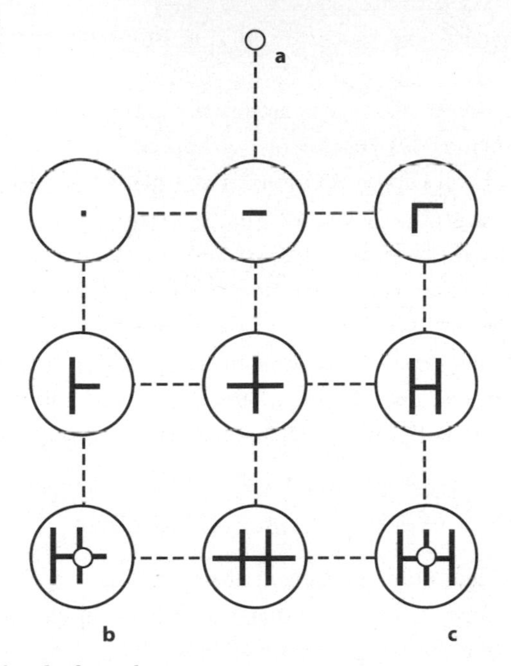

Figure 3. First level of number representation.

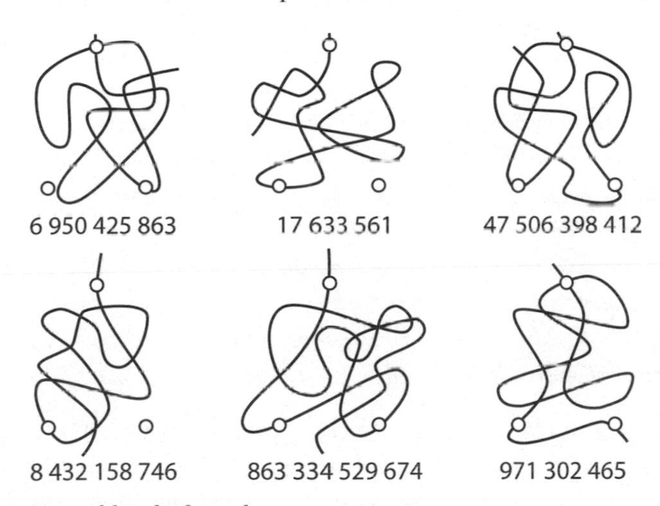

Figure 4. Second level of number representation.

and *c*. These points are sufficient to draw a unique regular grid (as in Fig. 3) that lies over the curve. Now proceed along the curve, starting in the upper point *a*, and list all the digits associated with vertices you pass through (you should list them from the right to the left). Eventually the generated numeral expresses the unique number represented by the curve.

I admit that the procedure is somewhat artificial and strange. Moreover, it has a serious drawback (for our approach): even if it is thinkable that the curve itself can be recognized at a glance (analogously to, e.g., human face recognition), the snag is the requirement of knowing its exact position with respect to three fixed reference points. Yet the proposed procedure has certain noticeable properties: first, one can in principle get along with visual imagery, second, there is no need of counting anything, and third, no standard arithmetical operations are used. Yet the procedure allows for handling considerably large numbers (as seen in Fig. 4) with only a modest increase of complexity.

Seeing Primes All at Once

As a matter of fact, quite a different alternative track of thought may be pursued, too. It is conceivable that number prodigies do not care so much about concrete numbers and their individual properties like primality. Let me quote in this respect a passage about a synesthetic subject from [RH03, pp. 56–7]:

> *If asked to visualize numbers the subject finds that they are arranged in a continuous line extending from one point in the visual field to another remote point—say from the top left corner to bottom right. The line does not have to be straight—it is sometimes curved or convoluted or even doubles back on itself. In one of our subjects the number line is centered around "world centered" coordinates—he can wander around the 3D landscape of numbers and "inspect" the numbers from novel non-canonical vantage points. Usually the earlier numbers are more crowded together on the line and often they are also coloured.*

Obviously a modified version of such a process may be hypothesized for the case of prime number savants like Sacks' twins. They may be miraculously gifted with a sort of direct access to a certain *whole set* of primes, small or large (certainly not to the set of *all* primes in the mathematical sense). They can perhaps visualize the entire set as a single, very complex geometrical object in which primes play a salient role, while nonprimes fill the space between or around them. A very simple illus-

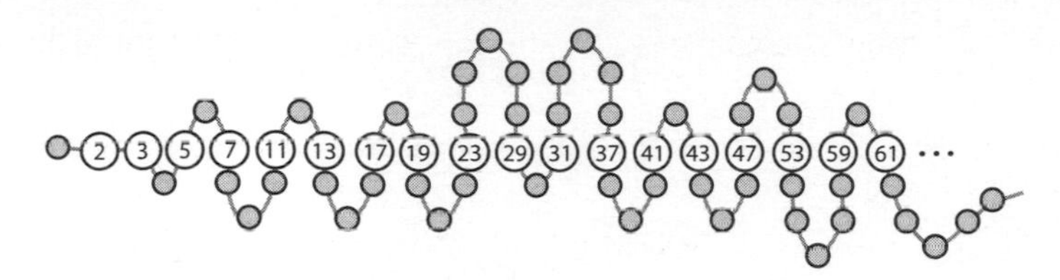

Figure 5. Prime number line and non-prime number line.

trative example of such an idea is depicted in Fig. 5 where the ordinary number line is curved around a straight "prime-number line".

Now is a good time to comment on the question concerning Sacks' prime twins who, even though severely retarded, paid special attention precisely to primes. Why was it just primes, out of all numbers, that made them so happy? A weird but plausible idea may be that the geometrically prominent positions of primes on or inside the complex geometrical object mentioned above would charge them with some emotionally strong, perhaps esthetic quality.

Another, easier guess may proceed from our earlier characteristic of primes: a prime number corresponds to such a count of elements of a group that wriggles out of all (nontrivial) rectangular formations. With a grain of aphorism: an entity that resists something as banal as a rectangular formation deserves a joyful welcome.

Conclusion

I outlined several purposefully simple but highly speculative ideas about mental number processing, motivated by some extraordinary cases of numerical savants. I admit that the ideas are far from being mutually compatible. The reader may have noticed, for instance, that there is no clear way to combine the curved-line pictorial representation of numbers with the procedure of primality testing. The remarkable powers of Sacks' twins and of other number prodigies remain, and probably will remain for a long time, a mystery.

However, my motivation for seeking various number representations and procedures did not consist in trying to suggest some explanatory tools and even less did it consist in proposing something of practical use. The only purpose was heuristic: to open possibilities of quite unusual lines of reasoning about cognition and arithmetic.

Dedication

This essay is dedicated to Petr Hájek, my teacher, colleague, and friend.

Notes

1. In this paper I expand on some of the ideas briefly presented in my earlier essays, e.g., [Hav08, Hav09b]. The work on this paper was funded by the Research Program CTS MSM 021620845.

2. Throughout this paper the term "number" always means "natural number."

3. What is characteristic of abstract numbers is that they do not emerge (or they are not constructed) individually but always together with some (or all) other numbers.

4. Let us live with this little ambiguity in terms—by the verbal form (*to count*) I refer to an active temporal process aimed at determining *the count*.

5. In this point I differ from the view of some scientists like Dehaene (quoted above) and Jean-Pierre Changeux. They associate the number sense with activation of specialized neuronal circuits in the brain (this applies also to non-human animals) so that human language and consciousness are not assumed to be a prerequisite for dealing with numerical quantities.

6. Perhaps even temporal sequences of events like sounds of a tolling bell.

7. For philosophical reasons I am not comfortable with the term "introspection" but the reader may happily make do with it.

8. The adjoining procedure may lead even to 2D or 3D figures.

9. The term synesthesia refers to perceptual experience in multiple modalities in response to stimulation in one modality. For some people, for example, letters or numbers (numerals) evoke vivid color sensations [RH03].

10. In response to Sacks' report, several researchers published various theoretical speculations. However, most of them (if not all) are based on ordinary arithmetical properties of primes (cf., e.g., [Yam09]).

11. In fact, the positional numeral system is hierarchical too, since each position represents different power of 10. In our case the word "hierarchical" should point to different representational strategy associated with each level.

Bibliography

[BB08] S. Beller and A. Bender. The limits of counting: Numerical cognition between evolution and culture. *Science*, 319:213–5, 2008.

[Cal09] E. Callaway. Animals that count: How numeracy evolved. *New Scientist*, 2009.

[Deh97] S. Dehaene. *The Number Sense. How the Mind Creates Mathematics.* Oxford University Press, New York, 1997.

[Deh01] S. Dehaene. Précis of the number sense. *Mind and Language*, 16(1):16–36, 2001.

[Deh02] S. Dehaene. Single-neuron arithmetic. *Science*, 297:1652–3, 2002.

[Hav08] I. M. Havel. Zjitřená mysl a kouzelný svět. *Vesmír*, 87:810, 2008. In Czech.

[Hav09a] I. M. Havel. Uniqueness of episodic experience. In *Kognice 2009*, pages 5–24, Hradec Králové, 2009. Gaudeamus, 2009.

[Hav09b] I. M. Havel. Vidět počty a čísla. *Vesmír*, 88:810, 2009. In Czech.

[Ifr94] G. Ifrah. *Histoire Universelle des Chiffres*, volume I et II. Robert Lafont, Paris, 1994.

[Jam07] W. James. *The Principles of Psychology*, volume II. Cosimo Inc., New York, 2007. Originally published in 1890 in Boston.

[Leh09] J. Lehrer. Inside the savant mind: Tips for thinking from an extraordinary thinker. *Scientific American*, 2009. http://www.scientificamerican.com/article.cfm?id=savants–cognition thinking.

[RH03] V. S. Ramachandran and E. M. Hubbard. The phenomenology of synaesthesia. *J. of Consciousness Studies*, 10(8):49–57, 2003.

[Sac85] O. Sacks. The twins. In *The Man Who Mistook His Wife for a Hat*, pages 185–203. Pan Books, London, 1985.

[Yam09] M. Yamaguchi. On the savant syndrome and prime numbers. *Dynamical Psychology*, 2009. Electronic journal, http://www.goertzel.org/dynapsyc/yamaguchi.htm.

Autism and Mathematical Talent

Ioan James

Autism is a developmental or personality disorder, not an illness, but autism can coexist with mental illnesses such as schizophrenia and manic-depression. It shows itself in early childhood and is present throughout life; sometimes it becomes milder in old age. Nowadays it is recognized as a wide spectrum of disorders, with classical autism, where the individual is wrapped up in his or her own private world, at one extreme. It is estimated that in the United Kingdom slightly under one percent of the population, about half a million people, have a disorder on the autism spectrum. The corresponding figure for other countries is not available, although it is unlikely to be very different. Autism is present in all cultures and, as far as we know, has existed for untold generations.

Hans Asperger, a Viennese psychiatrist, found that some of his patients had a mild form of autism, with distinctive symptoms that later became known as Asperger's syndrome. He was not the first to describe the syndrome but he may have been the first to recognize a connection with mathematical talent. As he observed (see Frith [13]):

> to our own amazement, we have seen that autistic individuals, as long as they are intellectually intact, can almost always achieve professional success, usually in highly specialized academic professions, often in very high positions, with a preference for abstract content. We found a large number of people whose mathematical ability determines their professions.

Later he wrote,

> It seems that for success in science or art a dash of autism is essential. For success the necessary ingredient may be an ability to turn away from the everyday world, from the simple practical, an ability to rethink a subject with originality so as to create in new untrodden ways, with all abilities canalised into the one speciality.

He went on to describe autistic intelligence—a kind of intelligence untouched by tradition and culture—unconventional, unorthodox, strangely pure, and original. The ability to immerse oneself wholeheartedly in work or thought is something that is seen time and time again in the Asperger genius.

Asperger's syndrome is not the only form of autism with this connection. The Irish psychiatrist Michael Fitzgerald, for example, tells me that virtually all the people he diagnoses as autistic have an interest in mathematics. Their greatest wish, he says, is to bring the world under the control of pure reason, to create order and meaning out of the chaos that they experience around them, particularly in the puzzling social domain. Such people are naturally attracted to science, especially to the mathematical sciences, since mathematicians tend to create order where previously chaos seemed to reign. He attributes this attraction to a feeling of security that they find in the rational world of mathematics, which compensates for their inability to make sense of the mysterious social world.

Much has been written about this, and the general public are now more aware of the presence of mildly autistic people in everyday life. Since I first wrote about autism in mathematicians in the *Intelligencer* [20] some years ago, more has been learned about the disorder and more has been published. In this follow-up article I begin by describing research that places the link between autism and mathematical talent on a firmer footing. Then I describe some of the more recent case studies of Asperger geniuses in mathematics and associated subjects.

Simon Baron-Cohen, director of the Autism Research Centre in Cambridge, has tried to put the connection on a more quantitative basis. For this purpose he devised a self-administered questionnaire for measuring the degree to which an adult with normal intelligence has the traits associated with the autistic spectrum. From the answers to the questions a number is obtained, which he calls the autistic-spectrum quotient, providing an estimate of where a given individual is situated on the continuum from normality to autism. (Anyone who wishes to take the AQ test will easily find it by Googling Simon Baron-Cohen.) When the questionnaire was administered to students at Cambridge University, interesting results were obtained. Briefly, scientists scored higher than nonscientists; and within the sciences, mathematicians, physical scientists, computer scientists, and engineers scored higher than the more human or life-centered sciences of medicine and biology. Full statistical details are provided in [4] and [5]. This research was taken a step further in [6], where among 378 undergraduates reading mathematics at Cambridge there were seven who reported a formally diagnosed autism spectrum condition, whereas there was only one among 414 students in a control group of Cambridge undergraduates reading medicine, law, or social science.

In the mathematical world, the establishment of a link between autism and mathematical talent will come as no surprise, but its recognition may have significant practical consequences for education and for choice of occupation. At school autism is regarded as a learning disability; its positive side should be recognized. Children with mild autism, who get on well in mathematics, may struggle with other subjects. They are likely to perform poorly at interviews, when they apply for a job, but they may be good at the right kind of work, for example in information technology, where their special abilities are appreciated. Although the disorder is a handicap in many ways, in others it is a great advantage. For the majority, life is a struggle, and only a minority make a success of it. There can be no doubt that gifted individuals with some degree of autism have contributed a great deal to research in mathematics. Not always, however; the tragic lives of Robert Amman [25] and William Sidis [28] show what can go wrong.

When combined with high intelligence, as it often is, autism is associated with outstanding creativity, particularly in the arts and sciences. An enormous capacity for curiosity and a compulsion to understand are evident in those who have the syndrome, as is a tendency to reject received wisdom and the opinions of experts. They often suffer from depression, and mathematical work can have an antidepressant effect. Work is a form of self-expression for the autistic who finds other forms of expression difficult; it boosts their often low self esteem.

The link with autism may throw fresh light on some aspects of mathematical creativity. More than a hundred years ago Henri Poincaré addressed a conference of psychologists in Paris on *Mathematical Creation* (translated by Halsted [17]). Poincaré's disciple Jacques Hadamard wrote a well-known monograph [16] on *The Psychology of Invention in the Mathematical Field*, which is mainly about mathematical creativity; a more recent discussion of this may be found in Changeux and Connes [7]. Much has been written about creativity in general, much of which applies to mathematical creativity, but Nettle [24] emphasizes that this differs from creativity in the arts. In a recent survey, comparing the psychology of a small sample of research mathematicians with poets and visual artists, Nettle finds that the cognitive style of the mathematicians was associated with convergent thinking and autism, whereas poetry and art are more associated with divergent thinking, schizophrenia, and affective disorders, such as manic-depression. (Divergent thinking means the ability to create new ideas based on a given topic; convergent thinking means the ability to find a simple principle behind a collection of information.)

In the history of mathematics it is not difficult to find people who may have had Asperger's syndrome, although without the right kind of biographical information we cannot say for sure whether each person had the syndrome or not. It is much less common among females than among males; it is difficult to find an example of an outstanding woman mathematician who was a clear case. It is not uncommon for individuals to have only a few features of the syndrome, not the full profile. Examples of well-known mathematicians who showed more than a trace of Asperger behavior, without necessarily meeting all the diagnostic criteria, are Paul Erdös, Ronald Fisher, G. H. Hardy, Alan Turing, André Weil, and Norbert Wiener. A detailed analysis for Srinavasa Ramanujan has been provided by Fitzgerald [10], for William Rowan Hamilton by Walker and Fitzgerald [27]. Some other cases are discussed by Fitzgerald and James [12]), whereas Baron-Cohen [1] describes one (who was, in fact, a Fields Medalist)

Sheehan and Thurber [26] have suggested that John Couch Adams had the disorder and that this lay behind both his success in identifying the unknown planet Neptune as the cause of anomalies in the orbit of Uranus and also his failure to persuade the Astronomer Royal to search for it in the orbit he had calculated. Most of those who encountered the mathematical physicist Paul Dirac have a story to tell about his eccentricity. His recent biography by Farmelo [8] describes his aloofness, defensiveness, determination, lack of social sensitivity, literal-mindedness, obsessions, physical ineptitude, rigid pattern of activities, shyness, verbal economy, and much else. Some features of his complex personality can be attributed to his strange upbringing but most of it goes with some form of autism.

Some people are critical of linking the syndrome with persons of genius. There is often strong resistance from the general public to any suggestion that a famous person might have had Asperger's, but this is generally because of the popular misunderstanding of the nature of the disorder. People who are otherwise well informed find it difficult to believe what some of those with this disorder may be capable of achieving. Attempts at diagnoses of individuals no longer alive often result in controversy when experts differ and amateurs also become involved. Unless one is absolutely sure, it is advisable to be careful, for example, to say that someone displayed autistic traits rather than that person was autistic, even when the case is a strong one, since otherwise the diagnosis is liable to be questioned. Some of the standard books on the subject, notably Frith [14], discuss the problems of historical diagnosis. On the one hand, to know that there have been outstanding Asperger mathematicians impresses the rest of us and enhances the self-esteem of gifted people with the syndrome. On the other hand, those who are not so gifted may feel depressed that they cannot aspire to mathematical fame.

References

[1] Baron-Cohen, S., *The Essential Difference: men, women and the extreme male brain*. Allen Lane, London, 2003.

[2] Baron-Cohen, S., et al., Does autism occur more often in families of physicists, engineers and mathematicians? *Autism* **2** (1998), 296–301.

[3] Baron-Cohen, S., et al., A mathematician, a physicist, and a computer scientist with Asperger syndrome: performance on folk psychology and folk physics test. *Neurocase* **5** (1999), 475–83.

[4] Baron-Cohen, S., et al.,The autism-spectrum quotient (AQ): evidence from Asperger syndrome/high-functioning autism, males and females, scientists and mathematicians. *J. of Autism and Developmental Disorders* **31** (2001), 5–17.

[5] Baron-Cohen, S., et al.,The systemizing quotient: an investigation of adults with Asperger syndrome or high-functioning autism, and normal sex differences. *Philosophical Transactions of the Royal Society, Series B (special issue on autism mind and brain)* **358** (2003), 361–740.

[6] Baron-Cohen, S., et al., Mathematical talent is linked to autism. *Human Nature* **18** (2007), 125–31.

[7] Changeux, J-P. and Connes, A., *Conversations on Mind, Matter and Mathematics*. Princeton University Press, Princeton N.J., 1995.

[8] Farmelo, Graham, *The Strangest Man: the Hidden Life of Paul Dirac Quantum Genius*. Faber and Feabre, London, 2009.

[9] Fitzgerald, M., Is the cognitive style of persons with Asperger's syndrome also a mathematical style? *J. of Autism and Developmental Disorders*, **30** (2000), 175–6.

[10] Fitzgerald, M., Asperger's disorder and mathematicians of genius. *J. of Autism and Developmental Disorders* **32** (2002), 59–60.

[11] Fitzgerald, M., *Autism and Creativity*. Brunner Routledge, Hove, 2004.

[12] Fitzgerald, M., and James, I.M., *The Mind of the Mathematician*. Johns Hopkins University Press, Baltimore, Md., 2007.

[13] Frith, Uta (Ed.), *Autism and Asperger Syndrome*. Cambridge University Press, Cambridge, 1991.

[14] Frith, Uta, *Autism: Explaining the Enigma*. Basil Blackwell, Oxford, 2003.

[15] Grandin, Temple, *Thinking in Pictures*. Vintage Books, New York, 1996.

[16] Hadamard, J., *The Psychology of Invention in the Mathematical Field*. Princeton University Press, Princeton, N.J., 1945.

[17] Halsted, G. B., *The Foundations of Science*. Science Press, Philadelphia, Penn., 1946.

[18] Hermelin, Beate, *Bright Splinters of the Mind*. Jessica Kingsley, London and Philadelphia, 2001.

[19] James, Ioan, Singular scientists. *J. Royal Society of Medicine* **96** (2003), 36–9.

[20] James, Ioan, Autism in Mathematicians. *Mathematical Intelligencer* **25**, No. 4 (2003), 62–5.

[21] James, Ioan, *On Mathematics, Music and Autism*. In Bridges London (Reza Sarhangi and John Sharp, Eds.), Tarquin Publications, London, 2006.

[22] Ledgin, Norm, *Diagnosing Jefferson: Evidence of a Condition that Guided his Beliefs, Behaviour and Personal Associations*. Future Horizons, Arlington, Tex., 2000.

[23] Ledgin, Norm, *Asperger's and Self-Esteem: Insight and Hope through Famous Role Models*: Future Horizons, Arlington, Tex., 2002.

[24] Nettle, Daniel, Schizotypy and mental health amongst poets, visual artists, and mathematicians. *Journal of Research in Personality* **40** (2006), 876–90.

[25] Senechal, M. The Mysterious Mr Ammann. *The Mathematical Intelligencer*, **26**(4) (2004), 10–21.

[26] Sheehan, W., and Thurber, S., John Couch Adams's Asperger syndrome and the British non-discovery of Neptune. *Notes Rec. R. Soc.* **61** (2007), 285–99.

[27] Walker, Antoinette, and Fitzgerald, Michael, *Unstoppable Brilliance*. Liberties Press, Dublin, 2006.

[28] Wallace, A., *The Prodigy*. E.P. Dutton, New York, 1986.

How Much Math Is Too Much Math?

Chris J. Budd and Rob Eastaway

What's It All About?

Mathematics is a difficult subject to communicate to the general public, for reasons that we will explore in this article. It also takes time and energy to communicate it well. So why do we bother communicating in the first place, and what do we hope to achieve when we attempt to communicate math to any audience, whether it is a primary school class, bouncing off the walls with enthusiasm, or a bored class of teenagers on the last lesson of the afternoon? We always have to tread a narrow line between boring our audience with technicalities at one end, and watering math down to the extent of dumbing down the message at the other. So, at the risk of stating the obvious, here are some good reasons for giving a math talk to a general audience:

- To communicate some real mathematics, maybe including a proof and/or an extended argument.
- To get the message across that math is important, fun, beautiful, powerful, challenging, all around us, and central to civilization.
- To entertain and inspire our audience.
- To leave the audience wanting to learn more math (and more about math) in the future, and not to be put off it for life.

That's a tall order, but is it possible? While the answer is certainly YES, there are a number of pitfalls to trap the unwary along the way.

So, What's the Problem?

Let's be honest, we do have a problem in conveying the joy and beauty of our subject. A lot of important math is built on concepts well beyond what a general audience has studied. Also mathematical notation can be completely baffling, even for other mathematicians working in a different field. Here for example is a short quote from a paper by CJB about the partial differential equations describing the (on the face of it very interesting) mathematics related to how things combust and then explode:

Let $-\Delta\phi = f(\phi)$. *A weak solution of this PDE satisfies the identity*

$$\int \nabla\phi\nabla\psi\, dx = \int f(\phi)\psi\, dx \quad \forall\psi \in H_0^1(\Omega).$$

Assume that $f(\phi)$ *grows sub-critically; it is clear from Sobolev embedding that* $\exists\phi \in H_0^1$.

This quote (which is actually saying that we can find a solution to the equations for combustion) is meaningless to any other than a highly specialist audience. Trying to talk about (say in this example) the detailed theory and processes involved in solving differential equations with an audience which (in general) doesn't know any calculus, is a waste of everyone's time and energy. As a result it is extremely easy to kill off even a quite knowledgeable audience when giving a math talk. At best this will bore them (and we have been to many boring math talks ourselves which we have not understood at all) and at worst it will positively put them off mathematics. So the challenge remains: how can you 'talk math' in a way that engages with, and entertains, an audience which either doesn't kill them off or (perhaps even worse) completely trivialises the nature, content and ideas which make up the subject of mathematics.

Is There a Solution?

As with all things there is no one solution to this problem, and many different math presenters have adopted different (and equally successful) styles. However some techniques that we have found have worked with many audiences (both young and old), including the following.

- *Start with an application* of math relevant to the lives of your audience, for example Google, iPods, crime, dancing (yes, dancing). Hook them with this and then show, and develop, the math involved (such as in the examples above, network theory, matrix theory, and group theory).

- *Be proud* not defensive of the subject. Math really DOES make a difference to the world. If we can't be proud of it then who will be?
- *Show the audience the surprise* and wonder of mathematics. It is the counter-intuitive side of math, often found in puzzles or "tricks," that often grabs attention and can be used to reveal some of the beauty of math. There are many links between math and magic; many good magic tricks are based on theorems (such as fixed point theorems in card shuffling and number theorems in mind reading tricks). Indeed a good mathematical theorem itself has many of the aspects of a magic trick about it, in that it is amazing, surprising, remarkable, and when the proof is revealed, you become part of the magic too.
- *Link math to real people.* Many of our potential audiences think that math either comes out of a book or was carved in stone somewhere. Nothing could be further from the truth. *All math at some point was created by a real person,* often with a lot of emotional struggle involved or with argument and passion. No one who has seen Andrew Wiles overcome with emotion at the start of the BBC film *Fermat's Last Theorem*, can fail to be moved when he describes the moment that he completed his proof. Also stories such as the life and violent death of Galois or even the famous punch up surrounding the solution of the cubic equation, cannot fail to move even the most stony faced of audiences.
- *Don't be afraid to show your audience a real equation*. Stephen Hawking famously claimed that the value of a math book diminishes with every formula. There are, however, many exceptions to this. Even an audience that lacks mathematical training can appreciate the elegance of a formula that can convey big ideas so concisely. Some formulae indeed have an eternal quality that very few other aspects of human endeavor can ever achieve. Mind you, it may be a good idea to warn your audience in advance that a formula is coming so that they can brace themselves. So here goes:

$$\frac{\pi}{4} = 1 - \frac{1}{3} + \frac{1}{5} - \frac{1}{7} + \frac{1}{9} - \frac{1}{11} + \frac{1}{13} - \cdots$$

Isn't that sheer magic? You can easily spend an entire lecture talking about that formula alone. When one of us (CJB) was asked to "define mathematics" that was his answer. And, if you want a little controversy, you can always wow your audience with Ramanujan's famous "formula":

$$1 + 2 + 3 + 4 + 5 + \cdots \text{“=”} -1/12.$$

That's one for the specialists, but that certainly will amuse, entertain, and inform a sixth form audience who think that they know it all. (Okay, it looks mad, but in a certain sense it is true and it leads to interesting results about the Riemann zeta function.)

- *Above all, be extremely enthusiastic*. If you enjoy yourself then there is a good chance that your audience will too. Look your audience in the eye, try always to imagine things from their perspective, and (if you are a man) make sure that your fly is up!

So, what can go wrong? Unfortunately, still lots. So, in the spirit of Goldilocks and the Three Bears we will now look at two sets of examples, the first of which has "too much math," the second of which has "too little math" before concluding with some cases where the math is "just right."

Too Much Math

We have already seen an example of where too much math in a talk can blow your audience away. It is incredibly easy to be too technical in a talk, to assume too much knowledge, and to fail to define our notation. We've all been there, either on the giving or the receiving end. The key to the level of mathematics to include is to find out about your audience in advance. In the case of school audiences this is relatively easy—knowing the year group and whether you are talking to top or bottom sets should give you a good idea of how much math they are likely to know. Yet too often we have seen speakers standing in front of a mixed GCSE [General Certificate of Secondary Education] group talking about topics like dot products and differentiation and assuming that these concepts will be familiar.

One of the ways of engaging audiences in math is by relating it to everyday life. This can, however, be taken too far. Taking a topic that is of general interest—romance, for example—and attempting to "mathematize" it in the hope that the interest of the topic will rub off on the math can backfire badly.

Much of the math that gets reported in the press is like this. Although we love the use of formulae when they are relevant, the use of irrelevant formulae in a talk or an article can make math appear trivial. One of us was once rung up by the press just before Christmas and asked for the "formula for the best way to stack a fridge for the Christmas dinner." The correct answer to this question is that there is no such formula, and an even better answer is that if anyone was able to come up with one they would (by the process of solving the NP-hard Knapsack problem) pocket $1,000,000 from the Clay Foundation. However the journalist concerned seemed disappointed with the answer. No such reluctance however got in the way of the person that came up with

$$K = \frac{F\left(T + C\right) - L}{S}$$

Which is apparently the formula for the perfect kiss. All we can say is: whatever you do, don't forget the brackets.

For the mathematician collaborating with the press this might seem like a great opportunity to get math into the public eye. To the journalist and the reading public, however, more often it is simply a chance to demonstrate the irrelevance of the work done by "boffins."

The public are not always hostile to the use of math to tackle everyday problems. Sometimes there is an understanding that math is essential, and that it can be effective. But while the public might accept the results, they rarely want to be presented with the detail, especially if that detail involves math that requires considerable thought. One example of this has been the invention by Frank Duckworth and Tony Lewis of an ingenious mathematical way of deciding the fair result of a cricket match that is interrupted by rain. Duckworth and Lewis have been frustrated by the reluctance of the media to give them an opportunity to explain their method. The reason for that reluctance, however, is that just as with a piece of technology, people want to know that it does work, but they don't generally have the time or inclination to know *how* it works.

This doesn't mean that communicators should avoid problems that involve difficult math. Sometimes a puzzle or problem can capture the imagination and have an intriguing answer, but an extremely complex solution. Fermat's last theorem is perhaps the most famous example, though it took great skill for Simon Singh to delve into the complexities of the proof without killing the excitement before getting to the punchline. Many math talks and articles go wrong because they lack ruthless editing, diving too deeply into the minutiae and losing their audience as a result.

Too Little Math

It is equally dangerous to put in too little math and to water down the mathematical content so that it becomes completely invisible, or (as is often the case) to talk only about arithmetic and to miss out math all together. Unfortunately in live media this can be hard to avoid, as both of us have found out to our cost when we have been asked to do tough mental arithmetic questions live on air. With a few notable exceptions, most producers and presenters in the media think that any math is too much math and that their audience cannot be expected to cope with it at all. But this only highlights the real challenge of presenting math in the media where time and production constraints make it very hard indeed to present a mathematical argument.

In his Royal Institution Christmas Lectures in 1978, Professor Christopher Zeeman spent 12 minutes proving that the square root of 2 was irrational. It is hard to think of any mainstream prime time broadcast today where a mathematical idea could be investigated in such depth. A couple of minutes would probably be the limit, far too short a time to build a proof. Perhaps at some point in the future this will change, but for the time being, math communicators have to accept that television is a very limited medium for dealing with many accessible mathematical ideas—though the good news is that there are other media that offer more scope—books, podcasts, and YouTube, for example.

More Math Than You Might Expect

We'll conclude with some examples of topics that contain a far higher level of math in them than might be anticipated. Part of the secret is that it is always possible to develop a mathematical argument live provided that you engage with your audience, acknowledge the audience's varying math levels, and show lots of enthusiasm.

Example 1: Aspergers

The following appeared in a popular book, which on the face of it had nothing to do with math:

> *A triangle with sides that can be written in the form* $n^2 + 1, n^2 - 1$ *and* $2n$ *(where* $n > 1$*) is right-angled.*
>
> *Show by means of a counter-example that the converse is false.*
>
> *[The full proof appears later in the book]*

The book in question was *The Curious Incident of the Dog in the Night-time*, which is a book about Aspergers written from a personal perspective. Millions of people have read this book, and many of these (who are not in any sense mathematicians) have read this part of it and have actually enjoyed, and learned something, from this.

The reason this worked was twofold. First, the math was put into the context of a human story, which made it easier for the reader to empathize with it. The second was that the author Mark Haddon used a clever device whereby he allowed the lead character to speak for math, while his friend spoke for the baffled unmathematical reader. As a result, Haddon (a keen mathematician) managed to sneak a lot of math into the book without coming across as a geek himself.

Example 2: Weather and Climate

One of the most important challenges facing the human race is that of climate change. It is described all the time in the media, and young people especially are very involved with issues related to it. From our perspective, climate change gives a perfect example of how powerful mathematics can be brought to bear on a vitally important problem, and in particular gives us a chance to talk about the way that differential equations can not only model the world, but are also used to make predictions about it. Much of the mathematical modeling process can be described and explained through the example of predicting the weather, with the audience led through the basic steps of

1. Making lots of observations of pressure, temperature, wind speed, etc.
2. Writing down differential equations which put the physical laws into a mathematical form and which then tell you precisely how these variables are related.

3. Solving the equations on a super computer, and the issues that this brings up in terms of prediction and the reliability of the prediction.
4. Constantly updating and checking the computer simulations with new data.

There are plenty of math/human elements to this story, starting from Euler's derivation of the first laws of fluid motion, the work of Navier and Stokes (the latter was a real character!), the pioneering work of Richardson (another great character) in numerical weather forecasting, and the modern day achievements and work of climate change scientists and meteorologists. However, the real climax of this talk should be the math itself, in particular the equations of the weather, which link the velocity u, pressure p, temperature T, density ρ, and moisture content q of the moist and solar heated air surrounding the rotating earth. Here they are in their full glory:

$$\frac{Du}{Dt} + f \times u + \frac{1}{\rho}\nabla p + g = \nu\nabla^2 u,$$

$$\frac{\partial\rho}{\partial t} + \nabla(\rho u) = 0,$$

$$C\frac{DT}{Dt} - \frac{RT}{\rho}\frac{D\rho}{Dt} = \kappa_h \nabla^2 T + S_h + LP,$$

$$\frac{Dq}{Dt} = \kappa_q \nabla^2 q + S_q - P,$$

$$p = \rho RT.$$

It may come as a surprise, but audiences generally like the "unveiling" of these equations despite not understanding them. A talk about mathematics can be exactly that, i.e., "about" mathematics. If the audience gains the impression that math is important, and that the world really can be described in terms of mathematical equations and that a lot of mathematics has to be (and still is being done) to make sense of these equations, then the talk to a certain extent has achieved its purpose. Talks on climate change often lead to a furious e-mail correspondence, particularly with sixth formers, who often ask many questions about the above equations, what they tell us about reality, how we solve them (or not) and whether they really do predict what the climate will be like in 50 years time.

Example 3: Imaging

Math saves lives. Many lives. One way to demonstrate this to any audience is to tell them that Florence Nightingale was actually a rather good statistician and that statistics plays a vital role in modern medicine. Another way is to talk about medical imaging. It is a fact both that modern medical imaging is an essential part of modern medicine and that medical imaging relies heavily on mathematics to work. This alone will interest almost any audience. The easiest way to explain imaging is to look at the example of computerized axial tomography (CAT), which uses X-rays. In CAT imaging a series of X-rays are shone through an object. It is a nice exercise to show that each X-ray has a prescribed angle θ relative to some base line and is at a distance ρ from a fixed point. Once this angle and distance are known then it is possible to measure the attenuation $R(\theta, \rho)$ of this particular X-ray as it passes through the body. The imaging problem can easily be motivated in terms of shadows. If you hold up an object to a strong light then it casts a shadow. As you change the orientation of the object, so it casts different shadows. The question is then, if we know all of the shadows, can we find the exact shape of the object. The answer to this question was found by the mathematician Radon in 1917 who showed, in the context of X-rays, that if $R(\theta, \rho)$ is known for enough angles and distances then the shape of the object (and in fact the details of its interior as well) can be determined exactly. Radon's formula can then be used to stun an audience. One of the problems with X-ray tomography is that to apply it the subject needs to be exposed to X-rays. Math again comes to the rescue. Modern mathematical techniques (in particular advances in the way that we invert the large matrices associated with tomography), now allow much more information to be extracted from the values of the X-ray attenuation values than before. This has two benefits. Firstly to get the same amount of information as before we can use much lower doses (and thus keep our patients more healthy). Secondly, for the same amount of X-rays as before we can find out much more detail about the object, and even plot it in real time. As an example of what is now possible (simply because of the use of good mathematics) the following figure shows, on the left a bee, on the right the gut of a bee, both obtained using a CAT scan of a beehive. (Both of these images were obtained by Mark Greco). The level of detail is astonishing and simply would not have been obtained without some very careful mathematics.

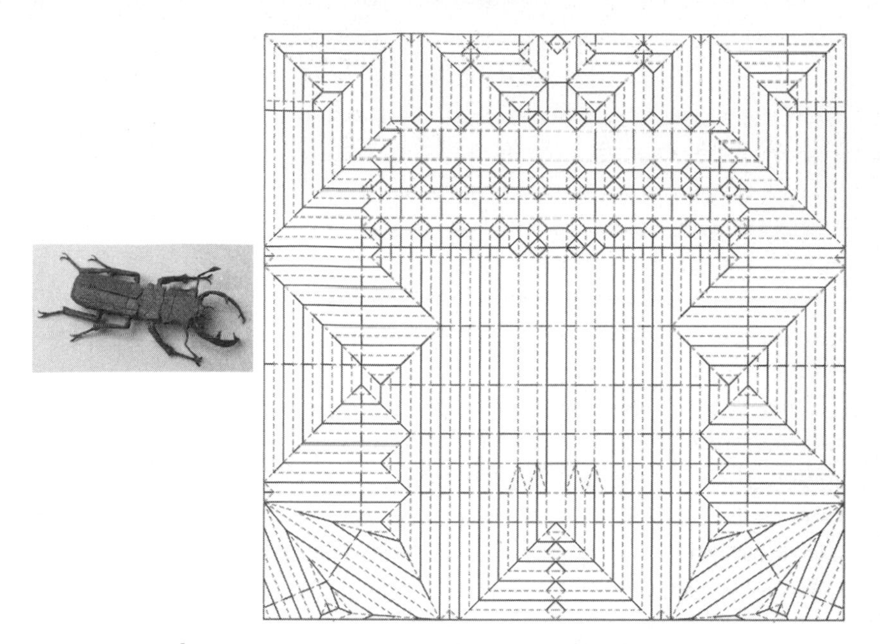

Figure 1. Beetle image. Images reprinted courtesy of Robert J. Lang, www.langorigami.com.

Example 4: Origami

As a final example, consider origami, a subject often considered to be a creative art. It is of course no coincidence that this should feature in a math presentation, as math is the most creative art that there is. Origami poses some nice mathematical questions. The most important being: given a particular shape, is it possible to find a folding pattern which will give that shape? Above is a model of a beetle folded from a single sheet of paper, and next to it the folding pattern of the paper. It is obvious from these images that a very significant amount of geometry is required to go from the model to the folding pattern. Huge advances in origami have been made in recent years by a number of highly gifted mathematical origamists, in particular Robert Lang. Their work has shown both how to go from a design to a folding pattern, and what sort of folding patterns correspond to real origami shapes. By doing this piece of mathematics it is now possible to design far more shapes than had ever been thought possible. It is great fun doing an origami workshop in which you both teach math through origami and origami through math. In fact it is possible to use origami to solve two of the great problems of the Greeks, namely tripling the angle and doubling the cube. Both are possible if you allow the paper to be folded. Explaining this is a very nice (but advanced lecture) for A-level students or undergraduates.

And Finally

Rocket Boys is the true story of a group of teenagers in West Virginia in the 1950s who are inspired by Sputnik and decide to develop rockets of their own.

Increasingly frustrated at their inability to project their missiles more than a modest distance, they begin to realize that they need some math to help them devise better technology. In one classic scene, they summon up the courage to approach their math teacher and ask him to teach them calculus, despite the fact that they are in the bottom set and barely able to comprehend simple algebra. After some resistance, the teacher relents.

This contrarian example of pupils wanting to do hard math shows that when they are sufficiently inspired and motivated to learn, people are prepared to take on math well beyond the level that might be expected of them.

As math communicators, our job is to foster that motivation, and to provide the inspiration.

Hidden Dimensions

Marianne Freiberger

That geometry should be relevant to physics is no surprise—after all, space is the arena in which physics happens. What is surprising, though, is the extent to which the geometry of space actually determines physics and just how exotic the geometric structure of our Universe appears to be.

One mathematician who's got first hand experience of the fascinating interplay between physics and geometry is *Shing-Tung Yau*. In a new book called *The Shape of Inner Space* (co-authored by Steve Nadis) Yau describes how the strange geometrical spaces he discovered turned out to be just what theoretical physicists needed in their attempt to build a theory of everything. *Plus Magazine* met up with Yau on his recent visit to London, to find out more.

Curvature and Gravity

An early indication that space is more than just a backdrop for physics came in 1915 when Albert Einstein formed his theory of general relativity. Einstein was working with a four-dimensional *spacetime*, made up of the three spatial dimensions we're used to and an extra dimension for time. His revolutionary insight was that gravity wasn't some invisible force that propagated through spacetime, but a result of massive bodies distorting the very fabric of spacetime. A famous analogy is that of a bowling ball sitting on a trampoline, which creates a dip that a nearby marble will roll into. According to general relativity, massive objects like stars and planets warp spacetime in a similar way, and thus "attract" other bodies that pass nearby.

Einstein's idea to unify space, time, matter, and gravity in this way was completely new—the physicist C. N. Yang has described it as an act of "pure creation." What wasn't new, however, was the mathematics

Figure 1. A doughnut mug.

Einstein used to describe the *curvature* of spacetime. This had been around since the 19th century, when the mathematician *Carl Friedrich Gauss* and after him his brilliant student *Bernhard Riemann* had come up with ways of measuring the curvature of an object from the "inside": they no longer needed to refer to a larger space the object might be sitting in. This intrinsic concept of curvature was just what Einstein needed.

"[At the time of Riemann] no-one believed that his new geometry would be useful," explains Yau. "But it turned out that it exactly suited Einstein's purpose. Without Riemann, Einstein would have taken much longer to develop general relativity. This then became an important reason for people to study geometry: geometry motivates physics and physics motivates geometry."

Gravity in a Vacuum?

When Yau first learned about general relativity he realized that it posed an interesting theoretical question: could there be a spacetime which contains no matter, a vacuum, in which there still is gravity? The spacetime we live in is not the only one that's possible in terms of general relativity. Einstein's field equations, which describe relativity, also permit other solutions, for example a spacetime without matter and without gravity, in which nothing happens at all. The question was whether a vacuum spacetime which still had some curvature and therefore gravity, was also possible. "In such a spacetime, gravity would be there because of the topology [the shape of the space], rather than because of matter," explains Yau.

Yau later realized that a geometric version of this question had been asked by the mathematician *Eugenio Calabi* in the 1950s. Calabi was interested in the interplay between the geometry of an object, that is precise features including size, angles, etc., and its *topology*. Topology is blind to exact measurements and only captures the overall form of an object. A sphere and a deflated football, for example, are very different geometrically, but they are topologically the same because one can be transformed into the other without any tearing or cutting. Similarly, topology regards a doughnut and a coffee cup as equivalent, because one can be morphed into the other. What differentiates the doughnut from a sphere is the fact that it has a hole.

An object with a given topology can be morphed into many different geometric shapes: a sphere into a deflated football, a pyramid, or a cube, etc. Calabi asked himself whether a shape with a certain kind of topology would admit a certain kind of geometry. And it turned out that if the answer was "yes," then the resulting object could be interpreted—in a general relativity setting—as a vacuum in which there was gravity.

Calabi's Question

There's no end to the variety of topological shapes you could think of, but topologists usually restrict their attention to what are called *manifolds*. These are objects that when viewed from up close, look like the ordinary "flat" space (called *Euclidean space*) we are used to. Spheres and doughnuts, for example, locally look like the flat 2D plane. If you're small enough, you won't notice their curvature, or whether there's a hole in them. You can easily draw a map of a patch of the sphere or doughnut on a flat piece of paper. So these are both 2D manifolds, also called surfaces.

Another thing the sphere and doughnut have in common is that they're *compact*: you'd only need a finite number of 2D maps to cover them. This means that they are finite in extent. Given a doughnut or sphere, you can always find a box to fit it into, even if it has to be a very big box.

But manifolds don't have to be two-dimensional. There are also 3D manifolds, which viewed from close up look like the familiar 3D Euclidean space given by three perpendicular coordinate axes. And since it's mathematically possible to think of Euclidean space in any dimension

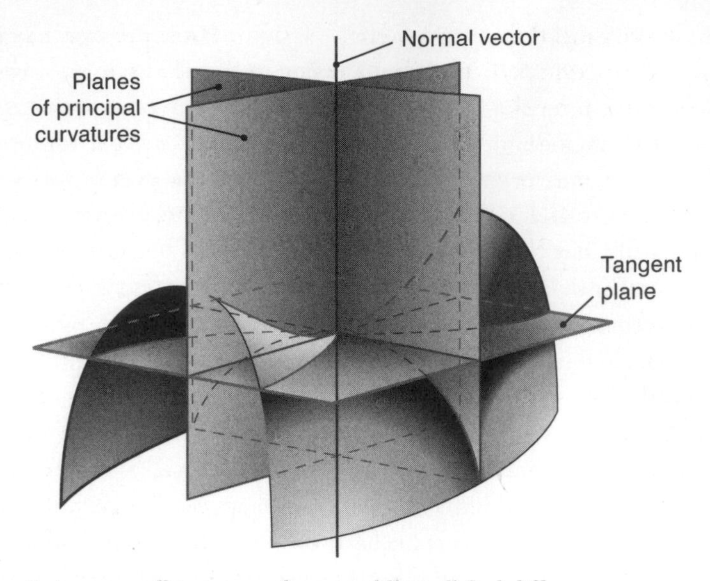

Figure 2. An ant walking around on a saddle will feel different curvature depending on its path. Image created by *Eric Gaba*.

you like (just use *n* coordinates rather than just three), there are manifolds of any dimension, too.

Calabi wanted to know what kind of geometry certain compact manifolds could exhibit; in particular, he was interested in curvature. Once you've given a topological manifold (say a sphere) a particular geometric structure (a deflated football), you can measure the curvature of the manifold at every point. This isn't entirely straightforward: an ant walking around on a saddle will feel upward curvature when it walks up the length of the saddle and downward curvature when it walks down the sides. In this example of a 2D manifold (which is what we consider the saddle to be), you can associate a notion of curvature to various 1D curves passing through a given point.

Similarly, in higher dimensions you can associate a notion of curvature to certain 2D surfaces that sit within the larger manifold and contain your point. Taking the average of all the curvatures associated to these 2D surfaces gives what's called the *Ricci curvature* at the point you're looking at. Since it's an average, Ricci curvature only captures one component of the full notion of curvature as defined by Riemann. This means that a manifold can have zero Ricci curvature at every point without being flat (or having zero Riemannian curvature) overall. In terms of physics, the component captured by Ricci curvature happens to be just the one that describes the curvature of spacetime that's induced by matter being present. So a space with zero Ricci curvature corresponds to a space with no matter—a vacuum in other words.

But Calabi was interested in Ricci curvature for purely geometrical reasons. The mathematician *Shiing-shen Chern* had shown in the 1940s that a manifold whose Ricci curvature is zero at every point can only have a certain kind of topology. In two dimensions, this corresponds to the rather boring topology of a doughnut. In higher dimensions the topology implied by zero Ricci curvature is a little harder to describe. Mathematicians say that manifolds which have that topology have a *vanishing first Chern class*.

Calabi turned Chern's question on its head: if you've got a compact manifold with a vanishing first Chern class, can you morph it into a geometric shape which has zero Ricci curvature at every point? Basically, what Calabi was asking is whether a certain type of topology guarantees that a certain type of geometry is possible. However, Calabi was not looking at any old manifold, but restricted his attention to so-called Kähler manifolds. These are easier to deal with because they deviate from Euclidean space in a limited way. They are also what's called *complex manifolds*: the maps that chart them preserve angles and the manifolds display a certain local symmetry. (The term *complex* refers to the fact that locally the manifolds look similar not just to plain old Euclidean space, but to something called complex space. In two dimensions this is just the complex plane you might be familiar with if you've studied complex numbers.) Being Kähler makes a manifold accessible to powerful mathematical machinery and also endows it with a special kind of symmetry.

Yau's Answer

When Yau first started working on this question in the early 1970s he was primarily motivated by geometry, though, as he tells us,

> it was always at the back of my mind that this would be interesting for physics: the construction of a closed universe [the compact manifold] that has no matter [since Ricci curvature is zero] but still has gravity [because of the curvature due to its topology]. But the existence of such a structure would also mean a lot to

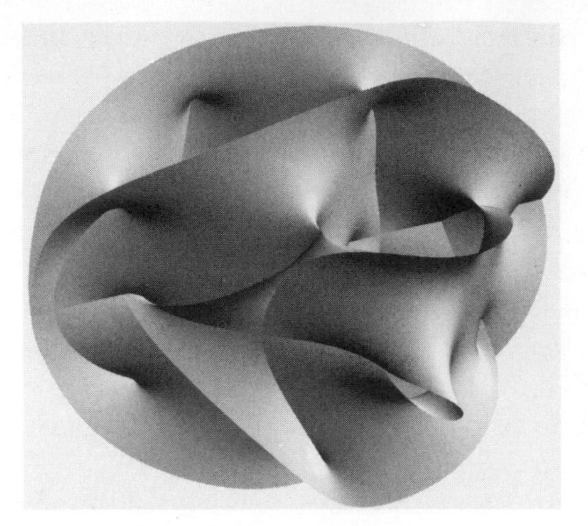

Figure 3. A 2D cross-section of a 6D Calabi–Yau manifold. Image created by *Lunch*.

geometry: Calabi's conjecture provided the clearest way to understand Ricci curvature.

At first Yau, like most other experts, believed that the answer to Calabi's question was "no." Since topology is a much looser concept than geometry, it seemed too much to expect that topology alone could guarantee such a special type of geometry. "For many years I tried to prove that the kind of manifold Calabi was after doesn't exist," he says. "But whatever I tried, I encountered difficulties. So I decided that nature cannot fool me so badly and that there must be something wrong with my reasoning."

In 1977 Yau finally proved that Calabi had been right. To state his result in its full glory, he proved that any compact Kähler manifold with a vanishing first Chern class could be endowed with a geometry with zero Ricci curvature. The kind of manifolds that fit this bill, and they exist in all dimensions, have since become known as *Calabi-Yau manifolds*.

Hiding Dimensions

In 1982 Yau received the Fields Medal, the highest honor in mathematics, for his resolution of Calabi's conjecture, which has had a major impact on geometry, as well as for other work. What he didn't know until a little later was that Calabi-Yau manifolds were just what some physicists were looking for. "I was in San Diego with my wife one day [in 1984], looking out at the beautiful ocean," he recalls. "The phone rang and it was my friends Andrew Strominger and Gary Horowitz. They were excited because string theorists were building up models of the Universe and needed to know whether [Calabi-Yau] manifolds really existed. I was happy to confirm that they did."

String theory is an attempt at a "theory of everything" which can explain all the physics in the Universe. Such a theory was, and still is, the holy grail of physics because the two major theories in existence, general relativity (which describes the macroscopic world) and quantum field theory (which describes the world at the sub-atomic scale) contradict each other. String theory resolves the mathematical contradictions by proclaiming that the smallest pieces of matter and energy aren't point-like particles, but tiny little strings. These strings can vibrate, just as guitar strings can vibrate, and the different types of vibration correspond to the fundamental particles and the physical forces we observe.

In the early 1980s string theory was in its infancy. One of its problems was that it needed ten dimensions to work in. Particles and forces were supposed to come from all the different ways in which the strings can vibrate. With less than ten dimensions there simply wouldn't be enough modes of vibration, not enough directions for a string to wriggle in, to produce all the physics we observe. With more than ten dimensions, on the other hand, string theory produced non-sensical predictions. So exactly ten dimensions it had to be. But then how come we can only perceive four of them, three for space and one for time?

String theory's answer to this riddle is that the six extra dimensions are rolled up tightly in tiny little spaces too small for us to perceive. "At each point of the 4D spacetime we observe there is in fact a tiny six-dimensional space," Yau explains. These tiny worlds that live at every single point in 4D spacetime are so small, we just can't see them. And what kind of six-dimensional geometrical structure can harbor this hidden world and also satisfy other requirements of string theory? You've guessed it: it has to be a six-dimensional Calabi-Yau manifold. "[Calabi-Yau manifolds] finally provided a concrete geometrical model for string theory," says Yau.

One reason that makes Calabi-Yau manifolds attractive for string theory is their compactness: the manifolds are extremely small, with a

Figure 4. Another cross section of a Calabi–Yau manifold. String theory claims that every point in spacetime is actually a tiny 6D world with the structure of a Calabi–Yau manifold.

diameter somewhere around 10^{-30}cm. That's more than a quadrillion times smaller than an electron. But there are other reasons too. To be consistent with the understanding of physics at the time, the manifolds harboring the hidden dimensions had to have zero Ricci curvature. What is more, string theory assumes a special kind of symmetry, called *supersymmetry*, which makes special demands of the geometry of spacetime. These demands make Calabi-Yau manifolds (with their special Kähler symmetry) excellent candidates for string theory, although we still don't know whether they are the only possible solution to the dimensional conundrum.

String Future

With the discovery of Calabi-Yau manifolds and other major advances, string theory experienced a landmark year in 1984. But the story didn't end there. A minor blow came in 1986 when it was discovered that string theory needed a slightly amended version of Calabi-Yau manifolds, whose

Ricci curvature wasn't zero, but only very nearly zero. What's more, there are many different 6D Calabi-Yau manifolds that could fit the string theory bill and, disappointingly, no one was able to work out which was the "right" one. All this somewhat undermined the manifolds' standing in physics. However, another boost came when it was discovered that pairs of different Calabi-Yau manifolds can give rise to a theoretical Universe with the same physics. This pairing of manifolds sparked new interest and gave rise to a new notion of symmetry, called *mirror symmetry* (rather misleadingly, since it's much more complicated that its name suggests).

The precise physical meaning of mirror symmetry is still a mystery, but, as Yau says, it led to "a spectacular new understanding of Calabi-Yau manifolds. It also had lots of rich mathematical consequences that were completely motivated by string theory intuition." In particular, the new notion of mirror symmetry provided a solution to a century-old problem in a nearly forgotten branch of geometry. We won't go into the problem here, but only say that it concerned counting the number of curves that live in particular geometric spaces. Mirror symmetry led to the formula which provided the answer, and its correctness was later proved by Yau and colleagues. (You can find out more about further developments in string theory and Calabi-Yau manifolds in Yau's book.)

Today, string theory is still far from complete. There are many physical quantities it can't yet describe and it has not, and currently cannot, be tested in the lab. Yau, however, believes that its sheer mathematical coherence means that it's not just a red herring.

> Mathematicians have been able to prove formulas that were motivated by the physical intuition from string theory. And there are many other spectacular contributions string theory has made to math. Because of string theory many apparently different areas of math have merged together in a smooth but totally unexpected manner. This means there must be some truth to string theory. Will it eventually lead to a fundamental theory of matter? It's too early to say, but we believe that there must be some truth to the intuition it provides.

Playing with Matches

Erica Klarreich

The mass entry of women into the American workforce in the 20th century has, proverbially, turned family life into a juggling act, with couples coordinating work schedules, childcare, and chores in days that never seem quite long enough.

Imagine, then, how difficult it is for professional couples—such as medical students and academics—who have to relocate just to find jobs in the first place. For these couples, solving the "two-body problem" can be far from trivial. All too often, one member of a couple has to decide about a job offer before the other has lined up any prospects. Do they grab the job and trust that a second job in the same city will turn up? Or do they hold out for a pair of jobs, hopefully in their dream city?

When the pair are medical students, they entrust their fate to a computer algorithm that matches them to hospital residency positions. Prior to Match Day, medical students submit preference lists of acceptable positions, and hospitals rank the students they are willing to hire. Couples are allowed to submit their preferences in pairs, to make sure they don't end up on opposite ends of the country. Then the computer comes up with a matching of hospitals and students that all the participants have committed to accept.

It would be nice if the computer could give every student, couple, and hospital its top choice, but in most cases this is not possible. In the absence of a match that will perfectly satisfy everyone, what constitutes an acceptable match? When the first resident matching algorithm was devised in the early 1950s, there was no consensus about what properties such a match should have. Then, in 1962, in a simple but groundbreaking paper, mathematicians David Gale and Lloyd Shapley came up with the idea of a "stable" matching. In the context of the market for medical residents, a matching is stable if no student and hospital (or couple and pair of hospitals) prefer each other over their computer-assigned matches. Such a matching may not result in everyone being overjoyed with their match, but it is stable in the sense that those participants who are disappointed with their match can't persuade any of their preferred choices to switch to them.

In markets that have no couples, it's always possible to find a stable matching, mathematicians have proved. Throw couples into the mix, however, and a stable matching may not exist. In fact, mathematicians have shown that in some cases, even computing whether a stable matching exists is so hard that it is, for all practical purposes, impossible (in technical terms, it is NP-complete).

Yet in the real world, the algorithm has been surprisingly successful at finding stable matchings in markets with couples, and economists have been wondering why. In the last decade, in several dozen annual matching markets—for medical students, psychologists, and a host of medical specialties—only a handful of markets have failed to find a stable matching.

"There's a sense in which we've been more successful than we had a right to be, based on the existing theorems," says Alvin Roth, an economist at Harvard University who was behind the design of one of the most widely used matching algorithms today.

Now, Roth and two other economists have proposed an explanation for why the matching markets seem to do so well. In a working paper, Roth—together with Fuhito Kojima of Stanford University and Parag Pathak of MIT—proves that in a large market, provided that there aren't "too many" couples (in a precise sense), the probability that a stable matching exists is very high. What's more, there is a simple algorithm that will successfully find a stable match with high probability.

The researchers also show that in these large couples markets, participants have strong incentives to tell the truth about their preferences. As we'll see, it's possible to construct examples of markets in which participants could successfully "game the system" by lying about their preferences. But as the size of the market grows, the researchers show, participants have less and less to gain from such chicanery.

"This is a first attempt to explain why the couples problem hasn't been such a bear," Roth says.

State of Residency

The first resident matching algorithm was brought on the scene in the early 1950s, in an attempt to clean up a market that had been messy and chaotic for decades.

Because medical residents—aka, cheap labor—were a valuable commodity, in the early 20th century hospitals would try to beat each other to the punch, locking in desirable medical students before other hospitals had made them offers. Gradually, offers to medical students crept earlier and earlier, until by the mid-1940s, jobs were being agreed upon almost two years before the students graduated.

Realizing that this trend was untenable, the medical schools decided jointly not to issue any student transcripts or reference letters before a fixed date in the senior year. This effort helped control the timing of the job market, but created a new problem for hospitals: often, if the first offer a hospital made got rejected, the student to whom it wanted to make its next offer had already accepted another job. Hospitals responded to this difficulty by issuing offers of residency positions at 12:01 AM on the earliest possible date, and giving students just a few hours to decide.

By the early 1950s, hospitals and students alike were frustrated enough to agree to a centralized market with jobs assigned by computer. After some trial and error, the National Resident Matching Program (NRMP) was born. Over the next few decades, the system spread to markets for dozens of medical specialties and related fields.

While the matching algorithm was originally designed in an ad hoc manner, over the next few decades mathematicians placed it on a solid theoretical footing. First, in 1962—motivated by a problem related to college admissions, rather than by the resident matching algorithm—Gale and Shapley published their seminal paper about the so-called "stable marriage" problem. Imagine, they wrote, a marriage market with *n* boys and *n* girls. Each boy ranks all the girls in order of preference, and each girl ranks all the boys. Is there any way to pair off the boys and girls, they asked, so that no boy and girl will want to ditch their assigned spouses and elope together? Gale and Shapley defined a match to be "stable" if it was elopement-proof.

Gale and Shapley designed a matching algorithm, called the deferred acceptance algorithm, and proved that it always produces a stable matching. In the "boy proposes" version of the algorithm, the boys all start by proposing to their favorite girls. Each girl looks through her offers (if she has any), holds on to the best one, and rejects the others. In the next round, each rejected suitor proposes to his number two girl; each girl holds her new best offer, possibly rejecting the suitor she held in round one. The algorithm proceeds in this way, with each rejected suitor in a given round proposing to the next girl on his list. (The algorithm could instead be run with the girls doing the proposing.)

To see that the algorithm eventually ends, instead of getting stuck in an infinite loop of proposals and rejections, note that the boys are working their way down their preference lists, so the proposals can't go on forever. When the algorithm ends, everyone is matched: it's impossible to have a leftover boy and girl, since the boy would eventually propose to that girl if no one else will have him, and she would accept him if she has no better offers.

And the matching the algorithm produces is stable. To see this, imagine that it has matched Alice to Bob and Carol to Dave, but Carol prefers Bob. Then Bob must never have proposed to Carol, since if he had, she would have accepted him and rejected Dave. Therefore, Alice must be higher than Carol on Bob's preference list, since he proposed to her before getting to Carol. So Bob will choose not to elope with Carol. A similar argument can be made when it's the boy who wants out of his match.

In 1984, Roth showed that the resident matching algorithm was equivalent to Gale and Shapley's deferred acceptance algorithm. (The hospitals and medical students in the NRMP don't actually go about proposing to and accepting or rejecting each other; the computer simply implements all the proposals and responses, based on the participants' preference lists.)

The resident matching market does have some important differences from Gale and Shapley's marriage scenario: Some hospitals may have more than one position, the number of positions may not match the number of applicants, and hospitals and students are not required to rank every single available option.

Despite these differences, Roth used arguments similar to Gale and Shapley's to show that the algorithm will always find a stable matching.

In contrast with the stable marriage scenario, the match will not necessarily produce a placement for every participant. The fact that there are (typically) more residents than available jobs pretty much guarantees this, but that's not all that is going on. The "leftovers" when the algorithm terminates will not necessarily be just the bottom-of-the-barrel medical students that no one wanted. There might also be students with glowing resumes, and there are typically also some empty hospital positions.

The reason that many participants get left out of the match stems from what happens before the algorithm does its work. To develop their preference lists, medical students send out resumes and get called for interviews. A given student can travel only to so many interviews, and a hospital can interview only so many applicants, before it's time to make preference lists. So, even though medical students could theoretically rank several thousand different programs, in practice they typically rank only 7 to 9. Thus, a talented student could get left out of the match if she foolishly listed only a few top programs at which she was outranked by other students. Similarly, a hospital could omit to include any "safe" students on its list, and miss out on a match. Or, a student or hospital might produce a well-balanced list but fail to get a match out of sheer bad luck.

In 2009, for example, about 12% of residency programs had unfilled seats after the official match. These positions got filled in an unstructured process affectionately dubbed the "Scramble," in which unmatched students and hospitals would work the phones until all the positions were filled.

The algorithm's inability to match up a substantial chunk of the participants seems at first glance like an unmitigated drawback. But there's a silver lining.

In their new work, Kojima, Pathak, and Roth show that this seeming failure is in fact the reason why the algorithm is successful in markets with couples. As we'll see, the large number of open positions provides just the wiggle room needed to integrate couples into the match. What's more, this wiggle room makes it close to impossible for participants in a couples market to successfully manipulate the market by lying about their preferences.

The Two-Body Problem

In the 1950s, when the centralized residents match was devised, virtually all medical students were men, and the idea of accommodating couples wasn't even on the radar. By the 1970s, however, enough women were graduating from medical school that couples were not unheard of, and today typically more than a thousand medical students look for positions as couples (in a market with about 25,000 to 30,000 participants).

In the 1970s, the NRMP made a first attempt at an algorithm that could accommodate couples. Couples had to be certified by their dean, and they had to specify one member of the couple as the "leading member." Each member of the couple made a separate preference list. The leading member went through the match as if single, and then the other member's list got edited to remove positions in a different community from the leading member's position.

This algorithm had some rather glaring drawbacks. For one thing, as Roth puts it, the algorithm violated an iron law of marriage: you can't be happier than your spouse. Furthermore, since the algorithm ignored many of the realities of couples' preferences, couples were often able to arrange better positions outside of the official match than the algorithm found for them. Gradually, over the 1970s and early 1980s, more and more couples started opting out of the match entirely.

It seems like the obvious solution is to allow couples to submit a single preference list consisting of pairs of positions—Alice at Massachusetts General Hospital and Bob at Boston Medical Center, for example. But doing so opens up a can of worms. In 1984, Roth showed that in such markets, a stable matching may not exist. Problems tend to arise, for example, when a couple gets placed in a pair of positions, and one hospital is happy but the other is not. (For a simple example of a couples market with no stable matchings, see the Appendix called Vicious Circle.)

Despite the potential for problems, in 1983 the NRMP adopted an algorithm that allowed couples to rank positions in pairs. Since 1998 it has used an algorithm designed by Roth and Elliott Peranson, of National Matching Services. This algorithm has now been adopted by more than 40 different labor matching markets.

The Roth-Peranson algorithm is based on the "sequential couples algorithm," which is in turn based on Gale and Shapley's deferred acceptance algorithm. The sequential couples algorithm starts by running Gale and Shapley's algorithm on the hospitals and single students, saving couples for later. Next, working one couple at a time, the algorithm places couples into positions (provisionally) by allowing them to apply to pairs of positions, working their way down their preference list until they find a pair of hospitals willing to accept them (potentially displacing two students who had previously been accepted by those two hospitals).

Once the algorithm has finished placing all the couples, the displaced students are brought back into the market by the computer and allowed to re-apply to hospitals, working their way down their preference lists.

The sequential couples algorithm can run into difficulties in a variety of ways. For example, once a couple has been placed (say, at hospitals A and B), suppose an individual student applies to hospital A and is preferred over the couple member. Then both members of the couple will be displaced, but only hospital A has improved its lot, not hospital B. Hospital B might now regret the students it had previously rejected in favor of the couple member. What's more, the couple will have to be placed in a new pair of positions, potentially displacing two new students and setting off an infinite cascade of placements, rejections, and dissatisfied hospitals.

Thus, the sequential couples algorithm is considered to succeed only if, once a couple gets placed, no one else applies to those two positions. If the algorithm succeeds, it's not hard to show that the match it produces is stable.

The Roth-Peranson algorithm starts with a process similar to the sequential couples algorithm, but if the algorithm falls into a hole along the way, it doesn't just declare failure, but tries to climb back out. In the above example—in which an individual student displaces two couple members, leaving a dissatisfied hospital—that hospital is put on an "unresolved" pile. After any displaced students have proposed down their lists, the unresolved hospital looks at the student currently assigned to it, and then all the students with whom it might be tempted to "elope" are thrown back into the mix and allowed to propose to hospitals anew, starting at the tops of their lists. In this way, the algorithm attempts to forestall each possible instability, one at a time.

The algorithm is also equipped with "loop-detectors" to keep it from cycling in infinite loops. If, for example, the algorithm detects two couples that keep displacing each other from the same pair of hospitals, it forces one of the couples to propose to a pair of hospitals lower down on its preference list.

The Roth-Peranson algorithm is guaranteed to terminate, but the matching it finds is by no means guaranteed to be stable, even if a stable matching does exist in the given market. And when an algorithm produces an unstable matching, it's not simply a matter for hand-wringing in the ivory tower. In two papers in the early 1990s, Roth studied a collection of different algorithms used by regions of the British National Health Service in the 1960s. He found that for the most part, the algorithms that produced stable matchings had succeeded and survived, while the algorithms that produced unstable matchings had gradually lost participants and support, eventually falling by the wayside.

Surprising Success

Yet so far, this has not been the fate of the Roth-Peranson algorithm. Somehow, it almost always seems to find a stable matching.

To understand what underlies this success, Kojima, Pathak, and Roth examined data from the first nine years that the clinical psychologists' market used the Roth-Peranson algorithm. In each year, the algorithm had successfully found a stable matching.

The market for clinical psychologists had several distinctive features, the researchers found. For one thing, couples were just a small fraction of participants, averaging about 1%. Also, as discussed above, participants typically ranked only a small number of the open positions available to them. The researchers also found that each year, about 17% of programs had unfilled positions after the match, necessitating an aftermarket like the NRMP's Scramble.

The researchers suspected that the large size of medical markets—usually numbering many thousands of participants—might be behind the Roth-Peranson algorithm's success, since large markets tend to work better than small ones in many respects. To test this theory, they decided to see what happens as the size of the market approaches infinity.

The researchers considered a sequence M_n of random markets with n hospitals, and made the following three assumptions (as well as some other regularity conditions):

1. The number of couples is small compared to the size of the market. (More precisely, the number of couples grows more slowly than a constant times the square root of n.)
2. The number of entries on the preference list of any student is bounded by a constant that does not depend on n.
3. The popularity of different hospitals on students' preference lists doesn't vary beyond a fixed ratio. In other words, not everyone is putting, say, the same few top-tier hospitals on their

lists. (It might seem at first glance as if everyone would want to put the top hospitals on their lists; but the students who didn't even get interviews at those hospitals would know that they don't stand a chance there, and focus on more attainable positions.)

The researchers found that as the size of the market approaches infinity, the probability that a stable matching exists tends to 1. What's more, with probability also tending to 1, the sequential couples algorithm will find a stable matching, without even having to call upon the Roth-Peranson algorithm's repair mechanisms.

The reason for this success lies in the fact—which the researchers prove—that in these large markets, at every stage of the matching algorithm's progress, a substantial number of programs have unfilled positions. This gives the algorithm a crucial degree of wiggle room for placing couples.

Recall that the sequential couples algorithm fails only if, after it places a couple, another participant wants to apply to one of the couples' two positions. In the researchers' model, they show that this is unlikely to happen: Participants are much more likely to apply to one of the many unfilled positions than one of the relatively few positions occupied by couples. And as long as no one ever displaces a couple, the algorithm is guaranteed to find a stable matching.

Honesty Is the Best Policy

The three researchers also show that as the size of the market approaches infinity, it becomes more and more difficult for any participants to successfully game the system by lying about their preferences.

In a 1982 paper, Roth proved that in a market with no couples, the side doing the proposing has nothing to gain from lying. But for individuals on the receiving end of the proposals, truth-telling may not be the best strategy. Roth later proved that in a market with couples, both sides may potentially benefit from lying.

To see how it might sometimes benefit market participants to lie, consider a market with two hospitals, Harvard Medical Center and Stanford Medical Center, and two students, Alice and Barbara. Suppose the algorithm is set up with the students doing the proposing, and the preference lists are as follows:

ALICE: Harvard, Stanford
BARBARA: Stanford, Harvard
HARVARD: Barbara, Alice
STANFORD: Alice, Barbara

If everyone tells the truth, then Alice will propose to Harvard and Barbara to Stanford; both proposals will be accepted, and the algorithm will end, with a stable match that is great for Alice and Barbara, but not so hot from the hospitals' point of view.

Harvard, for one, could do better for itself by lying and saying that it will accept only Barbara. In that case, the first round of the algorithm will end with Barbara temporarily accepted by Stanford, and Alice rejected by Harvard. A second round will occur, in which Alice proposes to her second choice, Stanford, and is accepted, displacing Barbara. Then Barbara proposes to Harvard in round three and is accepted, resulting in a stable match more to Harvard's liking than if it had been truthful.

In this example, Harvard's lie sets off a chain reaction of rejections and proposals that eventually results in Harvard receiving a proposal from a preferred student. But in a large market, the researchers show, it's unlikely that such a chain reaction would end back at Harvard, instead of landing on one of the many empty positions.

"In a large market, the opportunities to manipulate become increasingly rare," Roth says.

Design for Matching

The new work on couple matchings is the kind of thing market design is all about, says Paul Milgrom, an economist at Stanford University. "You have a genuine issue, like couples participating in a market, and the market design has to solve it somehow, even if it doesn't neatly institute the classical theory of Gale and Shapley."

While the group's result—that large markets with couples will usually have a stable matching—is a good start, ultimately economists will want to know just how large is large enough, Milgrom and Roth agree.

"Eventually we would like much better theorems, that say what the probability of finding a stable matching is for a market of a certain size, with a certain number of couples," Roth says. "I think it's definitely doable."

Market design is an appealing area of economics, Roth says, because mathematical theory informs the design of real markets, and the functioning of those markets in turn motivates the development of new mathematics.

"There's a nice back and forth between mathematics and the real world," Roth says. "The theory is turning out to be a powerful guide to what's going on in these markets, which is very gratifying."

Appendix: Vicious Circle

To see one way that a market with couples could fail to have any stable matching, consider a market with just two hospitals, Harvard Medical Center and Stanford Medical Center, and three students, Adam, Eve, and Bob. Suppose the preference lists are as follows:

Bob: Harvard, Stanford
Adam and Eve: Adam at Stanford, Eve at Harvard (they aren't interested in any other placement)
Harvard: Eve, Bob (not Adam)
Stanford: Bob, Adam (not Eve)

If the match places Adam at Stanford and Eve at Harvard, then Bob and Stanford will choose to "elope," since Stanford prefers Bob and Bob is available and willing to go to Stanford. So the match isn't stable.

If the match leaves Adam and Eve unmatched (the only other allowable option), then there are two possibilities:

If Bob gets Harvard, then Harvard and Eve, and Stanford and Adam will choose to elope, since Harvard prefers Eve to Bob and Stanford prefers Adam to no one.

If Bob gets Stanford (or nothing), then Bob and Harvard will choose to elope.

Thus, there is no stable matching.

Notable Texts

The following is a list of other notable texts on mathematics published during 2010 in periodicals, in collective volumes, or in the mass media. At one time or another during the preparation of this volume I considered including most of these texts but I decided against it due to constraints of space and/or copyright issues.

Antoy, Sergio, and Michael Hanus. "Functional Logic Programming." *Communications of the ACM* 53.4(2010): 74–85.

Barrow-Green, June. "The Dramatic Episode of Sundman." *Historia Mathematica* 37(2010): 164–203.

Bayley, Melanie. "Alice's Adventures in Algebra: Wonderland Solved." *New Scientist*, December 16, 2009.

Bialik, Carl. "Milestone Figures Grab Attention, but Their Impact Is Hazy." *Wall Street Journal*, February 12, 2010.

Cain, Alan. "Deus ex Machina and the Aesthetics of Proof." *Mathematical Intelligencer* 32.3(2010): 7–11.

Carey, Susan. "The Making of an Abstract Concept: Natural Number." In *The Making of Human Concepts*, edited by Denis Mareschal, Paul C. Quinn, and Stephen E. G. Lea. Oxford, U.K.: Oxford University Press, 2010.

Davis, Philip J. "The Prospects for Mathematics in a Multi-Media Civilization." In *Mathematics Everywhere*, edited by Martin Aigner and Ehrhard Behrends. Providence, R.I.: American Mathematical Society, 2010.

Davis, Philip J., and David Mumford. "What Should a Mathematical Professional Know About Mathematics?" *Svenska Matematikersamfundet Medlemsutskicket*, May 15, 2010, 5–14.

de Freitas, Elizabeth. "Making Mathematics Public: Aesthetics as the Distribution of the Sensible." *Educational Insights* 13.1(2010).

Dunham, Douglas. "Hamiltonian Paths and Hyperbolic Patterns." In *Communicating Mathematics*, edited by Timothy Y. Chow and Daniel C. Isaksen. Providence, R.I.: American Mathematical Society, 2009.

Erickson, Paul. "Mathematical Models, Rational Choice, and the Search for Cold War Culture." *Isis* 101.2(2010): 386–92.

Gaull, Marilyn. "From *Tristram Shandy* to Bertrand Russell: Fiction and Mathematics." *British Society for the History of Mathematics Bulletin* 25.1(2010): 81–91.

Gavalas, Dimitris. "On Number's Nature." In *Mathematics and Mathematical Logic: New Research*, edited by Peter Miloslav and Irene Ercegovaca, pp. 1–58. New York: Nova Science Publishers, 2010.

Glimm, James. "Reflections and Prospectives." *Bulletin of the American Mathematical Society* 47.1(2010): 127–36.

Hintikka, Jaakko. "How Can a Phenomenologist Have a Philosophy of Mathematics?" In *Phenomenology and Mathematics*, edited by Mirja Hartimo. Dordrecht, Germany: Springer Science+Business Media, 2010.

Kilpatrick, Jeremy. "Influences of Soviet Research in Mathematics Education." In *Russian Mathematics Education: History and World Significance*, edited by Alexander Karp and Bruce R. Vogeli. Hackensack, N.J.: World Scientific, 2010.

Koblitz, Neal, and Alfred Menezes. "The Brave New World of Bodacious Assumptions in Cryptography." *Notices of American Mathematical Society* 57.3(2010): 357–65.

Luecking, Stephen. "A Man and His Square: Kasimir Malevich and the Visualization of the Fourth Dimension." *Journal of Mathematics and the Arts* 4.2(2010): 87–100.

Mamolo, Ami. "Polysemy of Symbols: Signs of Ambiguity." *The Montana Mathematical Enthusiast* 7.2–3(2010): 247–62.

Peters, Ellen, Judith Hibbard, Paul Slovic, and Nathan Dieckmann. "Numeracy Skill and the Communication, Comprehension and Use of Risk–Benefit Information." In *The Feeling of Risk: New Perspectives on Risk Perception*, by Paul Slovic. Washington, D.C.: Earthscan, 2010.

Pimm, David. "'The Likeness of Unlike Things': Insight, Enlightenment, and the Metaphoric Way." *For the Learning of Mathematics* 30.1(2010): 20–3.

Rodgers, Joseph Lee. "The Epistemology of Mathematical and Statistical Modeling." *American Psychologist* 65.1(2010): 1–12.

Schachermayer, Walter. "The Role of Mathematics in the Financial Markets." In *Mathematics Everywhere*, edited by Martin Aigner and Ehrhard Behrends. Providence, R.I.: American Mathematical Society, 2010.

Sinclair, Nathalie, and David Pimm. "The Many and the Few: Mathematics, Democracy, and the Aesthetics." *Educational Insights* 13.1(2010).

Spelke, Elizabeth, Sang Ah Lee, and Véronique Izard. "Beyond Core Knowledge: Natural Geometry." *Cognitive Science* 34.5(2010): 863–84.

Strogatz, Steven. "Chances Are." *The New York Times Online*, April 25, 2010. This text concluded a series of approximately a dozen weekly articles written by Steven Strogatz for the newspaper.

Tall, David, and Juan Pablo Mejia-Ramos. "The Long-Term Cognitive Development of Reasoning and Proof." In *Explanation and Proof in Mathematics: Philosophical and Educational Perspectives*, edited by Gila Hanna, Hans Niels Jahnke, and Helmut Pulte. Dordrecht, Germany: Springer Science+Business Media, 2010.

van der Zande, Johan. "*Statistik* and History in the German Enlightenment." *Journal of the History of Ideas* 71.3(2010): 411–32.

Weinert, Friedel. "The Role of Probability Arguments in the History of Science." *Studies in the History and Philosophy of Science* 41.1(2010): 95–104.

Ye, Feng. "The Applicability of Mathematics as a Scientific and a Logical Problem." *Philosophia Mathematica* 18.2(2010): 144–65.

Contributors

Heather Anderson is an English teacher and team leader at Health Sciences High & Middle College in San Diego. She has her master's degree in curriculum and instruction from San Diego State University. Besides teaching English in high school, she teaches Spanish to health care workers. She can be reached at handerson@hshmc.org.

Chris J. Budd is a professor of applied mathematics at the University of Bath, where he does research into problems at the interface of mathematics, engineering, and industry. These problems include studying weather forecasting, microwave cooking, geology, and the optimal design of aircraft structures. He is interested in any application of mathematics to the real world, from the study of crowds to the design of folk dances. He is a passionate popularizer of mathematics and is heavily involved with many programs that make math fun and relevant to people's lives. This includes working as professor of mathematics at the Royal Institution, giving Saturday "mathematics master classes" to young people, and presenting many talks all around the world. He has recently been elected a Fellow of the British Science Association and has co-authored the popular math book *Mathematics Galore*. When not doing math he can be found in the mountains with his dog.

Martin Campbell-Kelly is professor emeritus in the Department of Computer Science at Warwick University, where he specializes in the history of computing. He has held visiting appointments at the London School of Economics, Manchester University, the Smithsonian Institution, and MIT. His publications include a history of the software industry *From Airline Reservations to Sonic the Hedgehog* and the popular textbook *Computer: A History of the Information Machine,* co-authored with William Aspray.

Barry A. Cipra is a freelance mathematics writer based in Northfield, Minnesota. He has been a contributing correspondent for *Science* magazine and a regular writer for *SIAM News*, the monthly newsletter of the Society for Industrial and Applied Mathematics. He is the author of *Misteaks . . . and How to Find Them Before the Teacher Does: A Calculus Supplement*, published by A.K. Peters, Ltd.

Peter J. Denning is a Distinguished Professor at the Naval Postgraduate School in Monterey, California. He chairs the Computer Science Department and directs the Cebrowski Institute, an interdisciplinary research center for information innovation. In his early career he was a pioneer in virtual memory and in the development of principles for operating systems. He co-founded CSNET, the computer science network of the National Science Foundation, and shared in the 2009 Internet Society Postel Award for this achievement; CSNET was the first community network bridge from the original ARPANET to the modern Internet. He was president of the Association for Computing Machinery from 1980 to 1982. He led the ACM Digital Library project that made ACM the first professional society to place all of its publications online. He has been a prolific author, publishing more than 345 articles in computer science and 10 books. His books include *The Invisible Future* (McGraw, 2001) and *Beyond Calculation* (Copernicus, 1997). His most recent book is *The Innovator's Way* (with Bob Dunham, MIT Press, 2010), about the essential practices of successful innovation. He holds twenty-six awards, including three honorary degrees, three professional society fellowships, six technical achievements, two best papers, three distinguished service awards, a hall of fame award, and several outstanding educator awards.

Underwood Dudley received his Ph.D. degree from the University of Michigan in 1965. He taught at the Ohio State University and at DePauw University, retiring in 2004. He is the author or editor of eight books, including *Mathematical Cranks* and *Numerology*. He has edited the *Pi Mu Epsilon Journal* and the *College Mathematics Journal* and is a recipient of the Trevor Evans Award for expository writing.

Freeman Dyson is a retired professor at the Institute for Advanced Study in Princeton, New Jersey. He began his career as a pure mathematician in England but switched to physics after moving to the United States. A volume of his selected papers in mathematics and physics was published by the American Mathematical Society in 1996.

Rob Eastaway is a U.K.-based author and speaker dedicated to the popularization of mathematics. He graduated from Cambridge University in 1985 with a degree in engineering science, specializing in operational research. His books include the best-selling *Why Do Buses Come in Threes?* and *Old Dogs, New Math*. He is the director of Mathematics Inspiration, a program of theatre-based lectures for teenagers in the United Kingdom, and was president of the U.K. Mathematical Association in 2007–2008. For several years he designed puzzles for *New Scientist* magazine.

Jordan Ellenberg is an associate professor of mathematics at the University of Wisconsin—Madison, specializing in number theory and arithmetic geometry. He is the author of *The Grasshopper King,* a novel.

Claire Ferguson has written extensively on Helaman Ferguson's work, including the Gold Ink and Ozzie award-winning book *Helaman Ferguson: Mathematics in Stone and Bronze*. She is a graduate of Smith College and an accomplished watercolor artist.

Helaman Ferguson loves math and creates art. Fortunately for him, his wife of nearly a half-century, Claire, and their seven offspring are similarly and sympathetically endowed. Helaman's current sculpture studio is in Baltimore, Maryland. Therein he has built a living room-sized, five-axes-gantry programmable industrial robot so he can quantitatively carve negative stone molds for casting positive silicon bronze.

Douglas Fisher is a professor in the college of education at San Diego State University and a teacher leader at Health Sciences High & Middle College. He is the co-author of *Better Learning Through Structured Teaching* and can be reached at dfisher@mail.sdsu.edu.

Marianne Freiberger is co-editor of *Plus* online magazine (http://plus.maths.org), a free online magazine about mathematics aimed at the general public. *Plus* is part of the Millennium Mathematics Project, based at the University of Cambridge. Before joining *Plus*, Marianne did research in holomorphic dynamics.

Nancy Frey is a professor in the college of education at San Diego State University and a teacher leader at Health Sciences High & Middle College. She is the co-author of *Guided Instruction: How to Develop Confident and Successful Learners* and can be reached at nfrey@mail.sdsu.edu.

Ian Hacking has retired from his chair at the Collège de France, Paris, and from a university professorship at the University of Toronto. He has written many books on many topics, of which his favorite is still *The Emergence of Probability* (1975, 2006). A companion work is *The Taming of Chance* (1990). He has published little on the philosophy of mathematics, but it was his first love. In 2010 he gave the Descartes lectures in Holland and the Howison lectures in Berkeley on this topic. A much extended version of these talks is in press from Cambridge University Press: *The Mathematical Animal: Philosophical Reflections on Proof, Necessity and Human Nature*. In 2009 he was awarded the Holberg International Memorial Prize given in Norway for "outstanding scholarly work in the fields of the arts and humanities, social sciences, law and theology."

James Hamlin is a software engineer in the DirectX Driver Group at NVIDIA. James received his M.S. in computer science from the University of California, Berkeley, where he published work on computer-aided design and visualization. He also holds a B.A. in computer science and philosophy from the University of California, Berkeley.

David J. Hand is Professor of Statistics at Imperial College, London, and Chief Scientific Advisor of Winton Capital Management, one of the world's leading CTA hedge funds. He served two terms as president of the Royal Statistical Society. He is a Fellow of the British Academy and an Honorary Fellow of the Institute of Actuaries and has been awarded the Guy Medal of the Royal Statistical Society.

Ivan M. Havel is a cognitive scientist and cofounder and past director of the Center for Theoretical Study, an international cross-disciplinary research institution affiliated with Charles University in Prague and the Academy of Sciences of the Czech Republic. He graduated in 1966 from Technical University in Prague and in 1971 earned his Ph.D. in computer science from the University of California, Berkeley. His current research focuses on human mind, consciousness, and related areas of philosophy and cognitive science, with a special interest in episodic experience. He teaches on artificial and natural thinking at the Faculty of Mathematics and Physics. He is editor in chief of the scientific journal *Vesmír*, a member of Academia Europaea, and serves on boards of several academic institutions and educational foundations.

Reuben Hersh is professor emeritus at the University of New Mexico. He is the co-author, with Vera John-Steiner, of *Loving and Hating Mathematics*. He is also the author of *What Is Mathematics, Really?* and, with Philip J. Davis, the co-author of *The Mathematical Experience*, which won a National Book Award in 1983.

Hans Niels Jahnke is Professor of Mathematics Education at the University of Duisburg-Essen, Germany. He is author of *Mathematik und Bildung in der Humboldtschen Reform* (1990), editor of *A History of Analysis* (AMS 2003), and co-editor of books on proof and on the history of science in the early nineteenth century. His research fields in education are proof and the inclusion of history of mathematics into mathematics teaching.

Ioan James is best known in the mathematical world for his contributions to homotopy theory. His academic career was mainly at Oxford University, where he held the Savilian Chair of Geometry. When he retired from this, he decided to reinvent himself as a writer. He had already written a number of books and many articles on mathematical subjects, but recently he has written about people with Asperger's syndrome, the psychology of mathematicians, and the lives of scientists. His latest book, *Driven to Innovate*, is about outstanding Jewish mathematicians and physicists of the nineteenth century. He is a Fellow of the Royal Society and an Honorary Fellow of two Oxford colleges. He was president of the London Mathematical Society. He was founding editor of the journal *Topology*.

Erica Klarreich is a mathematics and science writer based in Berkeley, California. She has a Ph.D. in mathematics from Stony Brook University and a certificate in science writing from the University of California, Santa Cruz. Her articles have appeared in *Nature*, *New Scientist*, *American Scientist*, *Science News*, and other publications.

When he was a child, ***Dana Mackenzie*** wanted to be a writer. However, he took a circuitous path. He majored in mathematics at Swarthmore College and received his Ph.D. in mathematics from Princeton University in 1983. After teaching at Duke University for six years and Kenyon College for seven, he decided to make his childhood dream come true. He completed the Science Communication Program at the University of California, Santa Cruz, in 1997, and since then he has made his living as a freelance writer. Mackenzie has written for such magazines as *Discover*, *Smithsonian*, *Science*, and *New Scientist*. In addition, he has both written and edited articles for *American Scientist*, where "A Tisket, a Tasket, an Apollonian Gasket" first appeared. His first book, *The Big Splat, or How Our Moon Came to Be*, was published by John Wiley & Sons and was named an Editor's Choice for the year 2003 by *Booklist* magazine. He has also co-authored a geology textbook for Wiley and he writes the ongoing series "What's Happening in the Mathematical Sciences" for the American Mathematical Society. A forthcoming book on the history of mathematics is scheduled for publication by Princeton University Press in 2012.

John Mason is emeritus professor at Open University in the United Kingdom. Previously he has held appointments at a number of other institutions, including at the University of British Columbia, University of Capetown, University of Alberta, University of Southern Queensland, Australian Catholic University, and Dalhousie University. Among his many books and other publications are *Thinking Mathematically*, *Questions and Prompts for Mathematical Thinking* (with Anne Watson), *Mathematics Teaching Practice*, *Fundamental Constructs in Mathematics Education* (with Sue Johnston-Wilder), and *Designing and Using Mathematical Tasks* (also with Sue Johnston-Wilder).

Melvyn B. Nathanson is professor of mathematics at the City University of New York (Lehman College and the Graduate Center), and the author of more than 150 research papers and books in mathematics. He has held visiting appointments at many universities, including Harvard, Princeton, Tel Aviv, and the Institute for Advanced Study.

Doris Schattschneider is professor emerita of mathematics at Moravian College in Bethlehem, Pennsylvania. She has studied the symmetry work of M. C. Escher for more than 35 years; her book *M.C. Escher:Visions of Symmetry* is the definitive publication on that topic. A past editor of *Mathematics Magazine*, she received from the Mathematical Association of America an Allendorfer Award for writing and a Haimo Award for distinguished college or university teaching. Her research interest is in discrete geometry (tiling problems) and visualization of geometric ideas; she assisted in developing the software *The Geometer's Sketchpad*.

Alan H. Schoenfeld is the Elizabeth and Edward Conner Professor of Education and Affiliated Professor of Mathematics at the University of California, Berkeley. He is a Fellow of the American Association for the Advancement of Science and of the American Educational Research Association and a Laureate of the education honor society Kappa Delta Pi. Alan has served as president of the American Educational Research Association and vice president of the National Academy of Education. In 2008 he was given the Senior Scholar Award by AERA's Special Interest Group for Research in Mathematics Education. After obtaining his Ph.D. in mathematics from Stanford in 1973, Schoenfeld turned his attention to issues of mathematical thinking, teaching, and learning. His work has focused on problem solving, assessment, teachers' decision-making, and issues of equity and diversity, with the goal of making meaningful mathematics truly accessible to *all* students. Alan has written, edited, or co-edited twenty-two books and almost 200 articles on thinking and learning. He has an ongoing interest in the development of productive mechanisms for systemic change and for deepening the connections between educational research and practice. His most recent book, *How We Think,* provides detailed models of human decision-making in complex situations, such as teaching.

Andrew Schultz is an assistant professor of mathematics at Wellesley College. After receiving his Ph.D. from Stanford University in 2007, Andrew was a J.L. Doob research assistant professor at the University of Illinois at Urbana–Champaign. He has authored numerous papers on module structures in Galois cohomology and for the mathematical community at large, and he has received teaching awards at both Stanford and Illinois.

Carlo H. Séquin is a professor of computer science at the University of California, Berkeley. He received his Ph.D. degree in experimental physics from the University of Basel, Switzerland, in 1969. From 1970 until 1976 he worked at Bell Telephone Laboratories on the design and investigation of charge-coupled devices for imaging and signal processing applications. At Bell Labs he was also introduced to the world of computer graphics in classes given by Ken Knowlton. In 1977 he joined the faculty in the Electrical Engineering and Computer Sciences Department at Berkeley. He started out by teaching courses on the subject of very large-scale integrated (VLSI) circuits, thereby building a bridge between the computer sciences division and the electrical engineering faculty. He concentrates on computer graphics, geometric modeling, and the development of computer-aided design (CAD) tools for circuit designers, architects, and mechanical engineers. Since the mid-1990s, Séquin's work in computer graphics and in geometric design have provided a bridge to the world of art. In collaboration with a few sculptors of abstract geometric art, in particular with Brent Collins of Gower, Missouri, Séquin has found yet another domain where new frontiers can be opened through the use of computer-aided tools. Dr. Séquin is a Fellow of the ACM, a Fellow of the IEEE, and has been elected to the Swiss Academy of Engineering Sciences. He has received the IEEE Technical Achievement Award for contributions to the development of computer-aided design tools, the Diane S. McEntyre Award for Excellence in Teaching, and an Outstanding Service Award from the University of California for Exceptional Leadership in the Conception, Design, and Realization of Soda Hall.

Francis Edward Su is a professor of mathematics at Harvey Mudd College and earned his Ph.D. from Harvard University. His research mixes geometry, topology, combinatorics, and applications to the social sciences. He has a passion for teaching and popularizing mathematics. From the Mathematical Association of America, he received the 2001 Merten M. Hasse Prize for expository writing and the 2004 Henry L. Alder Award for distinguished teaching. He authors the "Math Fun Facts" website, which receives more than a million hits each year, as well as the popular iPhone app by the same name.

Rik van Grol is a metagrobolist from the Netherlands; he studies, designs, makes, solves, and collects mechanical puzzles. Rik is chief editor of *CFF* (*Cubism For Fun*). *CFF* is the English newsletter from the Dutch Cube Club (http://cff.helm.lu), the longest surviving puzzle club. Rik has a Ph.D. in physics and works in the field of transport modeling. More information is available at http://www.significance.nl/rik_van_grol.html.

Feng Ye holds a doctoral degree in philosophy from Princeton University and is currently a professor of philosophy at Peking University, China. He is the author of *Strict Finitism and the Logic of Mathematical Applications* (*Synthese Library*, Vol. 355, Springer, 2011), *Philosophy of Mathematics in the 20th Century: A Naturalistic Commentary* (in Chinese, Peking University Press, 2010), and a number of articles on the philosophy of mathematics, constructive mathematics, and the philosophy of mind and language.

Acknowledgments

My thanks go first to the authors of the texts included in this anthology, for writing the pieces and for collaborating during the preparation and production phases, including (in some cases) by offering electronic files of the figures.

Once again I consulted closely with the in-house editor at Princeton, Vickie Kearn, whose overall guidance made it easier for me to decide on the best course to proceed at a few difficult junctures. Stefani Wexler obtained many reprint permissions and oversaw other matters related to copyright. Nathan Carr did an excellent job as production editor, for this and for the previous volume in the series—while Paula Berard, the copyeditor, spotted more typos in a few days than I could have done in several weeks. Thank you to all.

Reaching the final selection of texts was a daunting task. Gerald Alexanderson, Fernando Gouvêa, Ivars Petersen, and John Stillwell helped me to choose by conveying their preferences on texts initially belonging to a much larger pool. I especially appreciate Fernando Gouvêa's opinionated comments on all the pieces I advanced. I thank all my readers and ask for their understanding wherever I overrode their suggestions. The responsibility for the shortcomings in the final content of the book is entirely mine.

At Cornell University I benefited from a working environment conducive to my preoccupation with continuing this series. Thanks to David W. Henderson for a very flexible research schedule. Many thanks to Maria Terrell for assigning me an adequate teaching load. Thanks to Ravi Ramakrishna and Dan Barbasch for co-opting me into supplementary activities—as well as to Michelle Klinger and Stanley Seltzer for offering me additional teaching opportunities. Also thanks to the staff at the Cornell University Library, whose services I used intensively every day.

My friend Mihai Băileşteanu saved me from much trouble whenever my computers crashed over the past few years by recovering all my files intact—with sure hand and never-diminished wit.

Thanks to my daughter Ioana for being no less (and certainly no more!) patient with our library escapades than a child her age is expected to be. She showed all the understanding I needed when I carried books and piles of articles, even to beaches and on camping trips.

Georg Cantor wrote that the essence of mathematics lies in its freedom. Over the years, in wildly contrasting circumstances, I probed this statement and its opposites, trying to sort out their meanings. My high school mathematics teacher, Ioan Candrea, offered me the first such opportunities, with unparalleled trust, with insight, and with wisdom. I owe him the humble confidence that I can continue in the spirit of his good deeds, by treating my own students in kind. Foremost in his generosity toward me, during four crucial years of my life he often let me teach mathematics to my classmates, at a time when I was supposed to be only taught. I dedicate this volume to his memory.

Credits

"How Much Maths Is Too Much Maths?" by Chris J. Budd and Rob Eastaway. *Mathematics Today* (UK) 46.5(2010): 252–5. Reprinted by kind permission of the Institute of Mathematics and its Applications. This article was originally published in a Public Engagement Special Issue of *Mathematics Today*.

"Reflections on the Decline of Mathematical Tables," by Martin Campbell-Kelly. First published in *Communications of the ACM* 53.4(2010): 25–26. Reprinted by permission of the author.

"Lorenz System Offers Manifold Possibilities for Art," by Barry A. Cipra. First published in *SIAM News* 43.2(2010). Reprinted by permission of the Society for Industrial and Applied Mathematics.

"The Great Principles of Computing," by Peter J. Denning. First published in *American Scientist* 98.5(2010): 369–72. Copyright Peter J. Denning; reprinted with permission of the author.

"What Is Mathematics For?" by Underwood Dudley. First published in *Notices of the American Mathematical Society* 57.5(2010): 608–12. Reprinted with permission from the American Mathematical Society.

"Fill in the Blanks: Using Math to Turn Lo-Res Datasets into Hi-Res Samples," by Jordan Ellenberg. Originally printed in *Wired*. Copyright 2010 Condé Nast. All rights reserved. Reprinted by permission.

"Celebrating Mathematics in Stone and Bronze," by Helaman Ferguson and Claire Ferguson. First published in *Notices of the American Mathematical Society* 57.7(2010): 840–50. Reprinted with permission from the American Mathematical Society.

"Thinking and Comprehending in the Mathematics Classroom," by Douglas Fisher, Nancy Frey, and Heather Anderson. First published in *Comprehension Across the Curriculum: Perspectives and Practices*, ed. by Kathy Ganske and Douglas Fisher. New York: Guilford Press, 2010, pp. 146–59.

"Hidden Dimensions," by Marianne Freiberger. First published in *Plus Magazine*, December 23, 2010. http://plus.maths.org/content/node/5388. Reprinted by permission.

"What Makes Mathematics Mathematics?" by Ian Hacking. First published in *The Force of Argument: Essays in Honor of Timothy Smiley*, ed. by Jonathan Lear and Alex Oliver. New York: Routledge, 2010, 82–101. Reprinted by permission.

"Computer Generation of Ribbed Sculptures," by James Hamlin and Carlo H. Séquin. First published in *Journal of Mathematics and the Arts* 4.4(2010): 177–89. Copyright Taylor & Francis. Reprinted by permission of the publisher.

"Did Over-Reliance on Mathematical Models for Risk Assessment Create the Financial Crisis?" by David J. Hand. First published in *FST Journal* (www.foundation.org.uk), 20.2(Dec. 2009): 28–29. Reprinted by kind permission of the Foundation for Science and Technology.

"Seeing Numbers," by Ivan M. Havel. First published in *Witnessed Years: Essays in Honor of Petr Hájek*, ed. by Petr Cintula, Zuzana Haniková, and Vítězslav Švejdar. London: College Publications, 2009, pp. 71–86. Reprinted by permission.

"Under-represented then Over-represented: A Memoir of Jews in American Mathematics," by Reuben Hersh. First published in *College Mathematics Journal* 41.1(2010): 2–9. Copyright the Mathematical Association of America 2011. All rights reserved.

"The Conjoint Origin of Proof and Theoretical Physics," by Hans Niels Jahnke. First published in *Explanation and Proof in Mathematics: Philosophical and Educational Perspectives*, ed. by Gila Hanna, Hans Niels Jahnke, and Helmut Pulte. Dordrecht, Germany: Springer Science+Business Media, 2010, pp. 17–32. Reprinted with kind permission from Springer Science+Business Media B.V.

"Autism and Mathematical Talent," by Ioan James. First published in *Mathematical Intelligencer* 32.1(2010): 56–58. Reprinted with kind permission from Springer Science+Business Media B.V.

"Playing with Matches," by Erica Klarreich. Published by the Simons Foundation at https://simonsfoundation.org/mathematics-physical-sciences/news/-/asset_publisher/bo1E/content/playing-with-matches?redirect=/mathematics-physicalsciences/News. September 24, 2010. Reprinted by permission.

"A Tisket, a Tasket, an Apollonian Gasket," by Dana Mackenzie. First published in *American Scientist* 98.1(2010): 10–14. Reprinted by permission of the author.

"Mathematics Education: Theory, Practice, and Memories over 50 Years," by John Mason. First published in *For the Learning of Mathematics* 30.3(2010): 3–9. Reprinted by permission.

"One, Two, Many: Individuality and Collectivity in Mathematics," by Melvyn B. Nathanson. First published in *Mathematical Intelligencer* 33.1(2011): 5–8. Reprinted with kind permission from Springer Science+Business Media B.V.

"The Mathematical Side of M. C. Escher," by Doris Schattschneider. First published in *Notices of the American Mathematical Society* 57.6(2010): 706–18. Reprinted with permission from the American Mathematical Society.

"Reflections of an Accidental Theorist," by Alan H. Schoenfeld. First published in *Journal for Research in Mathematics Education* 41.2(2010): 104–16. Reprinted with permission from the National Council of Teachers of Mathematics. All rights reserved.

"Meta-Morphism: From Graduate Student to Networked Mathematician," by Andrew Schultz. First published in *Notices of the American Mathematical Society* 57.9(2010): 1132–5. Reprinted with permission from the American Mathematical Society.

"Teaching Research: Encouraging Discoveries," by Francis Edward Su. First published in *The American Mathematical Monthly* 117.9(2010): 759–69. Copyright the Mathematical Association of America 2011. All rights reserved.

"The Quest for God's Number," by Rik van Grol. First published in *Math Horizons,* November 2010: 10–13. Copyright the Mathematical Association of America 2011. All rights reserved.

"What Anti-Realism in Philosophy of Mathematics Must Offer," by Feng Ye. First published in *Synthese* 175.1(2010): 13–31. Reprinted with kind permission from Springer Science + Business Media B.V.